II International Scientific and Practical Conference: Materials Science, Engineering and Energy: Problems and Prospects of Development (MSEE 2020)

AA007033

IOP Conference Series: Materials Science and Engineering
Volume 1100

Online
20-21 November 2020

Editors:

A. Saprykin
D. Valuev
A. Sobachkin
V. Yakovlev

ISBN: 978-1-7138-4598-0
ISSN: 1757-8981

Printed from e-media with permission by:

Curran Associates, Inc.
57 Morehouse Lane
Red Hook, NY 12571

Some format issues inherent in the e-media version may also appear in this print version.

This work is licensed under a Creative Commons Attribution 3.0 International Licence.
Licence details: http://creativecommons.org/licenses/by/3.0/.

No changes have been made to the content of these proceedings. There may be changes to pagination and minor adjustments for aesthetics.

Printed with permission by Curran Associates, Inc. (2022)

For permission requests, please contact the Institute of Physics at the address below.

Institute of Physics
Dirac House, Temple Back
Bristol BS1 6BE UK

Phone: 44 1 17 929 7481
Fax: 44 1 17 920 0979

techtracking@iop.org

Additional copies of this publication are available from:

Curran Associates, Inc.
57 Morehouse Lane
Red Hook, NY 12571 USA
Phone: 845-758-0400
Fax: 845-758-2633
Email: curran@proceedings.com
Web: www.proceedings.com

TABLE OF CONTENTS

II International Scientific and Practical Conference «Materials Science, Engineering and Energy: Problems and Prospects of Development» (MSEE'2020), November 20–21, 2020, Barnaul, Russia (Online)

PREFACE

The International Conference on Materials Science, Engineering and Energy (MSEE'2020) was organized by the Polzunov Altai State Technical University and was successfully held online from November 20 to 21, 2020. The technical program included one plenary session with two plenary presentations and twenty-eight live online presentations.

The conference serves as a platform for discussion of advanced and innovative technical solutions in various fields of modernity. The aim of the conference is to present innovative approaches in the development of integration solutions in the field of engineering and technology, as well as in interdisciplinary research.

MSEE'2020 topics are based on sustainability, with close attention paid to all forms of engineering professions and research tools, including the following topics:

1. New materials and technologies for their production
2. Design, processing and quality control of machines and mechanisms
3. Problems of traditional and alternative energy
4. Actual problems of modeling and automation of technological processes in mechanical engineering
5. Additive technology
6. Methods for creating coatings and thin films
7. Breakthrough technologies and techniques for the development and development of underground space.

MSEE'2020 gathered more than 60 participants from 9 different countries. The key topics of the conference cover the priority areas of research in the field of technical and related sciences and take into account global development trends.

This volume contains materials from MSEE'2020 55 manuscripts, which, in accordance with the Journal's Peer Review Policy, have been strictly selected based on originality, relevance and contribution to this area after peer review.

Program Committee:

A. Svistula - Chairman of the Committee, Doctor of Technical Sciences, Professor, Barnaul, Russia

A. Sitnikov - Co-Chair, Doctor of Technical Sciences, Professor, Barnaul, Russia

A. Sviridov - Ph.D., St. Petersburg, Russia

M. Skakov - Doctor of Physical and Mathematical Sciences, Professor, Academician of KazNAEN, Kurchatov, Republic of Kazakhstan

S. Lebedev - Doctor of Technical Sciences, Klaipeda, Lithuania

V. Fedorov - Ph.D., Barnaul, Russia

Content from this work may be used under the terms of the Creative Commons Attribution 3.0 licence. Any further distribution of this work must maintain attribution to the author(s) and the title of the work, journal citation and DOI.
Published under licence by IOP Publishing Ltd

IOP Conf. Series: Materials Science and Engineering 1100 (2021) 011001 doi:10.1088/1757-899X/1100/1/011001

V. Yakovlev - Ph.D., Barnaul, Russia

A. Balashov - Ph.D., Associate Professor, Barnaul, Russia

V. Filimonov - Doctor of Physical and Mathematical Sciences, Professor, Barnaul, Russia

S. Leonov - Doctor of Technical Sciences, Professor, Barnaul, Russia

M. Loginova - Ph.D., Barnaul, Russia

A. Ikonnikov - Ph.D., Associate Professor, Barnaul, Russia

G. Leonov - Doctor of Physical and Mathematical Sciences, Professor, Barnaul, Russia

A. Gradoboev - Doctor of Technical Sciences, Tomsk, Russia

Organizing Committee:

A. Sobachkin - Ph.D., Barnaul, Russia

A. Myasnikov, Novosibirsk, Russia

A. Popova - Ph.D., Barnaul, Russia

A. Cherdantsev, Barnaul, Russia

N. Yakovleva, Barnaul, Russia

Editors:

Alexander Saprykin, Tomsk Polytechnic University, sapraa2020@mail.ru

Denis Valuev, Tomsk Polytechnic University, valuevden@rambler.ru

Alexey Sobachkin, Altai State Technical University, sobalvi@yandex.ru

Vladimir Yakovlev, Altai State Technical University, anicept@rambler.ru

Organizing Committee MSEE'2020

IOP Conf. Series: Materials Science and Engineering 1100 (2021) 011002 doi:10.1088/1757-899X/1100/1/011002

Peer review declaration

All papers published in this volume of *IOP Conference Series: Materials Science and Engineering* have been peer reviewed through processes administered by the Editors. Reviews were conducted by expert referees to the professional and scientific standards expected of a proceedings journal published by IOP Publishing.

- **Type of peer review: Double-blind**
  We had a blind review. 2 reviewers were assigned to one paper.

- **Conference submission management system:**
  https://en.altstu.ru/structure/unit/rcic/news/18705/
- **Number of submissions received:** 68

- **Number of submissions sent for review:** 63

- **Number of submissions accepted:** 62

- **Acceptance Rate (Number of Submissions Accepted / Number of Submissions Received X 100:** 91.2

- **Average number of reviews per paper:** 2

- **Total number of reviewers involved:** 4

- **Any additional info on review process:** During review process, all papers were checked for plagiarism, compliance with the topic of the journal, acceptable English language, and good quality of graphic images.

  The reviewers took into account the following key points regarding scientific content, quality:

Technical Criteria

• Scientific merit: notably scientific rigour, accuracy and correctness.

• Clarity of expression; communication of ideas; readability and discussion of concepts.

• Sufficient discussion of the context of the work, and suitable referencing.

Quality Criteria

• Originality: Is the work relevant and novel?

• Motivation: Does the problem considered have a sound motivation? All papers should clearly demonstrate the scientific interest of the results.

• Repetition: Have significant parts of the manuscript already been published?

• Length: Is the content of the work of sufficient scientific interest to justify its length?

- **Contact person for queries: Alexander A. Sitnikov,** Doctor of Technical Sciences, Professor, Director of ITC AltSTU, Barnaul. msee_astu@mail.ru

Content from this work may be used under the terms of the Creative Commons Attribution 3.0 licence. Any further distribution of this work must maintain attribution to the author(s) and the title of the work, journal citation and DOI.
Published under licence by IOP Publishing Ltd

IOP Conf. Series: Materials Science and Engineering 1100 (2021) 012001 doi:10.1088/1757-899X/1100/1/012001

Evaluation of the characteristics of a hybrid power nuclear mini-reactor fueled by super-heavy elements

V S Okunev

Department of Physics, Bauman Moscow State Technical University, 2-ya Baumanskaya ul. 5, Moscow, 105005, Russia.
E-mail: okunevvs@bmstu.ru

Abstract. The concept of a hybrid mini-reactor capable of operating in the mode of a nuclear fission reactor (on fast neurons), in the mode of a radionuclide source and simultaneously in both modes is considered. It is proposed to use the most long-lived isotopes of the 114th, 126th, and possibly 120th elements as fuel for such a mini-reactor. Such nuclides have not yet been obtained in laboratories around the world. The neutron effective cross sections and macroscopic cross sections for the nuclei of super-heavy elements are estimated. A simple and effective mathematical model of the neutron-physical calculation of a mini-reactor with fuel based on super-heavy nuclei is proposed. The model is based on the one-group diffusion approximation. The uncertainties of the cross sections are so great that the use of more accurate and high-precision calculation models can hardly be justified. The results of neutron-physical calculations of a mini-reactor with fuel based on the longest-lived isotopes of the 114th and 126th elements are presented. The critical mass of a ball (without a reflector) made from ^{114}Fl will be 1.492 kg; the critical radius will be less than 2.48 cm. (For element ^{342}126 1.474 kg and less than 2.33 cm respectively.)

1. Introduction

As the problems associated with technological development and safeties are solved, nuclear power in a number of countries of the world is gradually moving from the category of alternative energy to traditional energy. In the near future, there will be a need to use nuclear power sources in space: on spacecraft, orbital stations, space bases on the planets of the solar system being developed. For this reason, it is important to create compact mini-power plants with high energy density.

The principle of operation of new generation nuclear power plants can be based on fission reactions (nuclear reactors) and radioactive decays ("nuclear batteries") of super-heavy atomic nuclei. Traditionally, energy sources using heavy atomic nuclei as fuel are divided into fission reactors and radionuclide sources. Certain successes have been achieved in the world in obtaining new super-heavy elements of the islands of stability. Both types of installations can use the same fuel based on as yet undiscovered super-heavy nuclei. Such installations should be considered not as alternative systems used in our time, but as their further natural development. The high cost of the fuel used will limit the use of such installations. Their advantage is their small size and high energy density.

Projects of compact nuclear fission reactors with a thermal power of 1 ... 10 MW have been developed [1-3]. Such installations are characterized by a long service life (more than 10 ... 20 years without changing the fuel). In [4], the concept of a hybrid mini-reactor using fuel based on the nuclei of the islands of stability (the most long-lived isotopes of the 114th and 126th elements not yet obtained in laboratory conditions) is proposed. This concept unites two directions of development of energy sources:

Content from this work may be used under the terms of the Creative Commons Attribution 3.0 licence. Any further distribution of this work must maintain attribution to the author(s) and the title of the work, journal citation and DOI.
Published under licence by IOP Publishing Ltd

IOP Conf. Series: Materials Science and Engineering 1100 (2021) 012001 doi:10.1088/1757-899X/1100/1/012001

a nuclear fission reactor and a "nuclear battery" operating on the basis of radioactive decays of nuclei. The facility can operate in a nuclear reactor, "nuclear battery" and mixed mode.

2. Nuclear fuel

Obviously, nuclides can be used as fuel for a hybrid mini-reactor (fast reactor), the average lifetime of which will make it possible to manufacture nuclear fuel, load it into the reactor, and launch the reactor into outer space.

The average lifetime of the doubly magic nucleus of flerovium-298 (the number of protons is $Z = 114$, the number of neutrons is $N = 184$), which is the center of the first island of stability in the $N - Z$ diagram, is estimated at ~ 10^7 ... 10^8 years [5]. This is an underestimated estimate obtained using an elementary comparative analysis of the properties of heavy and super-heavy atomic nuclei. The estimate is consistent with the data of [6]. The most expected half-life of ^{298}Fl nuclei is much higher than 10^7 ... 10^8 years [5].

The doubly magic nucleus of the 126th element (^{310}126), which may be the center of the second island of stability, may not exist. On the $N - Z$ diagram, it is located near the boundary of proton stability or on it [5]. This nucleus can emit protons from the ground energy state, and its lifetime is only slightly longer than the characteristic time of nuclear interaction. The most long-lived isotope of the 126th element should be considered the nuclide with the number of neutrons $N = 216$ [5]. With this N / Z ratio in the nucleus, the forces of nuclear, electromagnetic and weak interactions are balanced. This nuclide can be considered the center of the second island of stability. The half-life of the 126th element isotope with mass number $A = 342$ can be ~ 10^6 ... 10^7 s (or 0.03 ... 0.33 years) [5].

Long-lived nuclides can be located in the “shallows” around the maximally deformed nucleus of element 120 ($N = 200$) with the maximum internal quadrupole electric moment in absolute value [5].

So, the nuclear fuel of compact energy sources of the future can be the nuclides of the islands of stability: flerovium-114, a nucleus with $N = 216$, $Z = 126$, a deformed nucleus with $N = 200$, $Z = 120$ (less preferable), and those close to them in the diagram $N - Z$ nuclides.

3. Operating modes of the installation

The installation can operate in the mode of a critical nuclear reactor on fast neutrons. It can be a super-temperature reactor with liquid or gaseous fuel, containing only one fuel element in a tungsten cladding with an inner diameter much smaller than 1.524 cm. The thickness of the fuel element cladding in conventional fast reactors is about 0.05 cm. The fuel element is cooled with liquid gallium or lead. A high effective fraction of delayed neutrons will make it easy to control the reactor. After stopping the reactor with an absorber introduced into the core, the fission of nuclear fuel will not completely stop. This is due to the presence of long-lived α-active nuclei in the fuel. Thus, when developing such power plants, it is necessary to provide for the possibility of efficient and reliable heat removal from the core after shutdown of the reactor.

The installation can operate in the mode of a radionuclide source. A subcritical system with an external neutron source can be used. When interacting with neutrons, at least a small number of fuel nuclei will be transmuted. Further (in the absence of an external source), the system begins to operate in the mode of a radionuclide energy source, i.e., to produce energy due to radioactive decays (first α-decays, then - spontaneous fission). Such a nuclear battery is in standby mode until it is "turned on" by an external neutron source. You cannot turn it off until the "battery" has exhausted its resource.

In theory, the burnup of fuel based on super-heavy elements can be almost 100%. Even after the termination of the fission chain reaction, the system will operate in the mode of a radionuclide source.

4. Mathematical model and results of neutron-physical calculation of a reactor with fuel based on the 114th and 126th elements

Let us estimate the critical radius of a sphere (without a reflector) consisting of ^{298}Fl or ^{342}126 nuclides in the simplest approximation. The uncertainties of the cross sections are so great that the use of high-precision calculation models can hardly be justified. In the one-group diffusion approximation:

$$\frac{k_\infty - 1}{M^2} = \left(\frac{\pi}{R_e^2}\right). \qquad (1)$$

Here k_∞ — infinity multiplication factor, M — the length of neutron migration, $M^2 = D / \Sigma_a$, R_e — extrapolated (in vacuum) radius of the sphere, D — diffusion coefficient, Σ_a — macroscopic absorption cross section. In the gas-kinetic neutron transport theory: $D = \frac{\nu_f \Sigma_f + \Sigma_s}{3\Sigma_{tot}^2}$, where Σ_f — macroscopic fission cross section, Σ_s — macroscopic scattering cross section, Σ_{tot} — total macroscopic cross section. In the one-group diffusion approximation $k_\infty = \frac{\nu_f \Sigma_f}{\Sigma_a}$.

Due to the neutron pairing effect in the nucleus, a heavy nuclide with odd N, located near the equilibrium curve of the forces of nuclear, electromagnetic and weak interactions on the $N-Z$ diagram, for example, the compound nucleus $^{343}126$, will be relatively short-lived even in the ground state. Any interaction of the $^{342}126$ nucleus with a neutron leading to the formation of a compound nucleus (that is, any inelastic interaction) will lead to the decay of the compound nucleus through the emission channel of one or the successive emission of several α-particles followed by fission, or immediately to fission. The decay of a compound nucleus through the fission channel is more probable, since the fission threshold is small, the neutron energy in a fast reactor is high, and the probability of emission of α-particles by a compound nucleus with odd N is small. Usually, the threshold for inelastic neutron scattering decreases with increasing mass number, i.e., inelastic scattering and capture of a neutron by a nucleus will ultimately lead to "instantaneous" (over the lifetime of a compound nucleus) or delayed (by one or several α-decays) nuclear fission. Elastic scattering of a neutron by a $^{342}126$ nucleus can initiate nuclear decay along a channel similar to direct nuclear interaction. Part of the neutron wave can penetrate into a large enough nuclei and excite energy levels.

When $\nu_f \Sigma_f >> \Sigma_a$, factor $D \approx \frac{\nu_f \Sigma_f}{3\Sigma_{tot}^2}$. Assuming $\Sigma_a \approx \Sigma_f$ get $M^2 \approx \frac{\nu_f}{3\Sigma_{tot}^2}$; $k_\infty \approx \nu_f$. In this case can be expressed as the ratio (1):

$$\frac{3\Sigma_{tot}^2(\nu_f - 1)}{\nu_f} \approx \left(\frac{\pi}{R_e}\right)^2. \qquad (2)$$

When $\nu_f >> 1$ ratio $\frac{\nu_f - 1}{\nu_f} \approx 1$ and the expression (2) become: $3\Sigma_{tot}^2 \approx \left(\frac{\pi}{R_e}\right)^2$ или

$$R_e \approx \frac{\pi}{\Sigma_{tot}\sqrt{3}} \approx \frac{1{,}814}{\Sigma_{tot}}. \qquad (3)$$

In the absence of a clear dominance of corpuscular or wave properties of neutrons, the total micro cross section is $\sigma_{tot} \approx 2\pi(R + ƛ)^2$, the cross section of inelastic processes is $\sigma_r \approx \pi(R + ƛ)^2$ [7]. Where R — radius of the nucleus, $ƛ = \frac{\lambda}{2\pi}$; λ — wavelength of the neutron. The cross section of elastic scattering

$$\sigma_{el} = \sigma_{tot} - \sigma_r \approx \sigma_r \approx \pi(R + ƛ)^2.$$

This formula is in good agreement with experimental data on actinides.

In the first approximation, the radius of the nucleus $R = r_0\sqrt[3]{A}$, where $r_0 \approx 1.3$ fm — nucleon constant radius, A — mass number. Radius of $^{342}126$ is 9.091 fm. Neutron wavelength

$$\lambda \sim \frac{h}{\sqrt{2mE}}.$$

The kinetic energy of the neutron $E = 2$ MeV is corresponds to $\lambda \approx 20.2$ fm or $\bar{\lambda} \approx 3.215$ fm. In this case $\sigma_{tot} \approx 9{,}515$ b; $\sigma_r \approx 4{,}758$ b; 1 b (barn) = 10^{-28} m^2. Radius of ^{298}Fl is equal 8.683 fm, $\sigma_{tot} \approx 8.895$ b; $\sigma_r \approx 4.447$ b.

Total macro cross section $\Sigma_{tot} = \sigma_{tot} C$, where C — concentration of nuclei. If we take the concentration of nuclei ^{298}Fl and ^{342}126 are equal $5 \cdot 10^{22}$ cm^{-3}, for the ^{298}Fl obtained: $\Sigma_{tot} = 0.445$ cm^{-1}, $\Sigma_{el} = \Sigma_{tot} - \Sigma_r \approx \Sigma_r = 0.222$ cm^{-1}. For the ^{342}126: $\Sigma_{tot} = 0.475$ cm^{-1}, $\Sigma_{el} \approx \Sigma_r = = 0.238$ cm^{-1}.

The extrapolated radius of the ball (3), made of the ^{298}Fl, is equal $R_e \approx 4{,}076$ cm.

The length of the linear extrapolation $d_e \approx 0{,}71 / \Sigma_{tot} = 1{,}596$ cm. The critical radius of the ball without reflector (in a vacuum) $R_e - d_e = 2.48$ cm, critical volume — 63.93 cm^3. The critical mass of the ball is equal 1492 g. For a sphere of ^{342}126: $R_e \approx 3.819$ cm, $d_e \approx 1.492$ cm, critical radius is 2.327 cm, critical volume is 52.781 cm^3, critical mass is 1474 g.

For a cylinder of infinite height ratio (3) takes the form

$$R_e \approx \frac{2{,}4048}{\Sigma_{tot}\sqrt{3}},$$

where 2.4048 — the first root of the Bessel function of zero order. The critical radius of the cylinder made of ^{298}Fl is equal 1.524 cm, made of ^{342}126 is equal 1.426.

Under section fission ^{298}Fl and ^{342}126 can be understood as the total cross section of inelastic processes. With the fast neutron flux density $\varphi \sim 10^{15}$ cm^{-2} s^{-1} (Typical for the fast reactor), $\Sigma_f \approx \Sigma_r \approx 0.238$ cm^{-1}, $E_f = 580$ MeV, specific volume energy release of fuel from the ^{342}126 is equal

$$q_V = \Sigma_f \Phi E_f = 22.105 \text{ кW/cm}^3$$

or about 1.167 MW in critically ball without reflector or 791.5 кW/кg. For ^{298}Fl-fuel: $q_V = 16.004$ кW/cm^3 or 1.023 MW in critically ball or 685.8 кW/кg. (Specific energy release accompanying the radioactive decay of nuclei ^{238}Pu is equal 0.56 кW/кg; for ^{232}U taking into account the complete decay chain is equal 8.27 кW/кg.) Energy release values are underestimated. The shell structure of fission products was not taken into account.

Elements with $Z \geq 112$ under normal conditions are gases [8], i.e., parameter C with two to three orders of magnitude smaller, but φ is much higher. As a result, q_V can be one or two orders of magnitude higher than the above estimates. There are projects of reactors with gas fuel [9, 10]. The fuel can be stationary or circulating. According to the developers' estimates, such reactors can be a powerful source of energy [9].

In the presence of a neutron reflector (made of any material other than super-heavy elements), the critical radius of the core can be noticeably smaller.

The estimates are made in the simplest approximation, which is poorly applicable for reactors of such small dimensions with a relatively high absorption cross section.

5. Discussion. Features of the fission of super-heavy nuclei

Super-heavy nuclear fission reactors are characterized by a good neutron balance (neutron yield) compared to traditional U or Pu-fueled reactors. This is due to the strong congestion of super-heavy nuclei with neutrons (with respect to $N = Z$). Since in the output channel of any nuclear reaction there is a high probability of finding doubly magic nuclei (if the neutron and proton balance of the system allows it), the most expected is asymmetric fission of super-heavy elements, and one of the fragments can be a doubly magic lead-208 nucleus with optimal N / Z or its isotope with a larger number of neutrons. An excited ^{208}Pb fragment is unlikely to emit a large number of neutrons. On the other hand, another fragment (with double fission of the nucleus ^{342}126 by a neutron) - ^{125}Ru (or the lighter isotope of ruthenium), heavily overloaded with neutrons, is located near the neutron stability boundary ($N = 83$). ^{100}Ru is stable. To move to the region of β-stable nuclei, ^{125}Ru must emit 25 neutrons or electrons. According to Glendenin's Law, fission fragments are equally overloaded with neutrons. But in the fission of Th, U, or Pu nuclei, the appearance of doubly magic fragments is unlikely. The ^{4}He yield upon fission of ^{235}U by a thermal neutron is 0.19%, and for ^{132}Sn - 0.70%. Nuclei of ^{16}O, ^{40}Ca, ^{48}Ca, ^{56}Ni, ^{208}Pb are absent from the fission products [11]. The yield of heavier tin isotopes than twice magic is small: ^{133}Sn - 0.16%; ^{134}Sn - 0.02%, and ^{135}Sn and ^{136}Sn - negligible [11]. Thus, Glendenin's law works well in the absence of doubly magic nuclei in the number of fragments, i.e. when describing the fission of super-heavy elements.

It is reasonable to assume that the yield of lead isotopes, heavier than twice magic, during fission of ^{342}126 (or ^{298}Fl) will also be small. In any case, during the fission of ^{342}126, theoretically, up to 25 neutrons can be emitted (the total congestion of the fragments with neutrons), and Glendenin's law can be violated due to the presence of doubly magic lead in the output channel.

In the fission of heavy nuclei, prompt and delayed neutrons are distinguished. The nuclei located near the neutron stability boundary are β^--active. They are characterized by a short half-life and high β^--decay energy. Such nuclei emit delayed neutrons [6]. As one approaches the neutron stability boundary, the probability of emission of two [12, 13] and three delayed neutrons [14] increases. Thus, the effective fraction of delayed neutrons produced during the fission of the ^{342}126 nucleus can exceed the corresponding values for the nuclei of thorium, uranium, and plutonium. This will make it possible to operate a nuclear reactor with super heavy fuel. The emission of neutrons and β^--decay of nuclei-fragments contributes to the approach to the condition of equilibrium of forces in the nucleus. Thus, when a single ^{235}U or ^{239}Pu nucleus is fission by a neutron with a kinetic energy of 2 MeV (the average energy of neutrons produced during fission), $\nu_f \approx 2.671$ or 3.147 neutrons, respectively, are produced on average, and fission fragments, on average, experience from 6 to 8 β^--decays (3–4 each of the fragments), while emitting from 6 to 8 electrons. The ratio of the number of β^--decays to the number of prompt neutrons produced is 1.905 ... 3.026. The total congestion of fission fragments with neutrons can be estimated at 7–11 (ν_f plus the total number of β^--decays of fragments). When ^{342}126 is fission by a neutron, the total congestion of the fission fragments with neutrons is 25, that is, the fragments can emit 25 neutrons or electrons. If the ratio of the number of electrons to the number of neutrons (1.905 ... 3.026) is extrapolated to the region of super-heavy elements (this is a rough extrapolation), then by fission of ^{342}126 by a neutron from 14 to 16 electrons and from 9 to 11 neutrons should be produced. This should be considered a lower estimate of ν_f, since fission fragments of super-heavy nuclei are closer to the neutron stability limit (than fragments of fission of uranium or plutonium), and can emit a larger number of prompt neutrons. On the other hand, such fragments have a much higher probability of the production of pairs and triples of delayed neutrons after β^--decays, which, in turn, will lead to a decrease in the number of β^--decays with an increase in the effective delayed neutron fraction. Thus, the number of neutrons produced by nuclear fission ^{342}126 will significantly exceed 9–11 and theoretically can reach 25.

A large number of excess neutrons (in relation to the maintenance of the fission chain reaction) opens up great opportunities for their utilization, for example, for the transmutation of radioactive waste (which is important, without preliminary separation of isotopes), placed in a blanket or in the core of such a reactor, for example, as a diluent fuel: long-lived minor actinides and fission products present in the spent nuclear fuel of conventional power reactors. Fission of minor actinides (Np, Am, Cm) in the hard neutron spectrum of a fast reactor will release additional energy - about 200 MeV per fission of one waste nucleus.

The even-even nuclides ^{298}Fl and ^{342}126 should formally be classified as fertile materials, not fissile. However, the threshold in the energy dependence of the fission cross section of these nuclei may turn out to be much less than the threshold for "traditional" fertile material nuclides (^{232}Th and ^{238}U), and with a large number of neutrons produced, the probability of fission of ^{298}Fl and ^{342}126 will increase significantly. The fission parameter Z^2 / A of the ^{342}126 nucleus is 46.4, and that of the ^{298}Fl nucleus is 43.6. According to the hydrodynamic model, nuclei with $Z^2 / A > 45 \dots 49$ should fission spontaneously: there is no fission threshold. The decay of these nuclei is prevented by the stabilization factors listed above.

6. Conclusion

The design of compact power plants with an abnormally high density of energy release, operating in the mode of a nuclear fission reactor or a radionuclide source "switched on" by external neutron radiation, will become possible only when relatively long-lived super-heavy elements are obtained. Nuclear reactions for their production are not yet known.

With the development of scientific and technological progress, developments that seem fiction are becoming reality. The synthesis of double-magic nuclei of flerovium-198 will allow in the near future considering fuel based on it as realistic. More distant prospects are associated with the use of fuel based on the nuclei of element 126. Hybrid mini-installations that implement the fission reaction and the radioactive decay of super-heavy elements can be in demand in space exploration. They can be used on spacecraft, orbital stations, and space bases on the planets being developed. The development of such mini-hybrid power plants with a small core size (less than 2.5 cm) and high energy density (from 0.7 to 70 MW / kg of fuel) is relevant for the development of the planets of the solar system. The high cost of nuclear fuel based on super-heavy elements is not conducive to the use of such facilities for generating energy on Earth.

The presented estimates of the minimum critical masses and radii of reactors with fuel based on super-heavy elements are somewhat overestimated and correspond to the upper estimate. Energy release values are underestimated.

References

[1] IAEA-TECDOC-1451 2005 *Innovative Small and Medium Sized Reactors: Design Features, Safety Approaches and R&D Trends* (IAEA, Vienna, Austria) https://www-pub.iaea.org/MTCD/Publications/PDF/TE_1451_web.pdf

[2] IAEA-TECDOC-1485 2006 *Status of Innovative Small and Medium Sized Reactor Designs 2005* (IAEA, Vienna, Austria) https://www-pub.iaea.org/MTCD/Publications/PDF/te_1536_web.pdf

[3] IAEA-TECDOC-1536 2007 Status of Small Reactor Designs Without On-site Refueling (IAEA, Vienna, Austria) https://www-pub.iaea.org/MTCD/Publications/PDF/te_1536_web.pdf

[4] Okunev V S 2019 *AIP Conference Proceedings* 2171, 040001; https://doi.org/10.1063/1.5133187

[5] Okunev V S 2018 *IOP Conf. Series: Materials Science and Engineering* **468** 012012

[6] Karnaukhov V A and Petrov L A 1981 *Nuclei Removed from Beta Stability Line* (Moscow: Energoizdat)

[7] Okunev V S 2015 *The basis of applied nuclear physics and introduction into physics of nuclear reactors. Ser.: Physics in Technical University* (2th ed., Bauman Moscow State Technical University)

[8] Pershina V 2014 *The Chemistry of Superheavy Elements* (In book: Theoretical Chemistry of the Heaviest Elements, ed M Schädel and D Shaughnessy, Second Edition, Springer: Heidelberg, New York, Dordrecht, London) pp 135-240

[9] Postnikov N S 2003 *Atomnaya Energia (Atomic energy)* **95** (**5**) 330-8

[10] Raevsky I I and Smirnov Yu V 1983 *Atomnaya Technika za Rubezhom (Atomic engineering abroad)* **2** 8-12

[11] The JEFF Nuclear Data Library 2011 http://www.oecd-nea.org/dbdata/jeff/

[12] Azuma R E et al 1979 *Phys. Rev. Lett.* **43** (**22**) 1652–54

[13] Gol'danskii V I 1961 Journal of Experimental and Theoretical Physics **12** (**2**) 348–51. http://www.jetp.ac.ru/cgi-bin/dn/e_012_02_0348.pdf

[14] Azuma R E et al 1980 *Physics Letters B* **96** (**1**, **2**) 31–4

Prospects for the development of the concept of a safe nuclear reactor BREST of maximum limiting power for nuclear energetics of the middle of the XXI century

V S Okunev

Department of Physics, Bauman Moscow State Technical University,
2-ya Baumanskaya ul. 5, Moscow, 105005, Russia.
E-mail: okunevvs@bmstu.ru

Abstract. In Russia (JSC NIKIET) a project of a fast reactor BREST-1200 is being developed. The article analyzes the safety of the BREST reactor with maximum limiting power (BREST-2000, -2400). In BREST reactors with an electric power of 1200 MW and more, the void reactivity effect of can be positive. The most dangerous scenario is associated with the entrainment of bubbles into the central part of the core (with a volume fraction of bubbles of 50 ... 60%). In this case, the magnitude of the void reactivity effect is several times higher than the effective fraction of delayed neutrons β. For the BREST-2400 reactor, the maximum void reactivity effect is about 7 β (for BREST-2000, about 6 β). When other ATWS are implemented against the background of erosion and corrosion of structural steel, the total introduced reactivity can also exceed β, which will lead to an accident. To ensure safety, it is necessary to adjust the design solutions. When used as a coolant, lead extracted from thorium ores (with a concentration of ^{208}Pb isotope from 75 to 98%) and tungsten coatings of fuel claddings, the void reactivity effect is negative even in the BREST-2400 reactor, and when switching to a new cermet fuel (0.8 MN - 0.2 U) it is easy to ensure the safety of the reactor.

1. Introduction

Promising energy sources capable of meeting the needs of mankind for the long term are based on the transmutation of atomic nuclei. Fusion of light nuclei is energetically beneficial. The principle of operation of fusion reactors is based on this. The decay of heavy atomic nuclei into lighter ones is also energetically beneficial. The principle of operation of fission reactors and radionuclide sources is based on this. Among the power plants based on nuclear transmutation, the most developed and industrialized are fission reactors. In addition to large-scale energy production, additional requirements are imposed on new generation energy sources. They are well known. These are economic efficiency (in the absence of competitors, this requirement can be formulated as the production of useful energy, that is, the production of energy must exceed the energy consumption of the installation), safety, and fuel availability for the long term. For sources based on the transmutation of atomic nuclei, the safety requirement includes nuclear and radiation safety (exclusion of unacceptable emissions of radioactive substances outside the nuclear power plant), non-proliferation of technologies and materials that can be used in the production of weapons of mass destruction, transmutation of radioactive waste.

Nuclear energetics is one of the youngest and fastest growing industries in the world economy. The development of nuclear energetics is caused by the growing demand for energy of mankind with limited non-renewable resources, uneven distribution of organic (hydrocarbon) fuel in the earth's crust.

Content from this work may be used under the terms of the Creative Commons Attribution 3.0 licence. Any further distribution of this work must maintain attribution to the author(s) and the title of the work, journal citation and DOI.
Published under licence by IOP Publishing Ltd

Compared to other energy sources, nuclear fuel has a millions of times greater energy concentration. The prospects for the development of energetics in a number of countries around the world provide for the gradual involvement and increase in the share of nuclear power plants in electricity production. (Electricity is the most convenient form of energy for the consumer.) In the world, under the auspices of the IAEA, the concepts of fourth-generation nuclear reactors ("Generation IV" or "GIF") are being developed that satisfy most of the listed requirements [1]. In 2006, Russia joined the Generation IV International Forum, whose members are developing several types of fourth-generation nuclear reactors. Among the concepts of fourth-generation reactors are fast reactors with sodium and heavy coolants (lead or eutectic lead-bismuth alloy). Fast reactors with sodium or lead coolant are considered as the main world concepts of liquid metal fast reactor [2-5].

For large-scale nuclear energetics, a lead coolant is more preferable due to the high cost of bismuth and the possibility of producing polonium-210 from bismuth-209. In Russia, the USA and France, the possibility of using an alloy of lead and bismuth as a coolant is being studied [6]. In total, there are 79 experimental installations for research of reactors with sodium coolant, 72 installations for research of reactors with lead coolant, and 14 installations for research of reactors with Pb-Bi coolant [6]. Belgium, the Czech Republic, the European Union as a whole, Italy, Spain and Sweden are showing interest in lead-cooled reactors. China, France, Germany, Japan, Republic of Korea, Latvia, Russia, and the USA are showing interest in both sodium and lead coolants [6].

2. Problem

Lead-cooled and mixed mononitride (MN) fueled fast reactors are attractive because they can meet all of the above requirements for the energy sources of the future. Projects of such reactors have been developed in Russia. This is a series of projects "BREST" ("fast reactor of natural safety") [7]. The technical design of the BREST-OD-300 reactor [8], the conceptual designs of the 1200 [9], and the technical proposal for the BREST-2400 [10, 11] have been developed. However, the BREST concept does not address all security issues. With an increase in the reactor power from 600 MW (e) and higher, the void reactivity effect (VRE) is positive and exceeds the effective delayed neutron fraction (β). This can lead to accidents and is unacceptable for new generation fast reactors. The implementation of a positive VRE is possible when the central part of the core is drained, for example, when bubbles appear (when the fuel-element cladding is depressurized). The involvement of bubbles in the core during depressurization of the steam generator tubes is unlikely, but it is not excluded in reactors with a high proportion of natural circulation of the coolant (BREST-2400).

The electric power of 2,400 MW can be considered the maximum limiting power for a power unit with a nuclear reactor. By limiting power we mean the smallest power at which the reactor operates stably and without damage. In other words, this is the maximum long-term power of the reactor. (This definition is borrowed from theory of acoustics.) Loss of stability is possible for LMFR electrical power of about 2000 MW. This value is close to the power limit. Moreover, further power increases may not be economically viable. The cost of produced energy decreases with increasing reactor power only up to a certain limit. This limit is about 2000 ... 2400 MW (e). When shutting down a high-power reactor, difficulties arise for energy consumers associated with the need to replace this power.

The main problem of the research is to analyze the safety of the high-power BREST reactor (BREST-2000 and -2400) and adjust the design solutions based on such an analysis. The traditional (cylindrical) geometric shape of the core was considered. The core of BREST-1200, -2000, -2400 consists of three profiling zones. The fuel assemblies and fuel pins of the BREST-OD-300 reactor are used. The diameter of the fuel pellet in the zones is 7.9; 8.3 and 9.0 mm, respectively. The height of the fuel column is 0.95 m. At the first stage of the research, the possibility of improving the safety of the BREST-1200 reactor was analyzed using a new cermet fuel (with the addition of uranium nanopowder up to 40% by weight [12]), a coolant based on lead extracted from thorium ores, and tungsten coatings of the fuel element cladding. Optimization of the core characteristics makes it possible to obtain a negative VRE not only when draining the entire BREST-1200 reactor, but also when draining the central part of the core. However, the most dangerous scenario for the implementation of VRE is associated with the entrainment of bubbles in the central part of the core (with a volume fraction of bubbles of 50 ... 60%). In this case, for BREST-1200 VRE is about 5 β. In what follows, we will consider such a scenario. The proposed innovations make it possible to ensure the internal safety of the BREST-1200 reactor. The reactor is ATWS resistant. VRE is negative (even in the most dangerous scenario of its implementation). The results of first stage have been published (see, for example, [12, 13] and other works of the author).

At the second stage, the possibility of achieving the internal safety of the BREST reactor of maximum power is analyzed. This article is devoted to this problem. Note that with a change in the lattice of fuel elements or fuel assemblies, the isotopic composition of the coolant, the use of uranium nanopowder additives to the fuel, the maximum dependence of the VRE on the fraction of bubbles in the central part of the core changes. For example, when using natural lead, the maximum corresponds to a 60% decrease in lead density. When using lead, thorium ores, the maximum is observed with a decrease in density by 20%.

3. Safety analysis: results and discussion

3.1. ATWS modes

Safety analysis involves the study of the behavior of the reactor in all emergency modes that are not deterministically excluded. First of all, emergency situations are considered, accompanied by a failure of emergency protection (anticipated transients without scram - ATWS modes). All ATWS modes can be combined into a small number of groups. The classification feature is a disturbance that initiates an emergency mode (change in flow rate, change in inlet coolant temperature, reactivity input, drainage of the core or its central part). There are only five groups of ATWS modes. Among them: LOF WS (loss of flow without scram), TOP WS (transient overpower without scram), LOHS WS (loss of heat sink without scram), OVC WS (overcooling accident without scram) and LOCA WS (loss of coolant accident without scram). This classification was first used by specialists from Idaho National Laboratory when analyzing ATWS for the EBR-II reactor (see, for example, [14] and other articles published in Nuclear Engineering and Design, 1987, no. 101).

The BREST-1200 reactor is considered as a basic option for the power industry of the future. Let's carry out a safety analysis for an abnormally high power reactor BREST-2400. The author uses the MCU code [15] and the many times modernized author's codes Dragon-M and FRISS-2D [12].

3.2. ATWS analysis

The approximate mathematical model of ATWS analysis is based on the solution of three groups of equations: a system of equations of point neutron kinetics with a six-group description of delayed neutrons (for the relationship between power and reactivity); non-stationary heat equation and coolant energy equation (to determine the relationship between power and temperature changes); conditions for the balance of reactivity (for the connection of reactivity with temperature changes) [12].

When simulating the LOF WS, it was assumed that all main circulation pumps were de-energized simultaneously. The LOHS WS was triggered by a failure of the secondary pumps. For a conservative assessment of internal self-protection, it is assumed that a coolant heated to the outlet temperature is supplied to the entrance to the core. The TOP WS is initiated by introducing the maximum possible positive reactivity (about β) linearly with time. It was assumed that the OVC WS could be caused by an increase in flow rate or a decrease in the coolant temperature at the inlet to the core. Both of these events do not pose a serious hazard to the reactor in question. The decrease in the flow rate in the LOF WS mode is determined by the difference in the levels of the ascending and descending lead sections in the nominal operating mode of the BREST reactor. Suppose that when all the main circulation pumps are de-energized, the coolant flow rate decreases to the G_{nc} level determined by natural circulation.

Figure 1 (a) shows the time dependence of the maximum fuel-element cladding temperature T_{cl} (t) in the LOF WS. After the lead transport time has elapsed along the primary loop, "hot" coolant will enter the core, which will lead to expansion of the base plate (collector) and, as a consequence, the introduction of negative reactivity, which in absolute value exceeds other components of the temperature coefficient of reactivity. In this case, the temperatures increase slightly (local maximum); however, the effect of negative feedback on the lead temperature at the inlet to the core ultimately leads to a decrease in the maximum temperatures in the reactor. The maximum fuel temperature does not exceed the permissible values. At temperatures above 1800 ... 2000 K, a significant increase in gas evolution in MN-fuel and fuel decomposition are possible. However, such temperatures are practically not reached in emergency modes. Fuel cladding is a critical element from the point of view of the possibility of ensuring internal self-protection.

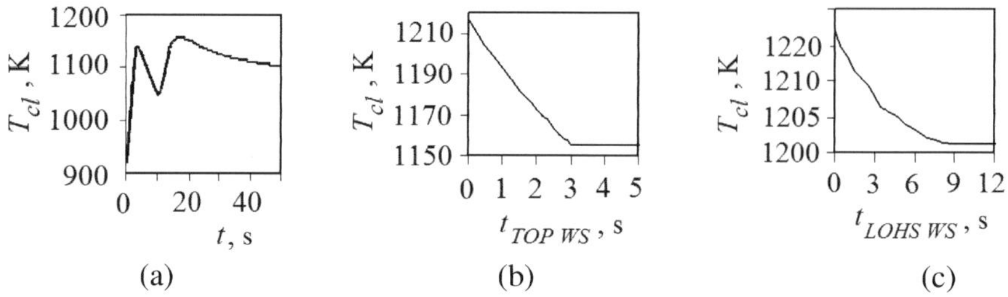

Figure 1. ATWS analysis

Studies of the dependence of the maximum temperatures on the time t_G of the decrease in lead consumption show that with an increase in t_G from 3 to 10 ... 15 s, the maximum temperatures decrease significantly, and the maximum shifts in time to the region of the steady state. A further increase in t_G has practically no effect on the safety functional (maximal temperatures and power of core). It should be noted that the scenario for changing the maximum fuel temperature in LOF WS is different for different G_{nc} values. (Depending on the G_{nc}, the role of the Doppler reactivity coefficient can be reversed.) In general, to reduce the fuel temperature in the transient mode at large G_{nc} (in BREST-2400 G_{nc} = 37% of the nominal), it is necessary to increase the negative Doppler reactivity coefficient in absolute value, at small G_{nc} (for example, in BREST-OD-300 G_{nc} = 6% of the nominal) it is necessary to reduce this coefficient in modulus. In the steady-state mode of natural circulation, the situation is the opposite: to reduce the temperature of the fuel at large G_{nc}, it is necessary to decrease the negative Doppler coefficient in absolute value; at small G_{nc}, it must be increased.

Elementary analysis shows that reactors of the BREST-1200 [12, 13], -2000 and -2400 types can achieve the properties of internal safety even in the absence of passive safety systems.

3.3. Analysis of the most dangerous combinations of emergency modes

Due to the mutual neutralization of disturbances that initiate emergency modes, the simultaneous superposition of the four modes LOF WS, LOHS WS, TOP WS and OVC WS is less dangerous than superposition any three of them. From the point of view of increasing the temperature of the fuel element cladding, the LOF WS mode is the most dangerous. The most dangerous combinations of emergency modes can be considered the superposition of three of the above, and one of them is LOF WS: (LOF + TOP + LOHS) WS or (LOF + TOP + OVC) WS. The uncertainty of the delay time of one or another process in the case of non-simultaneous superposition (LOF + TOP + LOHS) WS is illustrated in Figure 1 (b, c). Here $t_{\text{TOP WS}}$ and $t_{\text{LOHS WS}}$ are the latency times of the TOP WS and LOHS WS processes, respectively. When (LOF + TOP + LOHS) WS is applied, the sheath temperature is practically independent of the flow rate decrease time in the interval 2 … 7 s. Similar results were obtained for any combination of LOF WS, TOP WS, LOHS WS, and OVC WS.

3.4. VRE analysis

It is known that VRE is characterized by strong spatial dependence. For this reason, with an increase in the reactor power (due to an increase in the size of the core), these reactivity effects increase. For this reason, the analysis of VRE requires special attention.

As a rule, by minimizing VRE in the problem of optimizing the core of the BREST-1200, -2000 and -2400 reactors, it is possible to obtain an arrangement of core for which the VRE is negative even in the most dangerous scenario of its implementation (drainage of the central part of the core). The most dangerous is the decrease in the density of the lead coolant in the central part of the core by 50 ... 60%. In a reactor of an infinitely large radius, VRE (the absence of neutron leakage through the side surface is postulated) reaches 7 β (β is the effective delayed neutron fraction). In this scenario, for a reactor of the type BREST-2400 VRE is more than 6 β. In this case, the presence of bubbles (lead emulsion) in the core is postulated. If these bubbles contain hydrogen, then the VRE will be noticeably higher. (In cause of depressurization of the heat exchanger tubes in a two-circuit energy conversion scheme.) VRE becomes negative only when the density of the lead coolant decreases to 2 g / cm^3 and below (including with complete drainage: in the presence of a large bubble in the center of the core).

The use of a coolant based on lead extracted from thorium ores with a concentration of ^{208}Pb isotope allows one to reduce the VRE in a reactor of infinite radius to a value of about 1 β. The use of tungsten coatings of the cladding of fuel elements allows one to minimize the VRE to zero and negative values (depending on the thickness of the tungsten coating). The maximum VRE corresponds to a decrease in lead density by 20 ... 30%.

Thus, to improve the safety of the high-power BREST reactor, it is necessary to switch from natural lead to lead extracted from thorium ores.

3.5. "Corrosive" absorber

Erosion and corrosion of structural steels in liquid lead is one of the major problems. If in BREST-OD-300 the removal of steel only from the core is 39 kg / year, then for BREST-1200, -2000 and -2400 this value will increase by 4, 6.7 and 8 times, respectively. In terms of reliability and economics, the need to minimize the rate of erosion and corrosion is obvious. From a safety point of view, these processes, accompanied by the removal of structural steel (a "parasitic" neutron absorber) from the core and the deposition of erosion and corrosion products outside the core (in cold sections of the duct or in filters), is the input of positive reactivity. This phenomenon must be taken into account in neutron balance and reactivity balance. In BREST reactors, erosion and corrosion of structural steels in the core helps to minimize the positive reactivity margin for fuel burnup and, from this point of view, plays a positive role. (A "corrosive" absorber is, to a certain extent, an analogue of burnable absorbers in thermal reactors.) The problem of minimizing the reactivity margin for fuel burnup can be solved due to natural processes of corrosion and erosion of structural steels and their gradual removal from the reactor core [13].

At the same time, when steel is removed from the core, positive reactivity is introduced. Thus, the destruction of even a small part of the fuel claddings as a result of superposition (LOF + LOHS + TOP) WS can lead to an injection of reactivity greater than 1 β. (The density of steel is approximately 1.5 times less than that of lead; therefore, fragments of destroyed claddings will float even in the absence of natural circulation.) The introduction of reactivity greater than β can be excluded if the destruction of the cladding of the fuel elements is excluded, for example, when using tungsten coatings of the cladding and a denser and more heat-conducting fuel. For example, to reduce the maximum temperatures in ATWS, it is possible to propose the use of pellet fuel based on MN micro grains and uranium nanopowder (up to 20% by volume), filling the pores between micro grains [12].

4. Conclusion

For existing high-power BREST reactor designs, inherent safety is not fully ensured. This is due to the positive VRE, which is several times higher than the effective fraction of delayed neutrons (up to 7 β).

Erosion and corrosion of structural steel can lead to an additional reactivity input of more than 1 β and the maximum temperatures exceeding the permissible values in complex combinations of ATWS modes or even to a reactivity accident ("neutron burst").

To ensure safety, it is necessary to adjust the design solutions for high-power BREST reactors. It is associated with the transition to a coolant based on lead extracted from thorium ores, with the transition to fuel based on MN micro grains and uranium metal nanopowder placed between the grains, and with tungsten coatings of the fuel element cladding. Tungsten cladding of the fuel elements will make it possible to soften the requirements for the quality of the coolant. The use of lead with a high concentration of ^{208}Pb helps to reduce the risk of reactivity accidents due to a smaller neutron absorption cross section (the prompt neutron lifetime is maximum). With such a correction, the VRE is negative even for a reactor of infinite power (infinite radius of the core). The ATWS modes are terminated without failures even in the absence of passive safety systems for the BREST-2400 reactor.

With such an adjustment, reactors of the BREST-1200 [12], BREST-2000 and -2400 types can become the basis for safe large-scale nuclear power, capable of meeting humanity's energy needs in the long term. Thus, in the case of adjusting the design solutions, safety will not be a factor limiting the maximum power for BREST reactors.

References

[1] OECD 2020 *GIF 2019 Annual Report,* NEA **7527**, (OECD Nuclear Energy Agency for the Generation IV International Forum, Paris, France) https://www.gen-4.org/gif/jcms/c_119025/gif-2019-annual-report
[2] IAEA-TECDOC-1691 2013 *Status of Fast Reactor Research and Technology Development* (IAEA, Vienna, Austria) https://www-pub.iaea.org/MTCD/Publications/PDF/te_1691_web.pdf
[3] FR13 2015 *Fast Reactors and Related Fuel Cycles: Safe Technologies and Sustainable Scenarios* (IAEA, Vienna, Austria) https://www-pub.iaea.org/MTCD/Publications/PDF/Pub1665Vol1Web-68441300.pdf
[4] IAEA 2013 *Status of Innovative Fast Reactor Designs and Concepts, Information Booklet* (IAEA, Vienna, Austria)
[5] FR17 2017 *Fast Reactors and Related Fuel Cycles: Next Generation Nuclear Systems for Sustainable Development* (IAEA, Vienna, Austria) https://www-pub.iaea.org/MTCD/Publications/PDF/STIPUB1836web.pdf
[6] IAEA Nuclear Energy Series 2018 *Experimental Facilities in Support of Liquid Metal Cooled Fast Neutron Systems* NP-T-1.15 STI/PUB/1806 978-92-0-101018-6 (IAEA, Vienna, Austria) https://www-pub.iaea.org/MTCD/Publications/PDF/P1806_Web.pdf
[7] Orlov V V et al 1992 *Atomnaya Energia (Atomic energy)* **72** (**4**) 317–29
[8] Lemekhov V V et al 2018 *Innovative Design and Technologies of Nuclear Power* (Moscow, Russia: NIKIET) p 55
[9] Orlov V V et al 2003 11-th Int. Conf. on Nuclear Engineering: ICONE-11 (JSME/ASME, Shinjuku, Tokio, Japan, CD-ROM file)
[10] Okunev V S 2002 *Physical problems of effective and safe use of nuclear materials: 12th Seminar on Problems of Reactor Physics*. (Moscow, MEPhI) p 168–70
[11] Okunev V S 2002 *Izvestiya RAN. Energy* **6** 42–51
[12] Okunev V S 2019 *AIP Conference Proceedings* 2195 020012 https://doi.org/10.1063/1.5140112
[13] Okunev V S 2019 *Accident Tolerant Materials for LMFR* (IntechOpen, DOI: 10.5772/intechopen.90703) https://www.intechopen.com/online-first/accident-tolerant-materials-for-lmfr
[14] Lehto W K et al 1987 *Nucl. Eng. and Design* **101** 35–44. DOI: 10.1016/0029-5493(87) 90148-8
[15] Gomin E A et al 2000 *Advanced Monte Carlo on Radiation Physics, Particle Transport Simulation and Applications* (Monte Carlo, Lisbon, Portugal) p 2003–4

Efficiency evaluation of sustainable energy development financing mechanism

Vik A Borovkova[1], **Val A Borovkova**[2]

[1] International Banking Institute, 60 Nevsky pr., Saint-Petersburg, 191023, Russia
[2] Peter the Great St.Petersburg Polytechnic University, 29 Politechnicheskaya st., Saint Petersburg, 195251, Russia

E-mail: vaborovkova@mail.ru

Abstract. Transition to the sustainable energy development and modernization model requires adequate financing, and, consequently, generation of specific mechanisms, instruments, institutions, and sources of financing. All this predetermines the need to study methodological approaches to development and optimal implementation of financing mechanisms for sustainable energy development, and approaches to evaluation of their efficiency. This article clarifies the concept of the sustainable energy development financing mechanism (SEDFM), defines its specificities, proposes a system of indicators and methodology for efficiency evaluation of sustainable energy development financing mechanisms.

1. Introduction

Being a fundamental and life-supporting industry closely interconnected with all sectors of the economy and social sphere, the power industry plays an essential role in achieving the country's sustainable development goals. The power sector's contribution to the country's economy exceeds 3% of GDP [1]. Federal budget expenditures in 2018-2020 on the Energy Efficiency and Development Program amount almost to RUB 31 bln [2], investments in the domestic power sector are growing on the whole [3] and energy efficiency in the fuel and energy complex, energy saving, energy consumption efficiency is increasing [4].

Studies have shown that, in the presence of many papers by foreign and domestic scientists dedicated to sustainable development financing, there are only few studies investigating the issues of SEDFM use efficiency evaluation and there is no unified approach to SEDFM efficiency evaluation. As aggregate efficiency data, the efficiency level is most often used as compared to incurred expenditures determining the result that various approaches and methods of strategic analysis, ratio analysis etc. are applied to SEDFM evaluation. Despite their high theoretic and practical value, in terms of SEDFM efficiency evaluation, some of them require improvement, and often omit power industry peculiarities.

2. Results and Discussion

2.1. Peculiarities of SEDFM Efficiency Evaluation

Socio-economic, organizational and managerial, information- and communication-related, technical and technological, and other peculiarities of the power industry significantly impact the formation of mechanisms for its sustainable development financing and determine the specifics of their functional efficiency evaluation.

In addition, the financing of power sector enterprises has a number of peculiarities:

- Need for large investments;
- Diversified model for the industry's financial support;
- High capital intensity of power facilities;
- Long payback period of capital investments (normally, over 10-15 years);
- Presence of a natural monopoly;
- Continued government regulation;
- Diversification of investment risks by the industry's sectors.

Content from this work may be used under the terms of the Creative Commons Attribution 3.0 licence. Any further distribution of this work must maintain attribution to the author(s) and the title of the work, journal citation and DOI.
Published under licence by IOP Publishing Ltd

These specifics were taken into account by the authors in forming a system of indicators for SEDFM efficiency evaluation.

2.2. Integrated Approach to SEDFM Efficiency Evaluation

For the comprehensive evaluation of SEDFM efficiency, the authors suggest using an integrated approach characterized by: consistency; availability of background information; simplicity and low labor intensity; practical orientation; consideration of internal/external environment factors and current situation, available resources; consideration of industry specifics of SEDFM. The integrated approach unites systemic, structural, process-related, factorial, situational, resource and reframing approaches. In this approach there are several methodological approaches to SEDFM efficiency evaluation: general evaluation of SEDFM efficiency; efficiency evaluation of individual SEDFM elements; efficiency evaluation of the relationship between individual SEDFM elements, evaluation of SEDFM implementation results; evaluation of SEDFM implementation implications; SEDFM efficiency evaluation in relation to each project participant.

Here we should dwell on the methodology for general SEDFM efficiency evaluation, without division into subsystems (figure 1).

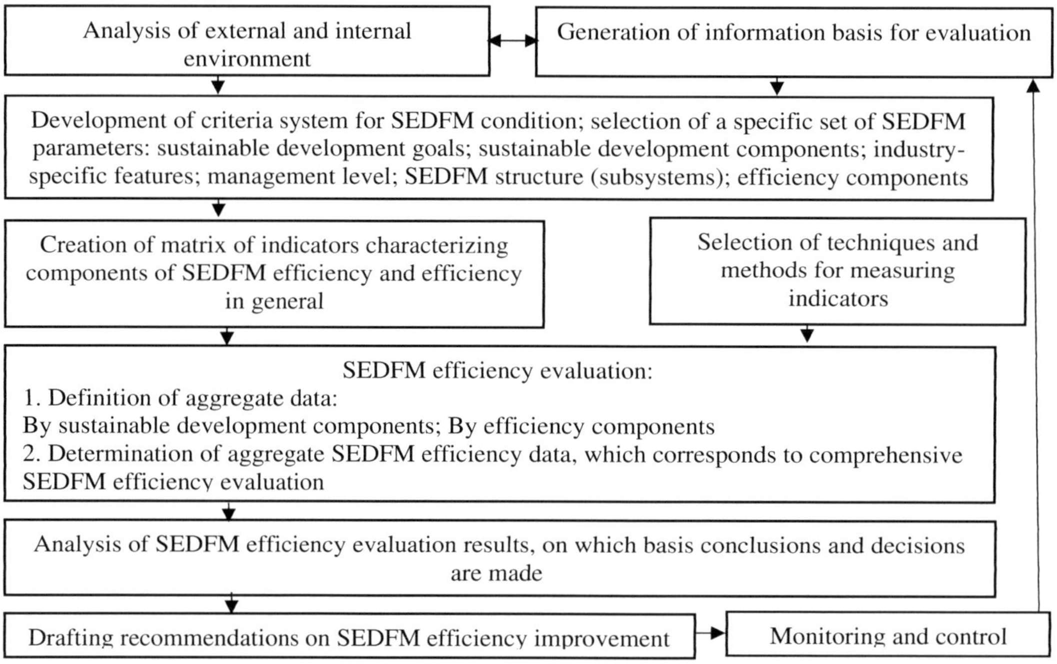

Figure 1. Algorithm for SEDFM efficiency evaluation.

In the said approach, the following provisions are essential.

1. The authors consider SEDFM as a system of economic, organizational and other forms, methods, tools, legislative norms, principles and methods for organization of financial relations, through which financial resources are generated, raised, transformed, accumulated, distributed and applied in order to achieve the power industry's sustainable development and competitiveness. SEDFM consists of three subsystems: the subsystem of financing tools and technologies, support systems and the ideological subsystem, each of them including a package of elements. In addition to financial benefits, SEDFM should deliver social and environmental benefits.
2. SEDFM efficiency should be evaluated taking into account national and industry characteristics, the specifics of each particular organization and particular project. One should take into account the level, at which SEDFM itself is generated and implemented (global, regional, national (federal, sub-federal, local), industry, organization or household), the type of SEDFM, the system of financial tools serving the basis for financing technology. The evaluation process must factor in the interests of all internal and external stakeholders (government, owners, investors, consumers, etc).

3. For evaluation purposes, the authors take into account 17 new global sustainable development goals (SDGs) [5], four sustainable development components: economic growth, environmental sustainability, social sustainability, digitalization. Peculiarities of the power sector and increased attention to the problems of introduction of new science-intensive digital technologies, environmental safety and social responsibility of business predetermine the need to include not only three classical SD components (economic growth, environment and society), but also a fourth one – digitalization – into the sustainable development model. The SEDFM system of indicators is relevant, maximally informative, measurable and compliant with SDGs, it is generated taking into account the principles of balance and reasonably limited evaluation.
4. The following factors are taken into account: the package of 244 global SDG indicators approved by the UN General Assembly; indicators laid down in 2035 Russian Federation Energy Strategy; indicators of national projects, the system of economic and energy indicators, and efficiency indicators.
5. Significant efficiency components to be included in calculation of the overall efficiency indicator, their content and evaluation are selected by experts, taking into account SEDFM specifics. It is proposed to evaluate overall SEDFM efficiency, its components and indicators using a 10-score scale (due to size variations of indicators): 10 – highly efficient SEDFM, 9 – very good SEDFM, 8 – good SEDFM efficiency, 7 – SEDFM efficiency above average, 6 – average SEDFM efficiency, 5 – SEDFM efficiency below average, 4 – low SEDFM efficiency, 3 – very low SEDFM efficiency, 2 – insignificant SEDFM efficiency, 1 – inefficient SEDFM, 0 – extremely inefficient SEDFM.
6. As the main criteria for comprehensive evaluation of overall SEDFM efficiency, the authors propose to take into account 7 efficiency components being the most significant for adequate presentation of SEDFM functioning efficiency at the country level: economic, environmental, social, technical and technological, organizational and managerial, risk-related, marketing and a package of indicators corresponding thereto.
7. SEDFM efficiency calculation comprises the following stages: 1) selection of SEDFM indicators; 2) rating evaluation of SEDFM indicators; each of selected indicators (K) is assigned certain points; 3) evaluation of each SD component based on individual SEDFM indicators (I_{er}, I_{ec}, I_s, I_d); 4) calculation of particular integrated indicators of efficiency components evaluation (I_{e1}, I_{e2}, I_{e3}, I_{e4}, I_{e5}, I_{e6}, I_{e7}); 5) generation of comprehensive and aggregate indicator of SEDFM efficiency evaluation (I_e) (table 1).

Table 1. Indicator System for SEDFM Efficiency Evaluation.

Efficiency	Sustainable development components			
	Economic growth	Environmental sustainability	Social sustainability	Digitalization
1. Economic	Share of energy products in GDP (K_{er11}) Energy export in total energy consumption (K_{er12}) Energy efficiency (K_{er13}) Organizations' debt on credits received from banks and granted loans (K_{er14}) Amounts owed by buyers and to organizations' suppliers (K_{er15}) Amount of budget subsidies (K_{er16})	GDP energy intensity (K_{ec11}) Volume of investments by energy organizations aimed at environmental protection and rational use of natural resources (K_{ec12}) Specific fuel consumption for electricity supply (K_{ec13}) APG utilization rate (K_{ec14}) Share of renewable energy resources' consumption (K_{ec15})	GDP per capita (K_{s11}) Energy consumption per capita (K_{s12}) Number of people employed in the power sector (K_{s13}) Organizations' wage arrears to employees (K_{s14}) Organizations' arrears of on payments to the national budget and national extra-budgetary funds (K_{s15})	Industry-specific research and development expenditure as a percentage of GDP (K_{d11}) Financing of science from the federal budget (K_{d12}) Internal R&D costs (K_{d13}) Grants, subsidies, competitive financing (K_{d14}) Research and development in the sphere of energy (K_{d15}) Rate of gasification with natural gas (K_{d16})
$I_{e1} = (I_{er1} \times I_{ec1} \times I_{s1} \times I_{d1})^{1/4}$	$I_{er1} = (K_{er11} \times K_{er12} \times K_{er13} \times ... \times K_{er1n})^{1/n}$	$I_{ec1} = (K_{ec11} \times K_{ec12} \times K_{ec13} \times ... \times K_{ec1n})^{1/n}$	$I_{s1} = (K_{s11} \times K_{s12} \times K_{s13} \times ... \times K_{s1n})^{1/n}$	$I_{d1} = (K_{d11} \times K_{d12} \times K_{d13} \times ... \times K_{d1n})^{1/n}$
2. Environmental	Share of coal-generated electricity (K_{er21}) Share of oil-generated electricity (K_{er22}) Level of the energy industry's environmental sustainability (K_{er23})	Greenhouse gas emissions by energy organizations (K_{ec21}) Volume of contaminated wastewater discharged by organizations (K_{ec22}) Rational use of land by organizations (K_{ec23}) Level of contamination with production and consumption waste by organizations (K_{ec24})	Level of economic, legal, digital and environmental culture of energy organizations (K_{s21}) Share of energy organizations that implement principles for responsible investment (K_{s22}) Share of population having access to electricity (K_{s23})	Share of organizations implementing environmental innovations (K_{d21}) Organizations' specific costs associated with environmental innovations (K_{d22}) Energy saving (K_{d23})

Efficiency	Sustainable development components			
	Economic growth	Environmental sustainability	Social sustainability	Digitalization
$I_{e2} = (I_{er2} \times I_{ec2} \times I_{s2} \times I_{d2})^{1/4}$	$I_{er2} = (K_{er21} \times K_{er22} \times K_{er23} \times \ldots \times K_{er2n})^{1/n}$	$I_{ec2} = (K_{ec21} \times K_{ec22} \times K_{ec23} \times \ldots \times K_{ec2n})^{1/n}$	$I_{s2} = (K_{s21} \times K_{s22} \times K_{s23} \times \ldots \times K_{s2n})^{1/n}$	$I_{d2} = (K_{d21} \times K_{d22} \times K_{d23} \times \ldots \times K_{d2n})^{1/n}$
3. Social	Average monetary income per capita (K_{er31}) Average cash expenditures per capita (K_{er32}) Profitability of sold goods, of the industry's products (works, services) (K_{er33}) Organizations' return on assets (K_{er34})	Environmental costs (K_{ec31}) Expenses for energy organizations' participation in social and charitable initiatives (K_{ec32}) Share of fines in energy organizations' profits (K_{ec33})	Real accrued wages of employees (K_{s31}) Active life expectancy (K_{s32}) Level of support to implementation of constitutional rights and freedoms (K_{s33}) Number of reported crimes in the industry (K_{s34})	Share of youth (adults) having information and communication technology (ICT) skills (K_{d31}) Share of population using the Internet (K_{d32}) Investments in modernization (K_{d33}) Innovation activity level (K_{d34})
$I_{e3} = (I_{er3} \times I_{ec3} \times I_{s3} \times I_{d3})^{1/4}$	$I_{er3} = (K_{er31} \times K_{er32} \times K_{er33} \times \ldots \times K_{er3n})^{1/n}$	$I_{ec3} = (K_{ec31} \times K_{ec32} \times K_{ec33} \times \ldots \times K_{ec3n})^{1/n}$	$I_{s3} = (K_{s31} \times K_{s32} \times K_{s33} \times \ldots \times K_{s3n})^{1/n}$	$I_{d3} = (K_{d31} \times K_{d32} \times K_{d33} \times \ldots \times K_{d3n})^{1/n}$
4. Organizational and managerial	Integrated indicator of efficiency of labor resources use (K_{er41}) Energy unit generation cost (K_{er42}) Investment activity level (K_{er43})	Share of organizations using an environment pollution control system (K_{ec41}) Integrated cost efficiency indicator (K_{ec42}) Share of environmental financing in total investments (K_{ec43})	HR stability level (K_{s41}) Share of women in management (K_{s42}) Organizational culture level (K_{s43})	Number of energy organizations involved in research and development (K_{d41}) Number of staff engaged in research and development (K_{d42}) ICT Development Index (K_{d43})
$I_{e4} = (I_{er4} \times I_{ec4} \times I_{s4} \times I_{d4})^{1/4}$	$I_{er4} = (K_{er41} \times K_{er42} \times K_{er43} \times \ldots \times K_{er4n})^{1/n}$	$I_{ec4} = (K_{ec41} \times K_{ec42} \times K_{ec43} \times \ldots \times K_{ec4n})^{1/n}$	$I_{s4} = (K_{s41} \times K_{s42} \times K_{s43} \times \ldots \times K_{s4n})^{1/n}$	$I_{d4} = (K_{d41} \times K_{d42} \times K_{d43} \times \ldots \times K_{d4n})^{1/n}$
5. Technological	Availability of fixed assets (K_{er51}) Investment in fixed assets (K_{er52}) Integrated indicator of efficiency of fixed assets' use (K_{er53}) Project payback period (K_{er54})	Depreciation of fixed assets (K_{ec51}) Fixed assets renewal ratio (K_{ec52}) Installed capacity of power plants (K_{ec53}) Power plants' installed capacity utilization ratio (K_{ec54})	Level of energy organizations' technical and technological infrastructure (K_{s51}) Industrial injuries per 100,000 employees (K_{s52}) Number of new jobs created (K_{s53})	Number of developed advanced manufacturing technologies (K_{d51}) Number of employed advanced manufacturing technologies (K_{d52}) Obtained patents (K_{d53}) Investments in gasification (K_{d54})
$I_{e5} = (I_{er5} \times I_{ec5} \times I_{s5} \times I_{d5})^{1/4}$	$I_{er5} = (K_{er51} \times K_{er52} \times K_{er53} \times \ldots \times K_{er5n})^{1/n}$	$I_{ec5} = (K_{ec51} \times K_{ec52} \times K_{ec53} \times \ldots \times K_{ec5n})^{1/n}$	$I_{s5} = (K_{s51} \times K_{s52} \times K_{s53} \times \ldots \times K_{s5n})^{1/n}$	$I_{d5} = (K_{d51} \times K_{d52} \times K_{d53} \times \ldots \times K_{d5n})^{1/n}$
6. Marketing	Availability of financial services (K_{er61}) State satisfaction level (K_{er62}) Owner satisfaction level (K_{er63}) Investor satisfaction level (K_{er64})	Customer satisfaction level (K_{ec61}) Compliance with the system of international and national standards (K_{ec62}) Public organizations' satisfaction level (K_{ec63})	Employee satisfaction level (K_{s61}) Counterparties' satisfaction level (K_{s62}) Infrastructure development level (K_{s63})	Digital consumption index (K_{d61}) Volume of organizations' innovative goods, works and services (K_{d62}) Technology trading with foreign countries (K_{d63})
$I_{e6} = (I_{er6} \times I_{ec6} \times I_{s6} \times I_{d6})^{1/4}$	$I_{er6} = (K_{er61} \times K_{er62} \times K_{er63} \times \ldots \times K_{er6n})^{1/n}$	$I_{ec6} = (K_{ec61} \times K_{ec62} \times K_{ec63} \times \ldots \times K_{ec6n})^{1/n}$	$I_{s6} = (K_{s61} \times K_{s62} \times K_{s63} \times \ldots \times K_{s6n})^{1/n}$	$I_{d6} = (K_{d61} \times K_{d62} \times K_{d63} \times \ldots \times K_{d6n})^{1/n}$
7. Risk-related	Risk resistance coefficient (K_{er71}) Risk intensity (K_{er72}) Risk ratio (K_{er73}) Autonomy ratio (K_{er74})	Environmental risk manageability level (K_{ec71}) Level of risk management system implementation (K_{ec72}) Risk management process continuity ratio (K_{ec73})	Risk management efficiency (K_{s71}) Risk management stability level (K_{s72}) Digital security index (K_{s73})	Share of risk management-related innovative technologies (K_{d71}) Share of organizations implementing innovations that ensure increased environmental safety during production of goods, works, services (K_{d72}) Share of organizations implementing innovations that ensure increased environmental safety as a result of consumers' use of innovative goods, works, services (K_{d73})

Efficiency	Sustainable development components			
	Economic growth	Environmental sustainability	Social sustainability	Digitalization
$I_{e7} = (I_{er7} \times I_{ec7} \times I_{s7} \times I_{d7})^{1/4}$	$I_{er7} = (K_{er71} \times K_{er72} \times K_{er73} \times ... \times K_{er7n})^{1/n}$	$I_{ec7} = (K_{ec71} \times K_{ec72} \times K_{ec73} \times ... \times K_{ec7n})^{1/n}$	$I_{s7} = (K_{s71} \times K_{s72} \times K_{s73} \times ... \times K_{s7n})^{1/n}$	$I_{d7} = (K_{d71} \times K_{d72} \times K_{d73} \times ... \times K_{d7n})^{1/n}$
Overall efficiency: $I_e = (I_{e1} \times I_{e2} \times I_{e3} \times I_{e4} \times I_{e5} \times I_{e6} \times I_{e7})^{1/7}$				

One can set the following intervals of the values of particular integral indicators for component efficiency evaluation and the aggregate efficiency indicator: 1-4; 4-7; 7-10, which correspond to the low, medium and high levels of SEDFM efficiency.

It should be noted that SEDFM efficiency is limited by a number of factors. They include lack of own financial resources, investment risks, high percentage of commercial loans, complex mechanism for obtaining loans for investment project implementation, price fluctuations in the global energy market, imperfect legal and regulatory framework governing investment processes, and existing tax treatment of investment activities [6].

The system of indicators should be regularly revised in connection with changes of priorities, emergence of new goals, tools, technologies, sources, financing institutions, and environmental changes.

The proposed methodology allows for an adequate evaluation of SEDFM efficiency level, to obtain a comprehensive and accurate picture of SEDFM condition, to reduce time and increase the efficiency of managerial decision-making and optimize risks, and can serve as an information basis for selection of the most promising ways to increase SEDFM efficiency. The methodology's distinctive feature is that it allows to use any number of the most significant indicators available, compare the achieved level of SEDFM efficiency with the expected one, dynamically, with other countries' indicators, and develop a step-by-step system for SEDFM adjustment to improve SDG implementation.

3. Conclusion

Thus, the goal of the study which is to develop an integrated approach, the system of indicators and the methodology for SEDFM efficiency evaluation has been achieved. Unlike the existing methodologies, the proposed one assumes an integrated, systematic and comprehensive evaluation of SEDFM efficiency by calculating a composite integral indicator based on particular integrated indicators of seven efficiency components, taking into account four components of sustainable development, power industry specifics and stakeholders' expectations. The practical application of the presented approach allows improving monitoring process efficiency, strengthening SEDFM control, identifying functional problems and directions for improvement of SEDFM, the industry and country economy as a whole, and also opens up great prospects for finding competitive advantages and new solutions to strategic issues.

References

[1] Power industry of Russia 2020 – URL: https://energoseti.ru/articles/energetika-rossii
[2] TASS 2018 – URL: https://tass.ru/ekonomika/4604123
[3] Russia's Fuel and Energy Complex 2019 (June 2020 Release) *Statistical Compilation Analytical Center at the Russian Federation Government* – URL: https://www.ac.gov.ru.
[4] Russia in Numbers 2020 *Brief Statistical Compilation* (Moscow: Rosstat) p 550
[5] 2030 Agenda for Sustainable Development 2015 (New York: UN) – URL: https://www.un.org/sustainabledevelopment/ru/summit/
[6] Investments in Russia 2019 *Statistical Compilation* (Moscow: Rosstat) p 228

IOP Conf. Series: Materials Science and Engineering 1100 (2021) 012004 doi:10.1088/1757-899X/1100/1/012004

Vented air heat recovery in ventilation and conditioning systems based on solid sorbent

A G Averkin, A I Eremkin and Yu. A Averkin

Penza State University of architecture and construction, German Titov street, 28, Penza, 440028, Russia

E-mail: algraw@mail.ru

Abstract. Method and equipment for the heat recovery of the air vented from the rooms with application of intermediate heat medium on the base of silica gel are presented. Analytical dependences for calculating the convective heat exchange in the apparatus with fixed layer, boiling bed and two-phase air flow applying granulated sorbents are specified.

1. Introduction

One of the main energy conservation measures in the ventilation and air conditioning systems is the heat recovery of the air vented from the rooms. This is done with the application of heat exchange units (heat recovery units) which are installed in the vent ducts [1].

According to their working principles and constructional features the heat recovery units are divided into recuperative, regenerative ones and units with intermediate heat medium. The most universal is the heat recovery system with the intermediate heat medium (Fig. 1).

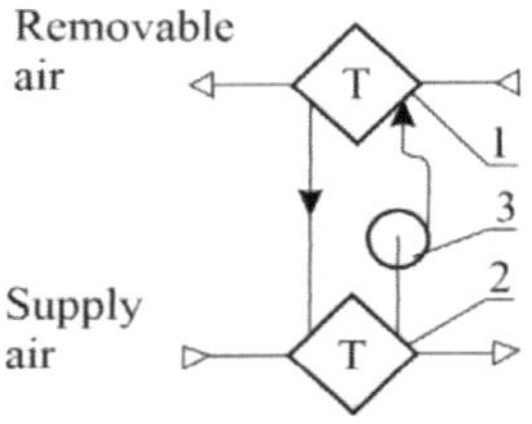

Figure 1. The basic diagram of heat recovery with the intermediate heat medium:
1 – heat exchange unit (heat recovery unit) heat extracting;
2 - heat exchange unit (heat recovery unit) exothermic;
3 – the pump for the intermediate heat medium (anti-freeze agent);

Extraction and emission of heat are ensured by the recirculation of liquid – intermediate heat medium (anti-freeze agent) between the heat exchange (heat recovery) units which are mounted in the extract 1 and supply 2 ducts. The heat exchange (heat recovery) units can be placed at significant distances from each other (dozens and hundreds of meters). Recirculation of anti-freeze agent is forced with the pump 3.

Methods of heat recovery integrated into the design of standard apparatus are based on the processes of heat transfer. They do not allow complete recovering of the main component of the air vented from the rooms – latent heat of the water vapor.

It is known [1] that the enthalpy of the of the vented air is determined by the total of components presenting the sensible heat (air + gaseous impurities) and the latent heat of vapor (mainly water vapor):

$$I = C_{B}t + (r + C_{П}t) \cdot d \cdot 10^{-3}, \qquad (1)$$

where I – enthalpy of moist air, kJ/kg; c_{B} – air specific heat, kJ/(kg·°C); t – air temperature, °C; r – latent heat of water vaporization, kJ/kg; $c_{П}$ – specific thermal capacity of water vapor, kJ/(kg·°C); d – moisture content, g/kg of dry air.

Total refluxing of water vapor in the air (total dehumidifying) is possible only when the air is cooled to absolute zero (on *I-d* diagram of humid air the curve of total saturation φ = 100% crosses the enthalpy axis I under t = - 273 °C which is practically unapproachable in the given conditions.

In [1] it is shown that the given disadvantage can be eliminated, the heat recovery unit becomes more efficient if its work is based on the sorption processes in the system humid air – solid sorbent. The air vented

Content from this work may be used under the terms of the Creative Commons Attribution 3.0 licence. Any further distribution of this work must maintain attribution to the author(s) and the title of the work, journal citation and DOI.
Published under licence by IOP Publishing Ltd

from the rooms is passed through a layer of granulated sorbent – a substance with well-developed microstructure which can absorb the water vapor intensively with formation of large amount of heat – absorption heat (latent heat). The total latent heat of water vapor absorption in the process of capillary condensation is 2930 kJ/kg.

2. Results and discussion

It is known [1, 2] that many sorbents may dry the air flow to zero humidity rate and release almost absolutely dry air into the atmosphere, i.e. recover all latent heat of water vapor and additionally sensible heat in the amount not less than that recovered in modern heat recovery units. Analysis of adsorbents produced by the domestic industry showed that silica gel is most efficient for adsorbing water vapor from the air flow. It has the following advantages: high selective adsorptive power for the water vapor and high hydrophilic properties; low reactivation temperature (110…200 °C) and, as a result, lower power consumption than that associated with reactivation of other industrial sorbents; possibility of synthesizing silica gels with a wide range of specified structural characteristics through the use of rather simple techniques; low cost-price under the large-scale industrial production; high mechanical strength with respect to abrasion and compression; incombustibility.

To improve the thermo technical efficiency in the recovery system with the intermediate heat medium it is suggested to equip the air-cooling unit with the functions of absorber and heat exchange unit (let us call it heat exchanger-absorber). For total extraction of the water vapor heat as well as for extraction of the sensible heat of the dried vented air the layer of the adsorbent is placed in the tube side of the shell-and-tube heat exchanger and the intermediate heat medium circulates in the intertubular space. The intermediate heat medium extracts the heat from the adsorbent layer, becomes warmer and transfers it via the exothermic heat exchanger to the outside (incoming) air.

We suggest using 20%... 30% water solution of ethylene glycol and propylene glycol having low freezing temperatures [1].

Standard constructions – pipe and flange contact apparatus which working principle is based on filtrating the gas through the layer of fixed granulated material (catalyst) – are used for the catalytical purification of gases [2].

We suggest completing the activation (reactivation) of adsorbent according to the method of replacing desorption method, i.e. by applying the element of heatless process of desorption which mechanism is considered in [1].

To ensure the process continuity two heat exchangers-absorbers are required: one works in the mode of recovering latent and sensible air heat, another works in the mode of sorbent activation (reactivation).

Through the layer of silica gel to be activated we filter the vented, completely dried air obtained in the nearby heat extracting heat exchanger before releasing it into the atmosphere.

Due to the large difference between the partial pressures of water vapor in the pore volume of the adsorbent and the air flow being blown through it the molecules of water are intensively diffused into the air flow.

As opposed to the mode of heatless reactivation of adsorbent all dried air flow will be used for blowing as dried air is not a marketable product in the given process and does not contain adsorption heat any more.

According to the equation [1]

$$G_d = G_a P_d / P_a, \quad (2)$$

under the equal flow rate of the air flow at adsorption and desorption stages $G_a = G_d$, the pressures of the air flows at the given stages will be equal in value $P_a = P_d$ and approximate to the atmospheric pressure.

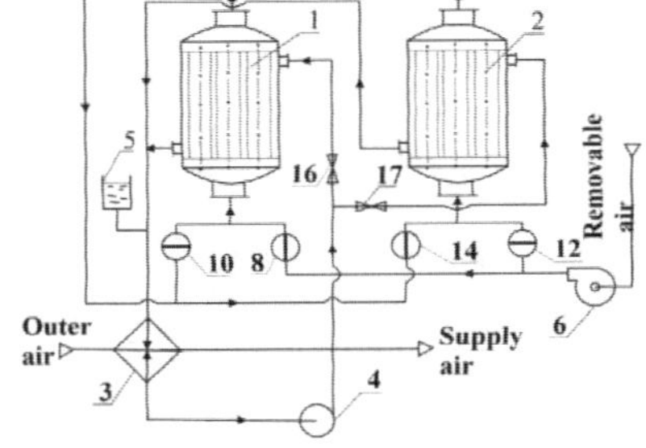

The principle diagram of the heat recovery of the air vented from the rooms on the base of a sorbent is given in Fig. 2, the principle diagram of heat exchanger-adsorber is given in Fig. 3.

Figure 2. The diagram of heat recovery with the intermediate heat medium:
1, 2 – heat exchangers-adsorbers (heat recovery units); 3 – heat exchanger-air heater for the outside air; 4 – the pump; 5 – expansion tank;
6 – air-supply ventilator;

7 – extraction ventilator;
8,9,10,11,12,13,14,15 – interception valves;
16, 17 – air valves;
18 – gate valve

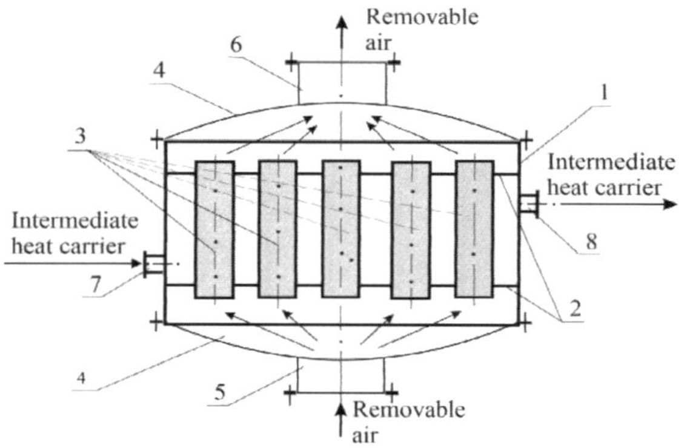

Figure 3. Diagram of heat exchanger-adsorber:
1 – case; 2 – tube sheet;
3 – tubular member with silica gel;
4 - covers; 5,6 – pipe pieces for the air;
7,8 – pipe pieces for the intermediate heat medium

Heat recovery of the vented air with application of the intermediate heat medium includes the following stages (modes).

The stage of water vapor adsorption and extraction of the latent and sensible heat of the vented air.

The air vented from the rooms is sent to the heat exchanger-adsorber 1 with the help of ventilators 6, 7. When it happens, valves 8, 9, 14, 15 are open, valves 10, 11, 12, 13 are closed.

In the heat exchanger-adsorber the air flow passes through the tube side filled with the layer of adsorbent – silica gel, dries due to water vapor sorption to the completely dry state (Fig. 2, Fig. 3). As this process goes on the heat of adsorption is emitted. Then the air is cooled with extraction of adsorption heat and partly of sensible heat with the help of the intermediate heat medium supplied into the pipe side of the apparatus by the pump 4 (Fig. 2). At this time air valve 16 is open, air valve 17 is closed.

The heated intermediate heat medium is transferred to the surface heat exchanger-air heater 3 and transmits the heat up to the cold outside air due to the process of heat conduction and then, being cooled down, is again supplied into the pipe side of adsorber-air cooler 1 to cool the air flow (Fig. 2, 3) and so on.

Stage of adsorbent activation. Ventilator 7 supplies the air dried to completely dry state and the cooled air from the heat exchanger-adsorber 1 to adsorber-air cooler 2 which is at the stage of adsorbent activation (Fig. 2). The air moves from the bottom upwards through the fixed layer of granulated adsorbent in the pipe side and causes desorption of water molecules into the air flow due to the difference of the partial pressures of the water vapor at the surface of silica gel capillary and in the air. The humidified air flow from apparatus 2 is released into the atmosphere through valve 15.

Air blowing (adsorbent activation) through the layer of adsorbent is stopped when the minimal residual humidity of silica gel is achieved. This can be monitored by measuring the moisture content of the released air. After that valves 14, 15 are closed and the gate valve 18 is opened and the dried air from the heat exchanger-adsorber 1 is released into the atmosphere by-passing heat exchanger 2.

After the silica gel in adsorber-air cooler achieves its equilibrium state (the duration of adsorption or the moisture content of the released air may serve as indirect indicators) the apparatus is switched to the stage of adsorbent activation and the air vented from the room is supplied to the heat exchanger-adsorber 2 for drying and cooling. To do this, valves 8, 9, 14, 15 must be closed and valves 12, 13, 10, 11 must be open. Air valve 17 must also be opened, while air valve 16 on the line of intermediate heat medium must be closed.

In the given cycle the air removed from the room moves through the layer of adsorbent from the bottom upwards in the heat exchanger-adsorber 2 driven by ventilators 6, 7. As it happens the air is dried emitting the heat of adsorption and also cooled due to the action of intermediate heat medium supplied into the shell side of the given apparatus.

Circulation of intermediate heat medium through apparatus 2 is ensured by pump 4 through the opened valve 17. The intermediate heat medium transports the heat extracted from the dried air to the heat exchanger-air heater 3 where the intermediate heat medium transfers the heat to the outside air and goes back for heating to the shell side of heat exchanger-adsorber 2 and so on.

Completely dried and cooled air flow from apparatus 2 is supplied to heat exchanger-adsorber 1 for adsorbent activation. Ventilator 7 supplies this air through valve 10 into the pipe side of heat exchanger-adsorber 1 where it encourages desorbing water molecules from the layer of adsorbent into the air. The humidified air from the top of the adsorber is released into the atmosphere through the opened valve 11 by ventilator 7.

To intensify the process of desorption it is possible to additionally install a vacuum pump to reduce the pressure in the heat exchanger-adsorber which is in the activation stage. The vacuum pump can be mounted in the general air line after valves 11 and 15 (the pump is not shown in Fig. 2).

This way, alternatively, one after another, the heat exchangers-adsorbers are in the stages of either adsorption or desorption ensuring the continuity of heat extraction and recovery from the air vented from the room.

The apparatus with the "boiling bed" of grainy adsorbent and cyclone-type apparatus can be also used as heat exchange-adsorber [1, 2]. The principle diagrams of the given apparatus are presented in Fig. 4.

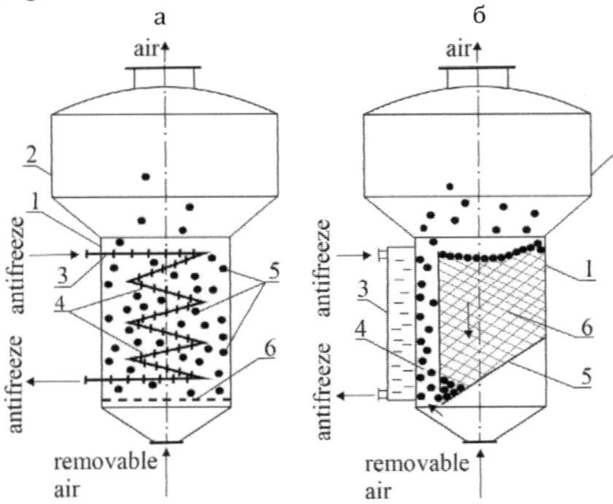

Figure 4. The diagrams of the apparatus with the moving bed of grainy adsorbent:
a – *apparatus with the "boiling bed" of adsorbent:* 1 – case; 2 – separator; 3 – pipe coil with external finning; 4 – bars for finning; 5 – "boiling bed" of adsorbent; 6 – grid;
b – *cyclone-type apparatus:* 1 – case; 2 – separator; 3 – coolant jacket; 4 – channel for the through flow (air + sorbent); 5 – channel; 6 – sorbent accumulator.

Changing the granular adsorbent from the fixed layer into the moving bed: into the "boiling bed" (Fig. 4a), into the through two-phase flow (Fig. 4b) – intensifies the heat and mass transfer [2]. At the same time, the linear dimensions of equipment can be significantly reduced and less production area will be required for the system of heat recovery.

To make the structural analysis of heat exchangers-adsorbers we have developed the methodology basing on the application of the number of transfer units. The number of transfer units is established with the help of the modified *I-d*-diagram of moist air [3].

To estimate the convective heat exchange in the system air flow – wall we can use the following analytic dependences [2]:

- in the apparatus with the fixed blown through grainy layer (Fig. 3)

$$\mathrm{Nu} = 0.31\,\mathrm{Re}^{0.5}[(1-\varepsilon)^{0.5}/\varepsilon], \quad (3)$$

$$\mathrm{Nu} = 0.1\mathrm{Re}^{0.8}[(1-\varepsilon)^{0.2}/\varepsilon]; \quad (4)$$

- in the apparatus with the "boiling bed" (Fig. 4a)

$$\mathrm{Nu}_{max} = 0.86\mathrm{Ar}^{0.2} \quad (5)$$

- in the apparatus with the through two-phase air flow (Fig. 4b) [1]:

$$\mathrm{Nu} = 0.3\mathrm{Re}^{0.54}\,Re_B^{0.38}\,\mu^{0.11}, \quad (6)$$

where Re – Reynolds number for the single-phase flow; ε – porosity;
$\mathrm{Re_B}$ – Reynolds number calculated using the weighing rate of the air flow
μ – weight concentration of sorbent particles in the air flow, kg/kg; Ar – Archimedes criterion.

Equation (3) is applicable under Re = 1.5...57; equation (4) – under Re = 57...1500; equation (6) –

under Re = 14000...36500; Re_B = 3.6...5.3; μ = (0.25...1.79) kg/kg.

In paper [1] it is shown that recovered heat gain ΔQ, kW in the suggested method in comparison to the traditional ones equals:

$$\Delta Q = 2.93 \cdot G \cdot d_y, \quad (7)$$

where G, d_y accordingly, consumption, kg/h, moisture content of the removed air, g/kg of dried air.

3. Conclusion

An efficient method of the heat recovery of the vented air with application of intermediate heat medium on the base of solid sorbent was developed. The advantage of the method is the opportunity of significant increase of the amount of the recovered heat from the air vented from the rooms due to complete extraction of the latent heat of water vapour contained in this air at the initial stage.

Разработан эффективный способ утилизации теплоты удаляемого воздуха с промежуточным теплоносителем на основе твердого сорбента. Достоинством способа является возможность значительного увеличения количества утилизируемой теплоты от удаляемого воздуха из помещения за счет полного отбора скрытой теплоты водяных паров, содержащихся в нем на начальной стадии.

References

[1] Averkin A G and Eremkin A I 2019 *Improvement of devices for heat and humidity treatment of air and methods for calculating climatic engineering* (Moscow Vologda: Infra-Engineering) p 168

[2] Gelperin N I 1981 *Basic processes and apparatuses of chemical technology* (Moscow: Chemistry Book one) p 384

[3] Averkin A G and Eremkin A I et al 2019 IOP *Conf. Series: Materials Sci. and Eng.* 665 012001

Evolution of the brass microstructure during rolling in relief and smooth rolls

A B Naizabekov[1], S N Lezhnev[1], I E Volokitina[1] and E A Panin[2]

[1]Rudny industrial institute, 38 50 let Oktyabrya Street, Rudny, 111500, Kazakhstan
[2]Karaganda industrial university, 30 Republic Ave., Temirtau, 101400, Kazakhstan

E-mail: sergey_legnev@mail.ru

Abstract. One of the most promising technologies for rolling thick-sheet metal, during which shear deformations are realized both in the longitudinal and transverse directions, is rolling in rolls with a relief surface in the form of annular channels that form trapezoidal protrusions and depressions that alternate one after another along the entire length of the barrel. At the same time, a relief with an unequal ratio of the protrusion to the depression is applied to the rolls. In this case, the gap between the rolls in all directions retains the value of the distance between the rolls, i.e. when deforming in these rolls, only the net shear scheme is implemented. This method of applying a relief surface to the rolls allows to save metal, reduce the required deformation force and labor and energy costs due to the reduced number of metal passes through the working stands. During the microstructure study of brass blanks rolled on existing and proposed technologies it was revealed that the original grain after the traditional deformation process is reduced from 22 μm to 9 μm, and after the proposed technology it is reduced from 22 μm to 3 μm.

1. Introduction

In the production of sheet and long rolled products from ferrous and non-ferrous metals and alloys, the main task is to improve its quality characteristics, both in geometry and mechanical properties. Most often, even at modern metallurgical enterprises for the production of rolled metal, the increase in mechanical properties is achieved by using special alloying additives in the smelting of alloys or by heat treatment of the finished product. However, it is also possible to achieve the required level of mechanical properties in metal products in a simpler and much more economical way, namely, by creating additional metal flow flows during rolling, i.e., by intensifying the developed shear deformations not only in the longitudinal but also in the transverse direction of the initial billet [1]. A review of scientific, technical and patent literature shows that at the moment there are already more ways to intensify shear deformations in the entire volume of the initial billet [2-10]. These methods also include a method that involves applying a system of alternating protrusions and depressions to the surface of the wide faces of the slab or to the working rolls. In this case, along with the deformation in the vertical and longitudinal directions, conditions are created for the deformation of the metal in the transverse direction, which has a positive effect on reducing the anisotropy of properties.

One of the most promising technologies for rolling thick-sheet metal, during which shear deformations are realized both in the longitudinal and transverse directions, is rolling in rolls with a relief surface in the form of annular channels that form trapezoidal protrusions and depressions that alternate one after another along the entire length of the barrel [11]. At the same time, a relief with an unequal ratio of the protrusion to the depression is applied to the rolls (Fig. 1a). In this case, the gap between the rolls in all directions retains the value of the distance between the rolls, i.e. when deforming in these rolls, only the net shear scheme is implemented (Fig. 1b). This method of applying a relief surface to the rolls allows to save metal, reduce the required deformation force and labor and energy costs due to the reduced number of metal passes through the working stands [12].

Content from this work may be used under the terms of the Creative Commons Attribution 3.0 licence. Any further distribution of this work must maintain attribution to the author(s) and the title of the work, journal citation and DOI.
Published under licence by IOP Publishing Ltd

(a) (b)

Figure 1. "ECAP-drawing» combined method: a – common view of rolls; b – scheme of the gap

The purpose of this work is to study the microstructure evolution of a brass alloy during rolling of thick-sheet billets using a technology that includes rolling at the first pass in relief rolls with underlying alignment in smooth rolls. At the same time, the task of research was to determine the possibility of relief rolls using in the production of rolled products at the Balkhash non-ferrous metal processing plant (Kazakhstan). For this purpose, preliminary negotiations were held with the plant management and the L80 brass alloy was jointly selected for rolling using the proposed technology in the laboratory. Experimental rolling was carried out at the laboratory mill "DUO-100" in the conditions of the Karaganda industrial university.

2. Experimental procedure

The current technology for rolling strips and rolls made of copper and its alloys at the Balkhash non-ferrous metal processing plant includes rolling on a universal 2-4-roll rolling mill. A special feature of the mill is that it operates both in 2-roll mode ("Duo") and in 4-roll mode ("quarto").

The mill consists of a working crate with replaceable cassettes, "quarto" and "Duo" cassettes, a device for handling rolls, hinged spindles with spring balancing, left and right winders, a roll feeder, a roll unwinder, a 5-roller regular machine with a strip start envelope, a centering device, a diverting roller. The power of the main drive motor is 500 kW. The diameter of the rolls when rolling in the "Duo" system is 630 mm, the width of the barrel is 650 mm, the maximum load on the rolls is 5.0 MN, the maximum rolling speed is 2.5 m/s, the maximum torque on the roll neck is 60 kN·m, the maximum current of the main motor is 2500 A. Initial dimensions of rolled strips after hot rolling and milling: thickness is 10-16 mm, width is 230-500 mm, length is 5500 mm.

For these studies, blanks with dimensions h x b x l = 10x100x300 mm from L80 brass alloy were made. Before starting the experiments, the samples in a muffle furnace at a temperature of 670°C for 1 hour were annealed. The deformation of brass alloy workpieces was carried out using the proposed technology in relief rolls with an unequal ratio of the protrusion to the depression and according to the current technology at the enterprise in smooth rolls. Rolling in relief rolls with an unequal ratio of the protrusion to the depression was carried out in five passes with 0.6 mm compression in each pass. After that, the workpieces were gradually deformed in smooth rolls to the final height of the workpiece equal to 5 mm. For this purpose, smooth rolls on the Duo-100 laboratory mill were installed.

The second batch of blanks was rolled at the Duo-100 laboratory mill using the current technology at the Balkhash non-ferrous metal processing plant with a compression mode of 10 mm → 8.5 mm → 7.2 mm → 6.3 mm → 5.8 mm → 5.4 mm → 5.0 mm.

After rolling, the samples in the initial (annealed) state and after two different types of rolling were examined for microanalysis of the structure and their hardness was determined.

3. Results and discussion

Figure 2a shows the microstructure of L80 brass in its initial, non-deformable state, after annealing at a temperature of 670°C. It is represented by conjoined equiaxed grains of the α-solid solution, between which there are rather thin layers of the β-phase, i.e., as a result of heating, a significant part of the CuZn intermetallic compound was dissolved in a copper-based solid solution (figure 2a). The α-phase is a solid solution of zinc substitution in copper with a HCC lattice, high plasticity, and low strength and hardness values; the β-phase is an ordered solid solution based on the intermetallic compound CuZn with a BCC crystal lattice. This phase is characterized by higher hardness than the α-phase and brittleness. Therefore, an increase in the α-phase leads to an increase in the ductility of the alloy and an increase in its ability to cold strain.

1) in the output channel of the matrix after the completion of the ECAP stage;
2) in the intermediate zone between the matrix and drawing die;
3) at the exit of the drawing die after the completion of the drawing stage.

Considering the microstructure in the shell of a bimetallic wire (figure 3), it was found that after passing the channels of the ECA-matrix, the initial grain size decreases to about 13 microns, while the shape of the grains becomes equiaxed, which corresponds to the ECAP effect.

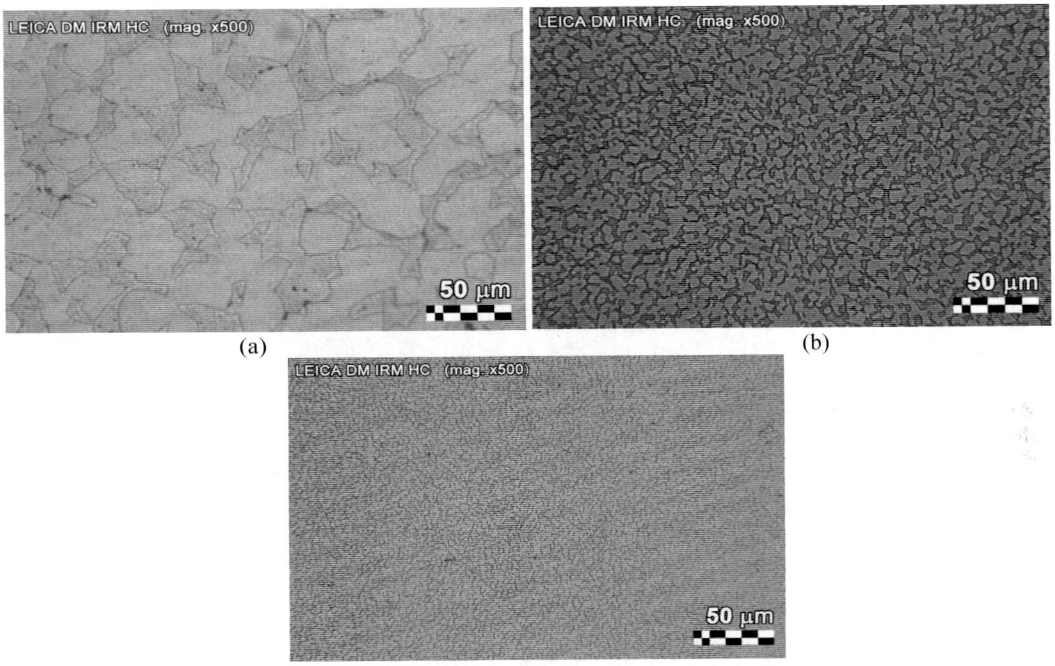

Figure 2. Structure of L80 brass in the initial state after annealing (a), after rolling according to the current (b) and proposed (c) technologies

During the microstructure study of brass blanks rolled on existing and proposed technologies it was revealed that the original grain after the traditional deformation process is reduced from 22 µm to 9 µm (figure 2b), and after the proposed technology it is reduced from 22 µm to 3 µm (figure 2c).

After deformation by both technologies, the proportion of grains of the β-phase, which surrounds the α-solid solution in the form of a thickened rim, increased in comparison with the initial state in brass. In this case, the overall orientation of the structure in the direction of rolling is enhanced, which is emphasized not only by the line arrangement of the main phases, but also by the impurities.

When rolling using the current technology, the increase in the amount of the β-phase is greater than with the proposed one and its expansion in directions perpendicular to the rolling direction indicates a predominant influence of tangential stresses on this process, in contrast to the change in the shape and location of the α-solid solution crystals. There was a peculiar redistribution of stresses between both phases. The deformation of the more plastic α-phase is provided by rigid normal stresses, and the less plastic β – phase is provided by soft tangential stresses. Their superposition leads to the formation of the line structure of the alloy.

In the microstructure of samples after the proposed technology, there are approximately the same probability of areas enriched with both the α-phase and the β-phase, which is squeezed out of the solid solution in local zones of intense deformation. The α-phase grains are pulled in the direction of the rolling tensile forces, and the β-phase germination is perpendicular. This self-consistent interaction of phases ensures the integrity of the structure and prevents the formation of cracks and pores.

Studies conducted on the JEOL JEM2100 electron transmission microscope have shown that the use of smooth rolls leads to uneven sliding of dislocations, since only the most favorably oriented sliding systems are involved, so there are more dislocations inside some grains than in others (figure 3 a-b). After rolling in relief rolls, this is not observed, the distribution of sliding lines inside individual grains is uniform (figure 3 c-d), but their location is at an angle to each other. Therefore, the slide goes along different planes and directions.

Figure 3. TEM images of brass microstructure: a) after rolling in smooth rolls, transverse direction; b) after rolling in smooth rolls, the longitudinal direction; c) after rolling in the relief rolls, transverse direction; d) after rolling in the relief rolls, longitudinal direction

4. Conclusions

The obtained results of microstructural studies indicate the advantages of the proposed technology for rolling thick-sheet blanks, including rolling in relief and smooth rolls, over the current technology for rolling blanks simply in smooth rolls in terms of obtaining an isotropic fine-grained structure of rolled metal, and therefore that the quality of thick-sheet rolled products will also increase. Based on the conducted laboratory studies, agreements were reached to conduct pilot testing of the proposed rolling technology at the universal 2-4-roll rolling mill of the Balkhash non-ferrous metal processing plant for rolling copper and its alloys.

References

[1] Kazeminezhad M 2012 Metal Forming: Process, Tools, Design (London: InTech Open)
[2] Traino A I, Polukhin V P and Nikolaev V A 2011 *Metallurgist* **55(5-6)** 358-63.
[3] Kotelkin A V, Berezin M V and Vorontsov V K 1983 *Notched ingot for rolling-with notches on one large edge on different sides of transverse cross-section axis relative to those on other large edge*

Patent SU942818-B
[4] Vorontsov V K, Brinza A V and Persiyanov S V 1983 *Ingot for rolling-has projections consisting of triangular prisms with side ribs* Patent SU948471-B
[5] Meshcheryakova A A, Konovalov Y V and Nabatov G I 1977 *Steel in the USSR* **7(10)** 571-3
[6] Dolzhenkov F E, Sunyaev A V, Ostapenko A L, Tishkov V Ya, Neustroev S L, Novik V I, Anokhina I Yu and Ponomarev V I 1984 *Steel in the USSR* **14(3)** 130-4
[7] Ostapenko A L, Soskovets O N, Anokhina I Y, Chepelyan I I and Romanika B P 1985 *Metallurgist* **29(5-6)**212-3
[8] Nukhov D 2019 *Materials Today: Proceedings* **19** 2562-5
[9] Bogatov A A, Nukhov D S and P'Yankov K P 2015 *Metallurgist* **59(1)** 113-8
[10] Bogatov A A, Nukhov D Sh and Leshchev I V 2016 *Chernye Metally* **10** 34-8
[11] Nayzabekov A B and Luzhnev S N 2007 *Roller used for rolling of thick-gage plate, includes upper and lower rollers with relief surface in the form of ring cavities, which form flanges and hollows of trapezoid form* Patent KZ14791-B
[12] Naizabekov A, Lezhnev S, Koinov T, Mazur I and Panin E 2016 *Journal of Chemical Technology and Metallurgy* **51(4)** 363-70

Methodology for estimating appropriate work parameters of motor vehicle and tractor maintenance and troubleshooting stations in farms of Kostanay Region Kazakhstan

G I Bondareva[1], S K Toigonbaev[2], V A Evgrafov[2] and O N Didmanidze[2]

[1] All-Russian Research Institute for Hydraulic Engineering and Land Reclamation (VNIIGiM), 44/2, Bolshaya Akademicheskaya street, Moscow, 127550
[2] Russian State Agrarian University-Moscow Agricultural Academy named after K.A. Timiryazev, 49, Timiryazev street, Moscow, 127550

E-mail: Boss2569@yandex.ru

Abstract. This paper presents a methodology for estimating an optimal zone of motor vehicle and tractor maintenance as well as an approach to calculate total discounted costs for maintenance and troubleshooting of motor vehicles and tractors in a production department. The proposed methodology allows substantiating an appropriate use of company motor vehicles and tractors for implementing mechanized operations in crop farming.

1. Introduction

This study represents a summary and a next step forward in the methodology to estimate optimal parameters of using facilities to maintain agricultural machinery, which we proposed in a series of works published from 2015 to 2018 [1,2,3,4].

A nine year research experience in estimating and substantiating optimal parameters for centers and stations has demonstrated it is necessary to precise optimization criteria and replace a hypothetical calculation unit of labor intensity – "a maintenance cycle" by a less complex unit "man-hour".

2. Materials and Methods

The developed maintenance and troubleshooting types and complexes for motor vehicles and tractors (stationary maintenance centers, mobile mechanized units, and tractor maintenance stations) are core elements of agricultural facilities and equipment. In most cases, maintenance operations are carried out by employees of production companies themselves with the help of mobile maintenance units at stationary maintenance centers or immediately where a machine is operated. Maintenance repair works of machines are pre-planned and carried out by employees of a production company and with certain equipment. The overhaul maintenance is implemented at specialized repair plants. When in service failures are possible because of faulty usage of equipment, poor quality of maintenance and repair, and aging of machine parts. As a rule, to eliminate failures of the 1st and 2nd complexity groups repair works are performed directly at a place where a machine is operated, either by a machine operator or with specialized units. Faults of the 3rd complexity group are rectified in a repair shop. An element in the preventive maintenance and repair system is troubleshooting carried out in both processes. There are certain types of troubleshooting, such as operational, production, complete, partial, planned, unplanned, specialized and combined ones. In operating conditions, machines are transported to places of work, maintenance and repair. They can move in self-propelled mode or are towed, transported on heavy-duty trailers, by rail and water transport. When in service a variety of factors interacting with each other affect a machine.

The specifics of motor vehicles operation in land reclamation and construction production companies can be represented as a counteraction of factors making a reliability assurance system less efficient (machine loading, machine aging, adverse environmental factors, etc.) and factors that improve the system functioning (incorrect operation of a motor vehicle and tractor facilities, organizational measures, etc.). Proper maintenance and troubleshooting of motor vehicle and tractor facilities and their functioning are related to certain

Content from this work may be used under the terms of the Creative Commons Attribution 3.0 licence. Any further distribution of this work must maintain attribution to the author(s) and the title of the work, journal citation and DOI.
Published under licence by IOP Publishing Ltd

challenges, such an efficiency analysis of a state-of-the-art motor vehicle and tractor maintenance station by available stationary and mobile means, developing a net of maintenance stations for machines, with production conditions are taken into consideration. To resolve these problems it is necessary to determine appropriate basic work parameters of motor vehicle and tractor maintenance station.

These parameters include the following:

1. an optimal zone of motor vehicle and tractor maintenance, characterized either by a correct average distance needed for tractors to be transported to maintaining or an optimal average distance of moving mechanized maintenance vehicles to tractors operating in fields (Lavopt.):
2. an optimal throughput of stationary maintenance centers, mobile vehicles, etc.,
3. optimal aggregate reduced costs for motor vehicle and tractor troubleshooting in a production subdivision.

In contrast to available methods for calculating a maintenance station, the proposed methodology is based on a more progressive criterion to conduct an economic assessment of agricultural processes – total reduced cost [5]. This criterion takes into account a deficit of capital investments and labor resources, allowing, therefore, a solution to the problems of interest from the standpoint of national economic efficiency.

3. Results and Discussion

In this regard, the function of specific aggregate reduced costs for maintaining and troubleshooting machinery used in a production subdivision, a mentioned above criterion taken into consideration, is written as follows:

$$C = \frac{Bb.\cdot(A\,rep.b.+En)+Be\cdot(A\,rep.e.+En)}{NL\cdot Lav.} + \frac{Fy.\cdot \Sigma i\,Cj}{Ny} + \frac{2\alpha\,av.\cdot m\cdot Lav.}{Ny} + \frac{El\cdot Mw{-}a\cdot\gamma\,w{-}a}{Nl\,Lav.} + \frac{dav\cdot Fy.}{Ny} + \frac{2d\,av.\cdot m\cdot Lav.}{Vav.t.\cdot Ny} \quad (1)$$

where Bb – a balance price (cost) of a stationary maintenance center or tractor maintenance workshop, ruble;
Be is a balance price (cost) of technological equipment used in a tractor maintenance center, ruble;
Arep.b., Arep.e. – a coefficient of annual expenditures on renovating buildings, repairing equipment, 1/year;
En – a standard coefficient of capital investments efficiency, 1/year. En = 0.15 1/year;
NL – a linear maintenance density (specific labor intensity of maintenance, per 1 square meter of an actual average distance tractors overcome for maintenance, man.h./year.km;
Lav. – an average distance of tractors overcome for maintenance (a "radius" of a maintenance zone), km;
Fy - an annual time for motor vehicle and tractor maintenance in a particular subdivision, h/year;
$\Sigma i\,Cj$ – a sum of standard specific elements in operating costs for motor vehicle and tractor maintenance in the subdivision, this amount includes an hourly tariff rate (with charges) of an adjusting worker, standards of hourly costs for repairing and maintaining buildings, equipment, for fuels and greases, materials for cleaning and other operations, fuel for flushing and technological needs in maintaining tractors and agricultural machines, as well as hourly costs for heating and electricity of mobile mechanized units or tractor maintenance stations, ruble/hour;
Ny. – an annual labor input of maintaining 1GGP unit, man. h/year;
αav. - transport aggregate reduced costs for idle driving of tractors for maintaining, ruble/km;
m is a number of trips per year of all tractors for maintenance in the subdivision 1/year;
Vav.tr. – an average speed of tractors during an idle run for maintaining, km/h;
El is an efficiency coefficient of using labor resources, ruble/person per year.
El = 1500 ruble/person year [7];
Mw-a – adjusting workers and their assistants (mechanics), people;
γw is a coefficient of using adjusting workers at a mobile mechanized unit, stationary tractor maintenance center (a share of annual working time adjusting workers need for their direct work in a mobile mechanized unit, stationary tractor maintenance station);
dav. – average losses for idle hours of tractors, ruble/hour.

This criterion can also be expressed in larger measurement units (rubles) per cycle (rub/cycle) via dividing annual labor intensity of maintaining tractor at the subdivision to labor intensity of maintaining any conventional tractor [6] and [7].

Taking a first derivative to L_{av}. from expression (1) and equating it to zero, we obtain a formula for estimating an optimal average distance of tractors driving to a stationary maintenance station, station of tractor maintenance:

$$L_{av.}^{opt.} = \left(\frac{Ny. \cdot [\,Bb \cdot (\,A\,rep.b.+En) + Be. \cdot (Arep.e.+En)] + Em \cdot Mw. \cdot \gamma w.}{2\,Nl \cdot m \cdot (\alpha\,av. + \frac{d\,av.}{Vav.t})} \right)^{1/2} \quad (2)$$

When using mobile means (mobile mechanized maintenance units), maintenance of tractors is carried out in the field. At the same time, a unit moves to tractor, and for a tractor $V_{av.tr.} = 0$. Therefore, using only first elements of function (1), similarly to expression (2), we obtain a formula for determining an optimal average distance of crossings (optimal service zone) of mobile mechanized maintenance equipment:

$$L_{av.}^{opt} = \left(\frac{Ny. \cdot [Bb. \cdot (\,A\,rep.b.+En) + Be. \cdot (Arep.e+En)] + Em \cdot Mw \cdot \gamma w}{2\,Nl \cdot m \cdot \alpha\,av} \right)^{1/2} \quad (3)$$

If we know $L_{av.}^{opt}$ we can determine an optimal (annual) throughput of stationary and mobile maintenance units (man.h/year).

$$W_{y\,t.y\,(y.t.)}^{opt} = Nl \cdot L_{av.st.(av.c)}^{opt} \quad (4)$$

An optimal throughput complies with a minimum value of specific aggregate, reduced costs for maintaining motor vehicle and tractor facilities of a production unit, which is determined by equation (1) with $L_{av} = L^{opt}_{av.st.(av.c)}$ (ruble/man.h):

$$C_{st.(c)} = \frac{Bb. \cdot (\,A\,rep.b.+En) + Be. \cdot (Arep.e+En)}{Nl \cdot L\,opt\,av.st.(av.c),} + \frac{Fy \cdot \Sigma i\,Cj}{Ny} + \frac{2\alpha\,av. \cdot m \cdot L\,opt\,av.st.(av.c)}{Ny} + \frac{Et \cdot Mw \cdot \gamma w}{Nl \cdot L\,opt\,av.st.(av.c),} + \frac{dav. \cdot Fy}{Ny} + \frac{2dav. \cdot m \cdot L\,opt\,av.st.(av.c)}{Vav.t \cdot Ny} \quad (5)$$

To use formulas (2), (3), (4), (5) in calculations, it is necessary to determine a number of initial values in the following order:

1. A number of maintenance works (M-1, M-2, M-3 and seasonal maintenance) per year for a given set of tractors in a subdivision is determined according to a well-known method and is calculated as a number of maintenance trips (m) per year of all tractors in the subdivision.
2. According to labor intensity standards for different maintenance operations annual the labor intensity of tractor maintenance in the subdivision (Ny) is estimated [8].
3. An annual time and a number of workers required are calculated according to the formula (people):

$$M_w = \frac{Ny}{Fy} \quad (6)$$

where N_y. - labor intensity of maintaining tractors in the subdivision, man.h/year.
F_y. working time resources for one adjusting worker [6, 50], h/year.

4. An actual average driving distance of a tractor for maintenance is determined by the formula (km):

$$L_{av.}^{f.} = 0.1 \cdot \varepsilon \cdot K_c \cdot F^{1/2} \quad (7)$$

where $\mathcal{E}$ is a coefficient depending on the subdivision territory configuration (values given in [5]);
Kc is an average coefficient of curvature of field roads [9];
F – a geographic area of the subdivision, ha

5. The linear density of the maintenance labor intensity is calculated (man.h / year.km):

$$Nl = \frac{Ny}{Lf.av} \quad (8)$$

When analyzing the existing maintenance organization, optimal parameters obtained as proposed above are compared with those actually available in this department. Designing a future maintenance station, the calculation of optimal parameters allows a reasonable choice of standard projects for stationary maintenance facilities, (stationary maintenance stations, and stationary tractor maintenance stations).

We carried out an analysis and found that every 10 years (since 1980) an average wholesale price of one domestically produced machine increased by 1.37-1.45 times, whereas its average capacity only by 1.1-1.4 times. Consequently, an average price of a machine increases faster than its use effect. Undoubtedly, some parameters of machines are improved: a degree of their automation and reliability increase.

The above methodology for estimating optimal parameters related to the work of maintaining and troubleshooting station of motor vehicles and tractors is simple in application for calculations and analysis. In 2010-2019 it was tested theoretically and practically at Faculty of Agricultural Engineering, Russian State Agrarian University – Moscow Agricultural Academy named after V. K.A. Timiryazev and in the farms of Kostanay Region, Kazakhstan.

4. Conclusion

The proposed methodology furthers resolving a wide range of problems related to optimal design of agricultural production operations and processes.

The method considered can be used in diverse agricultural production processes and individual operations. It allows substantiating an optimal use of motor vehicle and tractor facilities of an enterprise, grain cleaning and drying stations of post-harvest grain processing and can be successfully applied for optimization production complexes in crop production and similar optimization calculations.

References

[1] Leonov O A and Shkaruba N Zh 2019 A parametric failure model for the calculation of the fit tolerance of joints with clearance *Journal of Friction and Wear* **40** 4 332-336
[2] Leonov O A, Shkaruba N Zh and Vergazova Yu G 2019 Determining the tolerances in fitting for joints with interference *Russian Engineering Research* **39** 7 544-547
[3] McGregor B A, Kerven C and Toigonbaev S 2009 Sources of variation contributing to production and quality attributes of kyrgyz cashmere in osh and naryn provinces: implications for industry development *Small Ruminant Research* **84** 1-3 89-99
[4] Aldoshin N and Didmanidze O 2018 Harvesting lupinus albus axial rotary combine harvesters *Research in Agricultural Engineering* **64** 4 209-214
[5] Vinogradov O V, Moskvichev D A, Didmanidze O N and Parlyuk E P 2019 Methods of analyzing the structure of the modular car park and the intensity of its operation *Indo American Journal of Pharmaceutical Sciences* **6** 3 5289-5292
[6] McGregor B A, Kerven C and Toigonbaev S 2011 Sources of variation affecting cashmere grown in the Pamir mountain districts of Tajikistan and implications for industri development *Small Ruminant Research* **99** 1 7-15
[7] Ohrimenko O I, Vilisova M L, Rokotyanskaya V V, Bidova B B, Popovskaia A S 2019 Analysis and synthesis of complex technological systems *Journal of mechanics of continua and mathematical sciences* **14** 6 322-328
[8] Erokhin M N, Leonov O A, Shkaruba N ZH, Kataev YU V, Vergazova YU G 2020 Assessing the relative interchangeability in joints with preload *Russian engineering research* **40** 6 469-472
[9] Sergeyeva N V 2020 The cost planning and cash limits for repair and maintenance work in the AIC *Studies in systems, decision and control* **282** 585-594

IOP Conf. Series: Materials Science and Engineering 1100 (2021) 012007 doi:10.1088/1757-899X/1100/1/012007

Methodology for substantiating optimal location of oil storage, supply and transportation facilities in agricultural and industrial companies of Kostanay Region, Kazakhstan

G I Bondareva[1], S K Toigonbaev[2] and O N Didmanidze[2]

[1] All-Russian Research Institute for Hydraulic Engineering and Land Reclamation (VNIIGiM), 44/2, Bolshaya Akademicheskaya street, Moscow, 127550
[2] Russian State Agrarian University-Moscow Agricultural Academy named after K.A. Timiryazev, 49, Timiryazev street, Moscow, 127550

E-mail: Boss2569@yandex.ru

Abstract. This article proposes a methodology for substantiating an optimal location of oil storage facilities in agricultural and industrial companies, for estimating a consumption of fuel and greases, a placement of reserves and their operational distribution in the company. This approach allows reducing costs on providing consumers with petroleum products and saving fuel and oil when operating motor vehicles and tractors.

1. Introduction

In recent years, a mainstream of agriculture concentration and cooperation has been the establishment of inter-farms to mechanize agricultural works and regional agricultural and industrial companies responsible for the management of various farms, their motor vehicles and tractors, oil depots, oil storages, tank trucks and refueling facilities.

The concentration of agricultural machinery and material and technical facilities in inter-farms and regional companies makes a more efficient use of motor vehicles and tractors and allows their efficient management and controlling spare materials and technical means (petroleum products, spare parts, etc.).

A timely supply of high-quality petroleum products to motor vehicles and tractors of farms represents a burning issue which needs a conceptual solution. The latter is related to the development of maintenance programs; their management is based on unifying and coordinating information relevant for economic processes of a maintenance company. An analysis of processes results in a certain option for the development of a service product. In most cases, a decisive role in designing technical service programs is played by traditional methods of analysis focused on experience and intuition of maintenance company managers. In other words, a decision-making process, i.e. a choice of one of the possible actions hardly involves special research methods, since situations are very simple and management of maintenance companies has enough traditional knowledge and skills. However, sometimes situations are so difficult that there are doubts whether chosen solutions are correct. In such cases, it is necessary to apply techniques of formalized description and analysis of situations. Here is an approximate (and rather arbitrary) sequence of such an analysis:

- developing a model process or operation, e.g. as mathematical description;
- analysis of uncertainties and goal formalization, establishment of a target function and particular efficiency criteria;
- developing centralized management to process available or received information;
- solution to optimization related problems.

Any agricultural company should have oil storage, supply and transportation facilities to store, supply and transport petroleum products.

A rational structure of oil storage, supply and transportation facilities in an agricultural company is established depending on specific operating conditions of motor vehicles and tractors, state of roads, distance from the place where machines are operated to oil storage and filling points and other local features. Choosing a scheme for constructing and organizing oil storage, supply and transportation facilities, it is necessary to

Content from this work may be used under the terms of the Creative Commons Attribution 3.0 licence. Any further distribution of this work must maintain attribution to the author(s) and the title of the work, journal citation and DOI.
Published under licence by IOP Publishing Ltd

have a proper balance between delivery, storage and refueling capacities of oil products and maintenance means of motor vehicles and tractors to ensure an efficient use of equipment.

Centrally solved issues of delivery, storage, accounting the consumption of fuel and greases, placement of reserves and their operational distribution in the company are relevant for an entire administrative region, they allow reducing costs on providing consumers with petroleum products and saving fuel and oil when operating motor vehicles and tractors.

When agricultural and industrial companies and their subdivisions operate in such conditions, it is necessary to clarify the procedure of supplying fuel and greases to motor vehicles and tractors, locate oil storages and filling stations, taking into account bases of oil-supplying companies, an average distance to oil depots, oil storage facilities and oil-supplying companies. The problems above can be solved with the methodology described below intended to substantiate an optimal placement of oil facilities both in one company and in an administrative district on the whole.

Large volumes of consumed petroleum products, e.g. an average-sized company needs annually up to 600 tons of diesel, 400 tons of gasoline and 500 tons of diesel oil, as well as a significant share of expenditures on oil storage, supply and transportation (20%) in mechanized works, give grounds to claim the solution to this problem is of national economic relevance.

2. Materials and Methods

The proposed methodology for substantiating optimal placement of oil facilities is a step forward of methodological foundations developed in [2, 3, 4, 6] and focused on optimizing work parameters of an agricultural machinery maintenance station. The operation of oil facilities is similar to that of an agricultural machinery maintenance station. Therefore, basic work parameters of these branches are identical.
These include the following points:
1. An average distance of transporting oil products from storages and supply centers to farms and consumers (working groups, brigades, etc.) or to subdivisions in the administrative region.
2. An annual throughput of an oil storage or filling center, station, tanker truck and other stationary equipment, t/year, which makes it possible to determine a necessary oil container capacity in m^3.
3. Total, reduced monetary costs on storage of oil products, their transportation, filling and greasing operations per one ton of fuel – greasing materials, ruble/t.

3. Results and Discussion

In this connection, the target function to provide a solution to the considered problem from the standpoint of national economic efficiency is stated in the following expression suitable to the most generalized organization of oil facilities:

$$C = \frac{Bmech \cdot (Arep.mech + Ei) + Bg \cdot (A\,rep.e + Ei)}{Nl \cdot Lav} + \Sigma i \cdot Cj + 2\alpha av \cdot Lav + Lf \cdot \Sigma_m \cdot \mu_{l.f.g.} \cdot Pf + \Sigma_m \cdot \delta l.f \cdot Pf + \Sigma_m \cdot \frac{\Sigma_m \cdot Cf.t \cdot \rho_T \cdot Pf \cdot Ke.f.}{Kc.t \cdot \Sigma_m \cdot Cf.m \cdot \rho f} + \frac{Ms.w. \cdot Kf \cdot \beta av}{Nl \cdot Lav} \rightarrow min \quad (1)$$

where B_{mech} B_g – balance price (value) of construction, tank and other equipment of oil facilities, ruble.
$A_{rep.mech.}$, $A_{rep.e}$ – a coefficient of annual expenditures on renovation of constructions (buildings and structures) and equipment of oil facilities, 1/year;
Ei – a standard coefficient of capital investments efficiency, 1/year (Ei =0.15);
N_l – a linear density of fuel and greasing materials, t/km (an amount of oil products (ton) per one km of an average distance of their transport to consumers; it is calculated according to the formula (t/km):

$$Nl = \frac{Qy.d.}{L_{av.d.}} \quad (2)$$

where $Q_{y.d.}$ – a need of a department (other subdivision) in fuel and greasing materials, t/year;
$\Sigma i \cdot Cj$ – a sum of standardized operation costs of oil facilities (salaries of employees of oil storages, warehouses and centers with all charges, expenditures on repair of construction and installation units, repair and maintenance of reservoirs and equipment, electric power supply, etc.);
Σ_m – total reduced costs on transportation of oil products with a variety of technical means (tank trucks, refilling machines, etc.) per one unit of transport work, ruble/t.km;
$\mu_{l.f.g.}$ – a coefficient taking into account losses of fuel and greasing materials when transported per one ton-km, t/t-tm, (1/km);
P_f – price (cost) of one ton fuel and greasing materials, ruble/ton;

$\delta l.f$ – a share of fuel losses when loading and unloading per 1 ton;
$C_{f.m.}$ – capacity of m fuel, m^3;
ρf – a fuel density, t/m^3.
$K_{e.t,}$ – a fuel evaporation coefficient calculated as a share of evaporating fuel in a total annual throughput of oil facilities 1/year;
$K_{c.t.}$– a frequency of tanks circulation for a particular fuel, 1/year;
$M_{s.w.}$ – a number of service workers of oil storage, center or station, man.
K_l – an efficiency coefficient of labor resources, ruble/man year;
β_{av} – a coefficient purposeful employment of service workers within a year.

The function of specific total reduced costs for maintenance and troubleshooting of motor vehicles and tractors in a subdivision (ruble/ man hour):

$$C = \frac{Bb.\cdot(A\,rep.b.+En)+Bg.\cdot(Arep.e.+En)}{NL\cdot Lav.} + \frac{Fy.\cdot \Sigma i\,Cj}{Ny} + \frac{2\alpha\,av.\cdot m\cdot Lav}{Ny} + \frac{Ef\cdot Mw\cdot \gamma\,w}{Nl\cdot Lav.} + \frac{dav\cdot Fy.}{Ny} + \frac{2d\,av.\cdot m\cdot Lav}{Vav.t.\cdot Ny} \quad (3)$$

Taking a first derivative to $L_{av.}$ from expression (3) and equating it to zero, we obtain a formula for estimating a minimal average distance of oil products transport from oil storages to a consumer (km):

$$L_{av}^{opt} = \left(\frac{Bmech.\cdot(Arep.mech.+En)+Be\cdot(Arep.e+En)+Ms.w.\cdot Kf\cdot\beta av}{Nl\cdot(2dav+\cdot\Sigma_m\cdot\mu_{l.f.g.}\cdot Pf}\right)^{1/2} \quad (4)$$

If we know $L_{av.}^{opt}$ we can determine an optimal (annual) throughput of oil storage, center, station or mobile refuel capacity according to the formula (t/year):

$$Q_{year}^{opt} = N_l \cdot L_{av}^{opt} \quad (5)$$

Substituting Q_y^{opt} into (3), we obtain minimal (optimal) total reduced costs on storage, transport and loading and unloading works per 1 ton of fuel and greasing materials.

A volume of tanks for storing a production fuel stock is calculated in a percentage of annual demand, taking into account a various monthly fuel consumption and using the formula [5] (m^3):

$$G = \frac{Q\cdot E_{max}}{100\cdot\gamma\cdot m}, \quad (6)$$

where Q – a fuel consumption, t;
E_{max} – a maximal amount of fuel in % of annual consumption;
γ – a fuel density, t/m^3 (for diesel – 0.82) [5];
m – a coefficient of tanks filling, m = 0.95 [5].
E_{max} depends on the maximal coefficient of various monthly fuel consumption K_{max}.

$$Kmax. = \frac{Q_{max.m.}}{Q_{av.m.}}, \quad (7)$$

where $Q_{max.m.}$ – a maximal monthly demand of diesel, kg;
$Q_{av.Mm.}$ – an average monthly demand of diesel, kg.

An estimated volume is smoothed to values of standard tanks.

The described methodology makes it possible to analyze the existing organization of oil storage, supply and transportation at various production levels via comparing optimal work parameters with those really available in a company subdivision, as well as to select standard designs of oil depots in specific production conditions.

4. Conclusion

A distinctive feature of these techniques is the use of linear density of process labor intensity or an amount of fuel and grease or other units per one kilometer of an average transportation distance and characterizing the spatial difficulty degree of the process.

The use of linear density as a main design parameter simplifies optimization calculations, and improves their accuracy when taking into account specific field and road conditions of production units.

References

[1] McGregor B A, Kerven C and Toigonbaev S 2011 Sources of variation affecting cashmere grown in the Pamir mountain districts of Tajikistan and implications for industri development *Small Ruminant Research* **99** 1 7-15

[2] Aldoshin N and Didmanidze O 2018 Harvesting lupinus albus axial rotary combine harvesters *Research in Agricultural Engineering* **64** 4 209-214

[3] Kiryushin V I 2007 Assessment of land quality and soil fertility for planning farming systems and agrotechnologies *Eurasian Soil Science* **40** 7 785-791

[4] Borisov B A and Ganzhara N F 2008 Geographical features of the distribution and renewal of easily decomposable organic matter in virgin and arable zonal soils of European Russia *Eurasian Soil Science* **41** 9 946-952

[5] Zolotarevskaya D I 2011 Mathematical simulation and calculation of the soil compaction under dynamic loads *Eurasian Soil Science* **44** 4 407-416

[6] [Nikolsky Yu N, Castillo-Alvarez M, Bakhlaeva O S, Gama-Castro J and Landeros-Sanchez C 2010 Assessing the effect of possible global climate changes on the fertility of mexican soils and the prediction of crop yields *Eurasian Soil Science* **43** 9 985-992

[7] Dumnov A D, Khoruzhy L I, Kharitonova A E, Ukolova A V, Skachkova S A 2019 Eco-economic systems of russian agriculture: statistical analysis *Journal of reviews on global economics* **8** 362-372.

[8] Kerimov V E, Makhmudov A R, Khamidova L L, Khasanova S S 2019 Problems of using innovative potential in order to increase the efficiency of innovative activities in agricultural organizations *Espacios* **40** 20 4.

IOP Conf. Series: Materials Science and Engineering 1100(2021) 012008 doi:10.1088/1757-899X/1100/1/012008

The use of energy routers to improve reliability of power supply to non-traction consumers

Yu N Bulatov[1], A V Kryukov[2,3], A E Kryukov[2] and A V Cherepanov[2]

[1] Bratsk State University, 40, Makarenko st., Bratsk, Russia
[2] Irkutsk State Transport University, 15, Chernyshevsky st., Irkutsk, Russia
[3] Irkutsk National Research Technical University, 83, Lermontova st., Irkutsk, Russia

E-mail: bulatovyura@yandex.ru

Abstract. The purpose of the studies presented in the article was to determine the effectiveness of the use of energy routers (ER) in railway power supply systems to increase the uninterrupted power supply to non-traction consumers, reduce unbalance and nonsinusoidality, and reduce voltage dips in emergency modes. The research was carried out by means of computer modeling using the Matlab software system based on the developed models of power supply systems (PSS) equipped with ER with energy storage devices. The results obtained show that in standard circuits of PSS for linear consumers with short circuits in the supply network, voltage dips occur, the depth of which reaches 100%. The use of an ER and an energy storage device connected to DC buses allows maintaining voltage at consumer clamps near the nominal value in case of short circuits in the 10 kV feed line and 25 kV overhead contact system. The ER significantly limits the recharge of the short circuit location from the power storage. In addition, on the basis of the ER it is possible to provide the normative quality of electricity considering voltage deviations, asymmetry and harmonic distortions on the 0.4 kV buses of the substation that feeds non-traction consumers.

1. Introduction

In modern conditions, increased requirements for efficiency, quality of electric power and reliability of energy supply to consumers are imposed on power supply systems (PSS). It is possible to meet these requirements by upgrading PSS using smart grid (intellectual electric power supply network) technologies [1-5]. Smart grid networks allow one to implement a new generation of PSS, different from the existing ones by the following features:

- Uninterrupted and guaranteed supply of consumers with high quality electric power (EP);
- Flexible interaction of PSS segments through the use of active devices and smart controls;
- The availability of interfaces that connect EP storage devices and distributed generation (DG) plants that use, among other things, renewable sources.

A promising direction allowing the implementation of the listed properties consists in the use of energy routers (ER) [6-23], the central part of which are solid-state transformers (SST) [7, 8]. The ER can be classified as a cyber-physical system that provides management of energy flows, information exchange between active devices and electrical receivers, and integration of energy storage devices (ESD) and DG plants into PSS. The most significant effect of the introduction of ER equipped with ESD is a significant increase in the reliability of electric power supply in emergency and post-emergency PSS modes. In addition, it is necessary to note the possibilities of ER to improve the electric power quality, which is especially important for the power supply system of rail transport.

On the railway tracks there are transport facilities, which in terms of the degree of uninterrupted electricity supply belong to a special group and impose increased requirements for the continuity of electricity supply and the quality of the electric power supplied [24]. These consumers include telecommunications facilities, as well as signaling and interlocking arrangements (SIA). The safety of these trains depends on the reliable operation of such facilities.

Content from this work may be used under the terms of the Creative Commons Attribution 3.0 licence. Any further distribution of this work must maintain attribution to the author(s) and the title of the work, journal citation and DOI.
Published under licence by IOP Publishing Ltd

Below are the results of a simulation of power supply system modes designed to supply power to these facilities and the energy supply organization (ESO) connected to the network through an ER.

2. Energy router structure

The central link of the ER, which connects to medium voltage networks of 6-10 kV and can be used in PSS of telecommunication facilities or signaling and interlocking arrangements, is a solid-state transformer (Figure 1). This device is usually made in a single-unit design and includes the following segments: high-frequency power transformer; two active bridges made on the elements of power electronics.

The following positive results are achieved by having DC buses and high-frequency voltage conversions:

- the size and weight of the power transformer is significantly reduced;
- it is possible to connect DC- and AC-powered electric receivers, as well as storage devices in the form of accumulator batteries;
- an interface is formed to connect DG plants implemented on the bases of synchronous and asynchronous machines, DC generators, photovoltaic panels, and fuel cells.

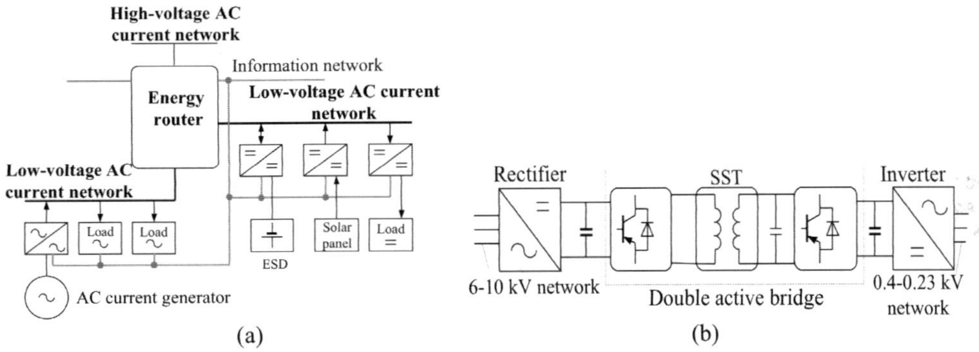

Figure 1. Energy router: (a) – power supply scheme with an energy router; (b) – a diagram of a router built on the basis of a solid-state transformer (SST).

3. Simulation results

To assess the effectiveness of the use of energy routers to organize highly reliable power supply for linear consumers of a special group, which include telecommunications facilities, as well as signalling and interlocking arrangements, modeling of normal and emergency modes was performed. The electricity system under study (Figure 2) included a traction substation, fragments of the overhead contact system, a 10 kV power transmission line mounted on separate supports, to which a linear consumer was connected through a 10/0.4 kV transformer to a specific group of the first category. The simulation was done with the Simulink and SimPowerSystems packages of the MATLAB system. The diagram of the developed computer model of energy router is shown in Figure 3.

When simulating the ER, a high-frequency 10/0.4 kV transformer with a capacity of 3000 kVA was used. The consumers' load on 0.4 kV buses was $1 + j0.5$ MVA. The simulation compared the quality of electricity in the PSS using an ER and a standard power transformer of the same power.

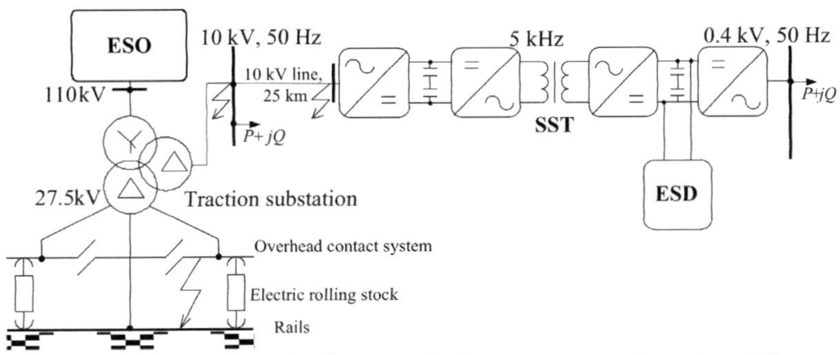

Figure 2. Power supply diagram of a linear consumer through an ER.

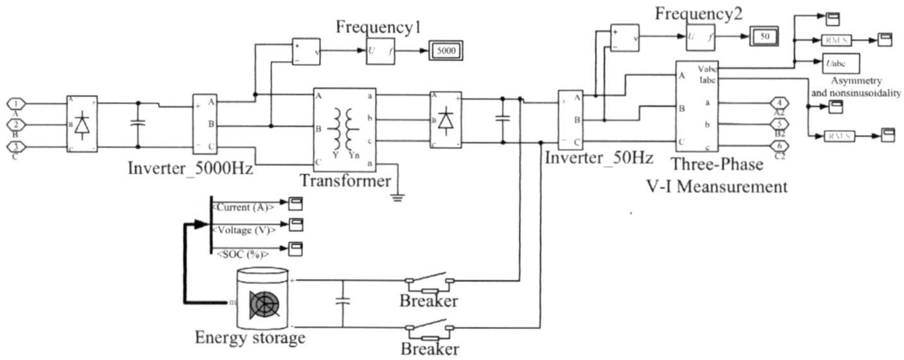

Figure 3. Schematic diagram of an ER model equipped with an energy storage.

As a result of the simulation, it was found that when the ER is switched on without a high harmonics filter, there are significant harmonic distortions in the 10 kV network on the high voltage side compared to the situation in which the standard transformer was considered: voltage harmonics ratios increase by about 3...4 times. This is due to the fact that a power router equipped with sophisticated power electronics generates additional odd harmonics into the network. The use of a passive filter on the 10 kV side allows one not to pass through these harmonics into the network. The results of the measurement of electric power quality indicators (EPQIs) on 10 kV TP buses, the receiving end of the 10 kV line and on 0.4 kV buses are presented in the table 1, and the corresponding time dependences of voltages – in Figures 4 and 5.

Table 1. The results of the measurement of electric power quality indicators.

A method of connecting consumers of signaling and interlocking arrangement	Electric power quality indicators measuring place	Voltage harmonic coefficients, %			Voltage unbalance ratio in reverse sequence, %
		k_{Ua}	k_{Ub}	k_{Uc}	k_{2U}
Through a standard power transformer	10 kV TP buses	6.9	7.2	5.2	5.2
	25 km of 10 kV after power transmission line	5.9	6.1	4.3	5.2
	On 0,4 kV buses	5.4	5.5	4.01	5.2
Through ER with harmonic filter	10 kV TP buses	6.6	6.6	4.5	5.3
	25 km of 10 kV after power transmission line	3.3	6.8	5.4	5.5
	On 0.4 kV buses	0.2	0.2	0.2	0.03

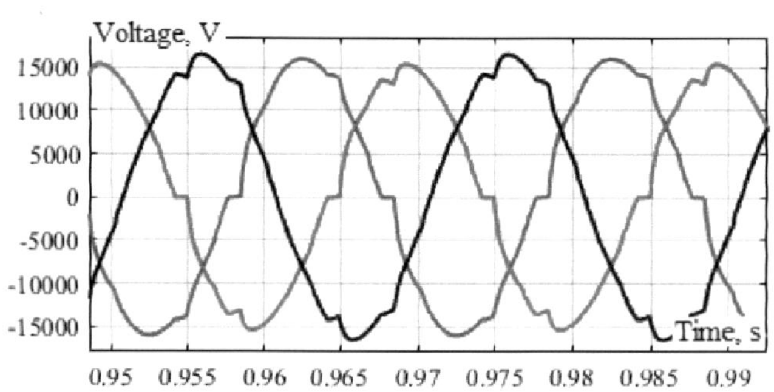

Figure 4. Time dependences of voltages on 10 kV TP buses.

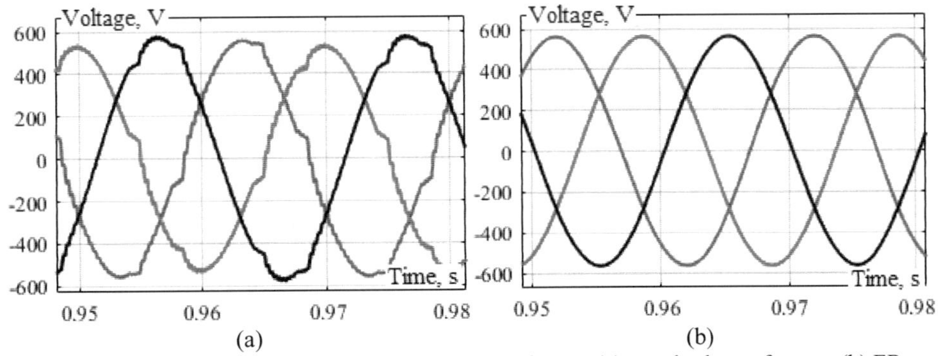

Figure 5. Time dependences of voltage on 0.4 kV buses: (a) standard transformer; (b) ER.

The results of computer modeling led to the following conclusions:

- to eliminate a decrease in the electric power quality in the supply network according to the criterion of sinusoidal voltage and current curves when using an ER, it is necessary to use harmonic filters;
- The use of an ER allows one to solve the problem of low quality of electric power in the network of 0.4 kV consumers, completely eliminating voltage unbalance and harmonic distortion.

With the help of an ER, one can connect power storage devices to the DC network without the use of additional equipment. In the experiments, we used a model of the ESD, operating on the basis of lithium-ion batteries, with a control system that monitors the consumers' voltage in order to connect the ESD when the power drops and turn it off if the set value is exceeded.

Below are the results of simulations of emergency modes caused by three-phase short circuits (SC) on 10 kV TP buses, the receiving end of the 10 kV line and the SC of the contact wire on the rail in the 25 kV network. Voltage dips with 0.4 kV consumers were investigated for the following situations: 1) a standard PSS without an ESD; 2) a standard PSS with an ESD; 3) a standard PSS with an ER without an ESD; 4) a standard PSS with an ER and with an ESD.

For situation 2, the ESD was connected through an inverter on 0.4 kV buses. The simulation results are shown in Figures 6-8 in the form of time dependences of voltages on 0.4 kV buses of a non-traction consumer (telecommunications facility or signaling and interlocking arrangement).

The simulation results of emergency modes made it possible to formulate the following conclusions:

- In a standard PSS designed to in-feed telecommunications facilities or signaling and interlocking arrangements, in the considered emergency modes significant voltage dips are observed (Figures 6-8), the depth of which reaches 100% of the nominal;
- The use of ESD in such circuits, connected through an inverter on the 0.4 kV side of the standard power energy storage, makes it possible to reduce the voltage dip by an average of 60% (Figures 6-8). However, in this case, there is a large recharge of the damage area with current from the ESD. The corresponding time dependences of the currents flowing on the 10 kV side of the feeder transformer are shown in Figure 9, a.
- The ER allows reducing the voltage dip up to 20% of the nominal value due to the presence of rectifiers, inverters and LC filters. The use of a ER and a controlled ESD connected to the DC buses allows maintaining the voltage close to the nominal value of the consumers in all the considered emergency modes; in this case, the feeding of the damage area by the current from the ESD is significantly limited. The corresponding time dependences of the currents flowing on the 10 kV side of the ER are shown in Figure 9, b.

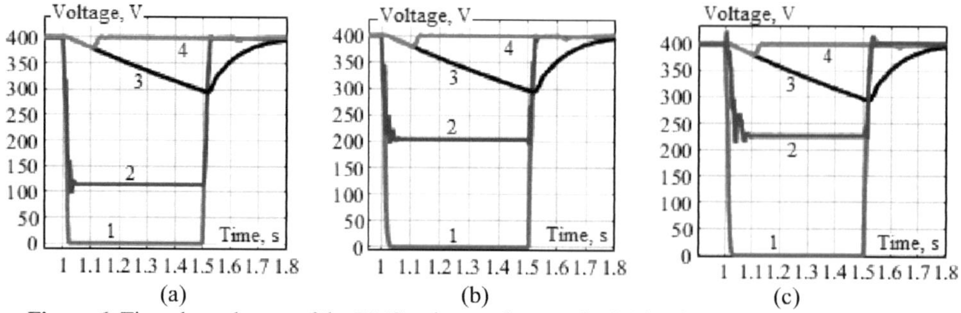

Figure 6. Time dependences of the RMS voltage values on the 0.4 kV buses: (a) – at a three-phase SC at the receiving end of a 10 kV line 25 km long; (b) – with three-phase SC on 10 kV TP buses; (c) – in case of a SC in a 25 kV contact network; 1 – a standard transformer without an ESD; 2 – a standard transformer with an ESD; 3 – an ER without an ESD; 4 – an ER with an ESD.

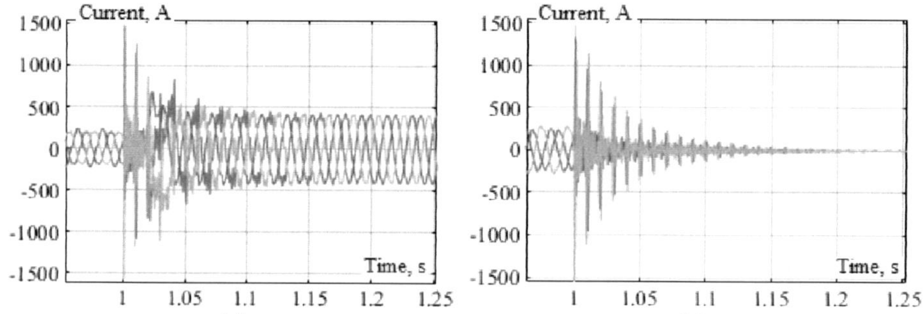

Figure 7. Currents flowing on the 10 kV side of the transformer with a SC at the receiving end of the 10 kV line: (a) – a standard 10/0.4 kV transformer was used; (b) – an ER was used.

4. Conclusion

An energy router built on the basis of an SST with a double active bridge can be used to form reliable power supply systems for critical non-traction consumers, which include telecommunication facilities and signaling and interlocking arrangements. This approach allows one to get high electric power quality indicators on the buses of the consumer's 0.4 kV substation by voltage deviations, unbalance (k_{2U} = 0.03%) and harmonic distortions (k_U = 0.2 %). In emergency modes caused by short circuits on the power transmission line or in the overhead contact system, the power supply system, equipped with an energy router with an energy storage, makes it possible to eliminate voltage dips on 0.4 kV buses.

Acknowledgments
The research was financially supported by a grant from the state task of the Ministry of science and higher education of Russia (project no. 0667-2020-0039).

References

[1] Buchholz B M and Styczynski Z A 2014 *Smart Grids – Fundamentals and Technologies in Electricity Networks* (Springer-Verlag Berlin Heidelberg) p 396
[2] Wang J, Huang A Q, Sung W, Liu Y and Baliga B J 2009 *IEEE Industrial Electronics Magazine* **2** 16-23
[3] Morzhin Yu I, Shakaryan Yu G, Kucherov Yu N et al. 2011 *Preprints of proceedings of IEEE PES Innovative Smart Grid Technologies Europe 2011* (Manchester, GB: IEEE, The University of Manchester) pp 1-5.
[4] Mohsen F N, Amin M S and Hashim H 2013 *IEEE 7th International Power Engineering and Optimization Conference (PEOCO)* (Langkawi, Malaysia) pp 427–31
[5] Torriti J 2012 *Energy Policy* **44** 199-206
[6] Juneja A and Bhattacharya S 2011 *IEEE International Conference on Smart Grid Communications (SmartGridComm)* pp 31–6

[7] Shilpakala G B and Prassad M J 2014 *Proceedings of 11th IRF International Conference* (Pune. India) pp 15-21
[8] Zhao T, Yang L, Wang J and Huang A Q 2007 *Electric Ship Technologies Symposium (ESTS)* (IEEE) pp145–9
[9] Rathod D K 2014 *Advance in Electronic and Electric Engineering* **4** 45-50
[10] Alzola R P, Gohil G V, Mathe L, Liserre M and Blaabjerg F 2013 *Proceedings of the IEEE Energy Conversion Congress and Exposition: ECCE* (IEEE Press) pp 380–7
[11] Wang K, Liu X, Zhao L, Zhou Y and Xu D 2019 *IEEE Power & Energy Society Innovative Smart Grid Technologies Conference (ISGT)* (Washington, DC, USA) pp 1-5
[12] Hambridge S, Huang A Q and Yu R 2015 *IEEE Energy Conversion Congress and Exposition (ECCE)* (Montreal, QC, Canada) pp 2355–60
[13] Zhang J, Wang W and Bhattacharya S 2012 *IEEE PES Innovative Smart Grid Technologies (ISGT)* (Washington DC, USA) pp 1-8
[14] Qin Y, Hua H and Cao J 2019 *IEEE International Conference on Energy Internet (ICEI)* (Nanjing, China) pp 161–6
[15] Ma Y, Wang X, Zhou X and Gao Z 2017 *29th Chinese Control And Decision Conference (CCDC)* (Chongqing, China) pp 4104–8
[16] Liu B, Wu W, Zhou C, Mao C, Wang D, Duan Q and Sha G 2019 *IEEE Access* 7 109069–82
[17] Wu R, Wang B, Zou Y, Fan B, Li L and Zhu Z 2017 *IECON 2017 - 43rd Annual Conference of the IEEE Industrial Electronics Society* (Beijing, China) pp 7771–6
[18] Han X, Yang F, Bai C, Xie G, Ren G, Hua H and Cao J 2017 *IEEE International Conference on Energy Internet (ICEI)* (Beijing, China) pp 320–5
[19] Sosnina E N, Shalukho A V, Shumsky N V and Kovtun S A 2017 *Science. Mysl' Publ.* **7** 22-27
[20] Sosnina E N, Shumsky N V and Shramko P A 2020 *All-Russia scientific-technical conference Information Technologies in Electrical Engineering and Electric Power Engineering (ITEE0)* (Cheboksary, Russia) pp 328–30
[21] Malysheva N N and Tretyakov E A 2019 *Bulletin of the Perm National Research Polytechnic University. Electrical engineering, information technology, control systems* **31** 22-34
[22] Bulatov Yu N, Kryukov A V and Arsentiev G O 2020 *Distributed generation and energy routers in railway power supply systems* (Moscow; Berlin: Direct Media Publ.) p 171
[23] Bulatov Yu N, Kryukov A V and Arsentiev G O 2018 *International Ural Conference on Green Energy (UralCon)* (Chelyabinsk, Russia) pp 143–8
[24] Kryukov A V and Zakaryukin V P 2020 *Power supply and in-feeding for non-traction consumers of railway transport* (Moscow; Berlin: Direct Media Publ.) p 293

Designing of reed switches for sensors and security alarm devices

M E Kalinkina[1], A G Korobeynikov[1,2], O I Pirozhnikova[1], N A Shmakov[1] and V L Tkalich[1]

National Research ITMO University[1], Federal State Budgetary Institution of Science Institute of Earth's magnetism, ionosphere and radio wave propagation. N.V. Pushkova Russian Academy of Sciences[2]

E-mail: mariia_kalinkina@mail.ru

Abstract. The article is devoted to the actual problem of designing a highly reliable domestic elemental reed switch base of devices designed to create sensors for monitoring fluid flow and detectors for car alarm systems. The developed method of analyzing the statics and dynamics of contact groups of reed switches is implemented when designing new types of magnetically controlled contacts. New designs of reed switches have successfully passed technical tests and are protected by patents in the Russian Federation.

1. Introduction

The development of modern road transport is steadily leading to an increase in technical and metrological requirements for detectors, distributors and reed switch devices, which are an integral part of fluid flow control systems and security alarm devices. It should be noted that important characteristics of magnetically controlled contacts (reed switches) such as speed, absence of chatter, mechanical stability indicators are ensured by a high quality of the elastic elements of the switching groups of magnetically controlled contacts included thereto [1-4].

A further increase in requirements of quality and safety of fire alarm and alarm devices leads to the need to develop new computational and experimental methods for studying the most important, in terms of reliability and metrological characteristics, elastic element base of reed switches.

The writings of the following domestic scientists were devoted to the analysis of statics and dynamics: Korsunov V.P., Andreeva L.E., Khazazova K.I., Shoffa V.N., Karabanova S.M., Mayzelsa R.M. other.

In modern models of automotive vehicles reed switches are widely used in various liquid level sensors (gasoline, antifreeze, water), as well as in security alarm systems, which makes them a highly sought-after element in the global market of magnetically controlled contacts. Automotive reed sensors comprise membranes and platelike sensitive elements. The reliability of these elastic sensing elements requires the improvement of methods for their calculation, as well as the creation of computer-aided design systems of the switching groups of magnetically controlled contacts in the ANSYS environment and the SolidWorks environment, which allows an automatic construction of 3D models of reed switches.

A combination of properties and profitability of reed detectors and security devices are in many ways better than semiconductor analogs, which leads to their widespread use in automotive vehicles.

Content from this work may be used under the terms of the Creative Commons Attribution 3.0 licence. Any further distribution of this work must maintain attribution to the author(s) and the title of the work, journal citation and DOI.
Published under licence by IOP Publishing Ltd

The development of models of shell and platelike switching groups of magnetically controlled contacts for the static mode can be carried out using deterministic or stochastic models.

In order to build mathematical models of shell and platelike switching groups of magnetically controlled contacts in a dynamic mode, it is necessary to analyze the amplitude and phase-frequency characteristics.

In this document new mathematical models of switching groups of magnetically controlled contacts in the devices of automotive vehicles for static and dynamic modes of operation are developed.

2. Formulation of the problem

Modern systems of physical protection of objects include security and fire alarm and alarm signaling detectors, namely magnetic contact and reed sensors comprising a modernized membrane and platelike elastic-deformed element base with a programmable strain value.

The authors of the work carried out the development and study of methods for analyzing thin wall sensitive elements of a multilayer structure. This goal is achieved due to the use of a multiparameter approach for the numerical simulation of dynamic processes occurring in the control and measuring sensors of the alarm system. In the course of numerical modeling of the processes of elastic deformation of the thermobimetallic membrane element of the microsensor, a number of design schemes were used, which ensured required operational characteristics and high reliability indicators of the physical protection systems of the objects.

Experimental verification of the developed theoretical models has been carried out.

A Russian patent application "heat fire detector" was prepared and a patent in the Russian Federation for a "temperature sensor" was received.

The development of a methodology for designing a membrane and platelike elastically deformable element base for fire alarm and alarm signaling detectors has been carried out.

3. Tasks to be solved

The goal solved by the authors is analyzing the statics and dynamics of the shell and platelike switching groups of magnetically controlled contacts devices of automotive vehicles. To achieve this goal the following tasks were solved:

- analysis of the patent and scientific literature dedicated to the problems of analysis of statics and dynamics of contact groups of reed switches was carried out;
- new elements were introduced into the classification of platelike and membrane contact sensitive reed switch groups;
- an analysis of the trends and prospects for the development of fire alarm and alarm sensors based on reed switches was carried out;
- analysis of the advantages and disadvantages of the existing methods of analysis of statics and dynamics, as well as the design and technological features of the switching groups of magnetically controlled contacts fire alarm and alarm signaling sensors were carried out;
- development of new mathematical models for the analysis of statics and dynamics of switching groups of magnetically controlled contacts was carried out;
- a library of finite elements of flat and membrane reed switches was created for the development of automated systems for designing sensitive elements of magnetically controlled contacts in the ANSYS environment;
- mathematical models of the attached and concentrated masses of the switching groups of magnetically controlled contacts were developed.

Obtained results are based on a systematic approach to the problem in question. They are based on numerical methods for calculating shell and platelike structures. For the construction of mathematical models of the added fluid masses, the methods of sections and electromechanical analogies were used. For the creation of mathematical models of the switching groups of magnetically controlled contacts when working in a static mode, methods of probability theory were applied. When calculating the amplitude-frequency characteristics of the elements of the switching groups of magnetically controlled contacts, the half-division method was used.

In the course of the work, an extended classification of modern types of switching groups of magnetically controlled contacts for liquid flow detectors and security devices of motor vehicles was carried out, the development of mathematical models of shell and platelike switching groups of magnetically controlled contacts for static operation was carried out, mathematical models of shell and platelike switching groups of magnetically controlled contacts during the dynamic mode were developed, derivation of the switching groups of magnetically controlled contacts equations for mercury wetted contacts was made; a finite element library has been developed for computer-aided design of switching groups of magnetically controlled contacts detectors and security alarm devices of motor vehicles.

4. Results

As a result of the work, the authors of the article obtained the following results:

1. Mathematical models for the analysis of static and dynamic processes in switching groups of magnetically controlled contacts detectors of fluid flow and alarm sensors in automotive vehicles were developed;

2. Equations for calculating added and concentrated masses to eliminate errors in estimating the dynamics of the switching groups of magnetically controlled contacts are derived;

3. Criteria are derived and conditions for the stability of membrane and platelike switching groups of magnetically controlled contacts liquid flow detectors and security alarm sensors in automotive vehicles are specified.

This study is of great importance due to the ever-growing requirements for alarm systems, namely their reliability characteristics, since reliability is determined by the quality level of the elastic elements of the switching groups of magnetic contacts included in their composition.

A new reed switch model has been developed and patented with improved reliability indicators for use in liquid flow detectors and security alarm sensors in automotive vehicles.

Thus, the use of modified elastically deformable elements in magnetic contact and heat detectors starts a new line in the design of modern fire alarm and alarm signaling devices, which once again confirms the relevance of the problem under consideration.

The resulting models can be used in computer-aided design of reed switches for the case of mass production or individual production.

The correctness and adequacy of the results obtained in the work was confirmed in the course of an analytical comparison with the data of experimental studies and results presented in other scientific sources.

References

[1] Labkovskaya R Ya, Pirozhnikova O I and Tkalich V L 2014 Condition and stability criterion of elastic sensitive elements of reed switches *News of higher educational institutions. Instrument Engineering* **57** (10) 34-37

[2] Labkovskaya R Ya, Tkalich V L, Kozlov A S and Pirozhnikova O I 2016 Analysis of the dynamics of multilink sensitive elements of reed switches of control systems *Almanac of scientific works of young scientists of ITMO University* **3** 88-90

[3] Tkalich V L, Labkovskaya R Ya and Pirozhnikova O I 2012 Analysis of the attached masses of elastic sensitive elements of mercury reed switches *News of higher educational institutions. Instrumentation* **55** (7) 32-35

[4] Kozlov A S, Shmakov N A, Tkalich V L, Labkovskaia R I, Kalinkina M E and Pirozhnikova O I 2018 Development library of finite elements for computer-aided design system of reed sensors *Journal of Physics: Conference Series* **1015** 032080

[5] Lepeshkin O M , Kopytov V V and Beetle A P 2009 Integrated security and technical equipment for fire alarm *Helios* M ARV 288
[6] Rude I I 2012 Alarm systems *Technical means of detection* M 220
[7] Kalinkina M.E., Kozlov A.S., Labkovskaia R.I., Pirozhnikova O.I., Tkalich V.L., Shmakov N.A. Development of computer-aided design system of elastic sensitive elements of automatic metering devices // Journal of Physics: Conference Series - 2018, Vol. 1015, pp. 052018
[8] Klaassen C 2008 The basics of measurements Sensors and electronic devices *Intellect* M 352
[9] Hashemian H M 2008 Sensors of technological processes. Characteristics and methods for improving reliability *Binom* M 336
[10] Kozlov A.S., Shmakov N.A., Tkalich V.L., Labkovskaia R.I., Kalinkina M.E., Pirozhnikova O.I. Development library of finite elements for computer-aided design system of reed sensors // Journal of Physics: Conference Series - 2018, Vol. 1015, pp. 032080
[11] Kashkarov A P 2017 Microelectromechanical systems and elements *DMK Press* M 114
[12] Raspopov V Ya 2007 Micromechanical devices *Mechanical Engineering* M 400
[13] Tkalich V.L., Labkovskaia R.I., Pirozhnikova O.I., Kalinkina M.E., Kozlov A.S. Analysis of errors in micromechanical devices // 14th International Scientific-Technical Conference on Actual Problems of Electronic Instrument Engineering, APEIE 2018: Proceedings - 2018, pp. 272-276
[14] Rude I I 2012 Alarm systems *Technical means of detection* M Solon-Press 220

Structure and mechanical properties of a wedge-rolled steel skelp

V G Pachurin[1], V V Galkin[1], G V Pachurin[1], M V Mukhina[2], Zh V Smirnova[2] and EA Chelnokova[2]

[1]Nizhny Novgorod State Technical University named after R.A. Alekseev, Minin St., 24, Nizhny Novgorod, 603950, Russia
[2]Minin Nizhny Novgorod State Pedagogical University,1, Ulyanova str., Nizhny Novgorod, 603002, Russia

E-mail: mariyamuhina@yandex.ru

Abstract. Using the example of steel 50CrMnV, we present the regular changes in the structure and mechanical properties of the metal product with a wedge-shaped longitudinal section obtained by hot rolling of a skelp with the highest reduction rate of 40% and data on the structure of its deformation zone.
Keywords: spring steel, structure, mechanical properties, skelp, hot rolling, reduction rate

1. Introduction

Many metal products are exposed to cyclic loads during operation [1-3]. Due to the fact that the process of crack generation and propagation is local, the critical factor in fatigue failure is not the average resistance to deformation and failure under static loading, but local characteristics and their combinations [4,5]. The process of die forging is accompanied by uneven deformation, which leads to heterogeneous distribution of structure and mechanical properties over the product [6-8]. The effect of plastic deformation on fatigue strength is carried out both at the micro level, when it changes the density and structure of lattice imperfections and at the macro-level when retained stress occur and remain due to nonuniform deformation of the blanks. It is known [9-13] that the effect of homogeneous prior deformation on fatigue strength is not unambiguous [14-17]. Thus, plastic deformation in the zone of homogeneous deformations leads, as a rule, to an increased cycle life of metal products. However, hardening at degrees smaller and greater than homogeneous plastic deformation can negatively affect fatigue resistance. During operation, springs are exposed to cyclic loads, both in the air and in the corrosive environment. There is little information on the assessment of fatigue strength of specific products [9,18]. This is due to the fact that the study of nonuniform deformation in a specific technological process of metal forming, until recently, had significant difficulties [19-21]. Therefore, the study of the structure and mechanical properties of the obtained spring sheets is of great practical importance.

In this paper, using the example of steel 50CrMnV, the structure changes and mechanical properties of the metal product with a wedge-shaped longitudinal section obtained by hot rolling of a skelp into a wedge are shown.

2. Research methods

The spring sheet manufacturing technology includes induction heating of a blank (hot-rolled skelp of rectangular cross-section) and forming operations: punching the central hole, rolling both ends of the blank to a wedge with non-drive rolls (rolling speed 200 mm/sec) and wrapping the lugs. The cycle time is 17-19 seconds. Then the skelp is heated, bent and heat-treated according to the factory technology (oil quenching and tempering), followed by shot peening of its surface. The use of hot-rolled skelps of cross-section to manufacture springs has a number of features. Rolling the blank into a wedge is carried out in one pass, which means a different degree of deformation along the length of the rolled sheet; the production is

Content from this work may be used under the terms of the Creative Commons Attribution 3.0 licence. Any further distribution of this work must maintain attribution to the author(s) and the title of the work, journal citation and DOI.
Published under licence by IOP Publishing Ltd

computer-aided, which leads to a significant post-deformation pause and the possibile recrystallization processes in the metal; long products that are not intended for conversion are used.

To assess the quality of the metal sheet obtained by rolling a rectangular section blank with rounded edges of long steel 50CrMnV on a wedge, the following work was carried out: assessment of strength characteristics and ductility index in the hot-rolled and heat-treated state of rectangular section of 13x75 blanks and their comparison with the original metal; assessment of the structure influence on the mechanical properties of the heat-treated sheet; comparison of the performance properties of a skelp made from rolled products. Three zones of the rolled spring plate were selected that differ in the degree of deformation: small deformations ≤ 4%; medium ones ≤ 23% and large ones 40%. The material of the blanks was taken after rolling and final heat treatment. The samples were cut out in the selected areas for mechanical testing and structural survey. Mechanical tests included hardness measurements, uniaxial tension of the samples, the impact hardness of metal in the hot-rolled and heat-treated state. Metallographic testing was performed using an "Epitup" microscope at magnification >300. The zones of cross-sections after rolling and heat treatment were studied. After picric etching the grain size was measured, the size changed depending on the reduction degree (along the rolled plate). The study of hardness was carried out in the cross sections of the selected zones according to the Rockwell method according to GOST 9015 in three levels: in the middle of the rolled thickness, in two layers separated from the surfaces by 2 mm. Along the width of the skelp, five punctures were made, 75 mm in size, with an interval of 15 mm and side margins of 7.5 mm. Stretching was performed in accordance with GOST 1497-84 on the all-purpose machine UME-10TM (with a strain gauge), on flat samples 2 mm thick, made of longitudinal blanks of selected zones in three layers (the upper one – in contact with a stationary plate; the middle one –the middle thickness of the rolled blank and the lower one - rolled by a roller). The deformation rate is 5 mm/min. Fatigue tests of flat samples (GOST 25.502-73) were carried out according to the cantilever bending with a symmetrical cycle with a frequency of 1500 cycles/min at room temperature according to a "rigid" loading mode. Impact tests were carried out on samples with a U-shaped incision in accordance with GOST 9454, made from transverse blanks of the selected zones. To study changes in the fine structure (substructure) of rolled metal, the method of x-ray diffraction analysis was used. The sizes of mosaic blocks (coherent scattering zones) and the magnitude of microstresses on the diffractometer DRON-2 were determined. Fatigue tests of springs were carried out with a loading frequency of 0.6 Hz on the spring stand, before the fracture of any sheet.

3.Results and discussions

Products with a wedge-shaped surface obtained by forming are widely used. At the same time, obtaining a wedge-shaped surface in the workpiece is performed by two main technology processes: stretching and rolling. In the first case, hammer forging is used, in the second one – rolling on special equipment: forge rolling machines and rolling mills. The process of rolling a rectangular-section skelp into a wedge is a periodic process. However, unlike longitudinal rolling, it has a constantly changing zone of deformation, due to the fact that the rule of forming is set by a moving tool – a roll, which, in addition to translation movement, goes perpendicular to the rolled surface of the workpiece. The results of studies of the rolled blanks of spring sheet made of steel 50CrMnV showed uneven thickness of layers of the rolled blanks, depending on the reduction rate with a certain regularity. From the start of rolling to the degree of deformation ε 20%, the layers in contact with the tool (roll) have the greatest deformation intensity, and then it shifts to the layers in contact with the stationary plate of the rolling mill.

In order to study the structure of the actual deformation zone in the strip of rectangular cross-section blanks of spring steel sheet 50CrMnV, the uneven strain state of a workpiece thickness was evaluated, and in particular the zone of maximum intensity (deformation zone), depending on the reduction rate, as well as microstructure of the actual deformation zone.

As for the first aspect, the mechanical properties of the material were studied by the thickness of the rolled skelp. To do this, the skelp was cut out of blanks by EDM process on plates 2 mm thick.

Speaking about the second one, the structure of the actual deformation zone was analyzed on a section of the longitudinal section of the skelp at the point of its contact with the roll, fixed by stopping the process, in particular at a maximum reduction rate of 40%. The assessment was carried out by identifying zones of different strain intensity. In addition, mathematical modeling of the process by the finite element method (FEM) was performed using the DEFORM software package.

It is established that if during cold processing, as the deformation rate increases, there is an increase in the deformation resistance (hardening), then during hot processing, the processes of hardening and softening occur simultaneously. The recrystallization rate depends on the deformation rate. Thus, the faster recrystallization proceeds, (it is determined by an increase in ductility and a decrease in strength properties), the greater the deformation rate preceded. In other words, the deformation is associated with a significant metal softening. The strip temperature was monitored during rolling using an infrared pyrometer "Micron MSOP" (USA). The measurement results showed that there was no temperature drop, and rolling was performed at a constant temperature. The analysis of the yield strength and hardness values showed uneven softening, both in thickness and in the cross section of the rolled strip at different degrees of reduction.

Thus, in the longitudinal section, the greatest softening, and as a result, the greatest degree of deformation correspond to:

- at reduction rate up to 23% - metal layers on the side of the skelp, while the difference is in the range of 10-15%;
- at reduction rates from 23% to 40% - layers that are adjacent to the stationary plate of the rolling mill;

in cross section:

- with the reduction rate $\varepsilon \approx 23\%$ at the edge of the skelp, the yield strength was in the range of 542...745 MPa, in the middle – 622...719 MPa;
- with the reduction rate $\varepsilon \approx 40\%$ along the edge of the strip, the yield strength was in the range of 434...647 MPa, in the middle-581...674 MPa.

The obtained results are confirmed by data on the "crown shape" of the rolled blank, depending on the reduction rate. For small reduction rates $\varepsilon \approx 4\%$, the greatest broadening is in the layers of the strip on the rolling side. With a subsequent increase in the reduction rate, the greatest broadening goes to the layers near the stationary plate of the rolling mill.

Strength characteristics: yield strength and temporary failure resistance confirm the above-mentioned nature of uneven changes in hardness along the length of the hot-rolled blank. At the same time, some discrepancy in the results can be explained by the fact that the samples for the tensile test were made from layers of hot-rolled metal that have uneven mechanical properties both in width and in thickness. For small deformations, the rolled layer has the greatest hardening.

Changes in the ductility parameters: percentage extension and contraction, and impact hardness are similar to changes in strength characteristics. The difference is that they have an extreme minimum. At the same time, the changes in the strength and ductility of the hot-rolled and heat-treated metal in deformation are almost identical [9]. However, the following differences occur in a heat-treated blank: increase in strength characteristics: hardness by 1.5 times, strength and yield strength by 1.52 times; decrease in ductility by 2.5 times, impact strength by 1.25 times.

The results of fatigue tests in general confirmed the data in strained condition in terms of mechanical properties. Metal layers that have a high intensity of recrystallization processes, and as a result, better healing of structural defects obtained during deformation, had a longer service life [22].

4.Summary

Along a hot-rolled blank, with increasing degree of deformation, the strength increase, and the ductility index decrease, passing through an extremum in the range of degrees of deformation of 15-20%, depending on the metal layer.

The hardening obtained in the process of hot rolling is partially preserved after the final treatment, while the nature of strength characteristics and ductility of the metal of the rolled blank and the blank after the final heat treatment in deformation is the same.

The studies of hot-rolled and heat-treated metal indicate different conditions for the formation of a hot-deformed structure and its influence on the final mechanical properties of the spring sheet. Layers with the most intensive recrystallization have a higher cycle life in a wedge-rolled blank, which causes better healing of structural defects during deformation.

References

[1] Pachurin G.V., Goncharova D.A., Filippov A.A., Shevchenko S.M., Mukhina M.V. et al., Development of fatigue test technology of sheet automobile materials, Eastern-European journal of enterprise technologies, ISSN 1729-3774, Vol 5, No. 12 (95), 2018, pp. 31-37, Materials science, UDC 620.169.1. DOI: 10.15587/1729-4061.2018.144524.

[2] Guslyakova G.P., Zhbannikov S.I., Pachurin G.V., Fatigue failure resistance of deformed structural steels, Materials Science. 2 Vol. 28 (1993), 182-185.

[3] Pachurin G.V., Guschin A.N., Increase of operational durability of metal products by technological methods, Vestnik mashinostroeniya, 6 (2007), 62-65.

[4] Pachurin G.V., Fatigue failure of pre-deformed alloys at normal temperature, Metallurgy and heat treatment of metals. 10 (1990) 35-38.

[5] Pachurin G.V., Kinetics of fatigue failure of copper M1 and brass L63, Nonferrous metallurgy. Proceedings of higher institutions of the USSR. 1 (1989) 96-101.
[6] Pachurin G.V., Ruggedness of structural material and working life of metal components, Steel in Translation. 3 Vol. 38 (2008) 217-220.
[7] Pachurin G.V., Life of Plastically Deformed Corrosion-Resistant Steel, Russian Engineering Research, Vol. 32, Nos. 9–10 (2012) 661–664.
[8] Pachurin G.V., Durability of sheet die stamped materials in air and in corrosive environment, Materials Science. 7 (2003) 29-32.
[9] Pachurin G.V., Shevchenko S.M., Filippov A.A., Mukhina M.V., Kuzmin N.A., Defining rolled metal performance for cold bolt upsetting (bolt head). IOP Conference Series: Materials Science and Engineering, 2018, 327, 032040 doi: 10.1088 / 1757–899X/327/3/032040. DOI: 10.1088/1757-899x/327/3/032040
[10] Bachurin G. V., Goncharova D. A., Filippov A. A., Shevchenko S. M., Mukhina M. V., etc. Development of fatigue testing technology for sheet automotive materials. Eastern European journal of corporate technologies ISSN 1729-3774. - Vol. 5, № 12 (95). – 2018. – Pp. 31-37: materials Science. UDC 620.169.1 DOI: 10.15587/1729-4061. 2018. 144524.
[11] Pachurin G.V. Ruggedness of structural material and working life of metal components // Steel in Translation. 2008. №3. T. 38. P. 217-220.
[12] Goncharova D.A., Pachurin G.V., Filippov A.A., Kuzmin N.A.. Fatigue destruction of automobile construction materials // Journal of Physics: Conference Series 2019. - C. 012025. DOI: 10.1088/1742-6596/1177/1/012025
[13] Pachurin G. V., Goncharova D. A., Filippov A. A., Nuzhdina T. V., Deev V. B.. Investigation of the fatigue process of automotive materials // Izvestiya vuzov. Ferrous metallurgy. - 2019. - No. 9. - Pp. 732-738. DOI: 10.17073/0368-2019-9-732-738.
[14] Pachurin G. V., Goncharova D. A., Kuzmin N. A., Filippov A. A., Nuzhdina T. V.. Process of fatigue damage of sheet automobile steels / / Technology of metals. - 2019. - No. 8. - Pp. 16-22. DOI: 10.31044 / 1684-2499-2019-8-0-16-22
[15] Kuzmin A.N., Filippov A.A., Pachurin G.V., Goncharova D.A.. Optimization of perlite steels mechanical properties for car fasteners stamping. AAE IOP Conf. Series: Materials Science and Engineering 695 (2019) 012030. IOP Publishing DOI:10.1088/1757-899X/695/1/012030
[16] Kuzmin N.A., Goncharova D.A., Pachurin G.V., Filippov A.A., Korotunova I.A. Improvement of car repairability based on the material destruction assessment during its operation. AAE IOP Conf. Series: Materials Science and Engineering 695 (2019) 012029. IOP Publishing. DOI:10.1088/1757-899X/695/1/012029
[17] Pachurin G.V., Goncharova D.A., Gevorgyan G.A., Filippov A.A., Mukhina M.V., Trunova I.G., Konyukhova N.S.. Fatigue of the automobile structural materials in corrosive environment. E3S Web of Confer-ences. Topical Problems of Green Architecture, Civil and Environmen-tal Engineering, TPACEE 2019. 2020. C. 03039.
[18] Pachurin G. V. Assessment of damage to structural materials and improvement of service life of metal products / / Izv. VUZ. Ferrous metallurgy. 2008. No. 3. P. 55-58.
[19] Pachurin G.V, Vlasov V.A., Mechanical properties of sheet structural steels at operating temperatures, Metal Science and Heat Treatment, 56, Nos. 3–4. (2014) 219–223. DOI: 10.1007/s11041-014-9735-8
[20] Pachurin G.V. Life of Plastically Deformed Corrosion_Resistant Steel // Russian Engineering Research, 2012, Vol. 32, No. 9–10. P. 661–664.
[21] Filippov A.A., Pachurin G.V., Kuzmin N.A., Matveyev Yu.I., Deyev V.B.. Evaluation of the quality of rolled steel for cold die forging, Proceedings of higher educational institutions. Ferrous metallurgy, 7, Vol. 61 (2018), 551-556. DOI: 10.17073/0368-0797-2018-7-551-556
[22] Bachurin V. G., Galkin V. V., Pachurin G. V., Filippov A. A.. Method for predicting the cycle life of metals. Patent RU 2 619 480 C1 G01N; Appl. 10.05.2016; Publ. 16.05.2017, Bul. No. 14.

High-pressure cylinders of gas-fuel system made of composite materials

V B Markin[1] and N A Myagkova

Polzunov Altai State Technical University
Altai Territory, Barnaul, 46, Lenina Avenue, 656038, Russia

[1]E-mail: mvb1942@mail.ru

Abstract. Use of gas fuel systems will allow to carry out the ecological tasks facing the Russian automotive industry, it is essential to cut down expenses of consumers on fuel and to increase an engine lifetime. As a basis of such modernization it is necessary to consider development, calculation, a design binding and technological security of production of cylinders of pressure as capacities for the compressed or liquefied natural gas. For the purpose of weight reduction of cylinders the preference is given to the polymeric composite materials using methods of winding of the rein

1. Introduction

Russian automotive industry is being developed in several directions: assembly of well-known brands of foreign vehicles and production of new Russian vehicles, for example Lada series (Lada Kalina, Lada Priora, Lada Granta, Lada Vesta, Lada Largus). One of the ways of increasing competition of Russian vehicles is the improvement of ecological characteristics and decrease fuel expenses. This direction can be developed without high expenses on purchasing new technologies and equipment, increase of sales volume can be reached by transition to new consumer appeals and fulfillment of discharge consent of dangerous gaseous products. Directions of improvement that were mentioned above are possible due to the usage of compressed gas vehicles equipment with pressure balloons made of composite materials.

2. Advantages of gas fuel

Gas fuel burns fully than petrol, so the concentration of carbonic oxide and other gases in car emission is much lower than in petrolic cars (figure 1).

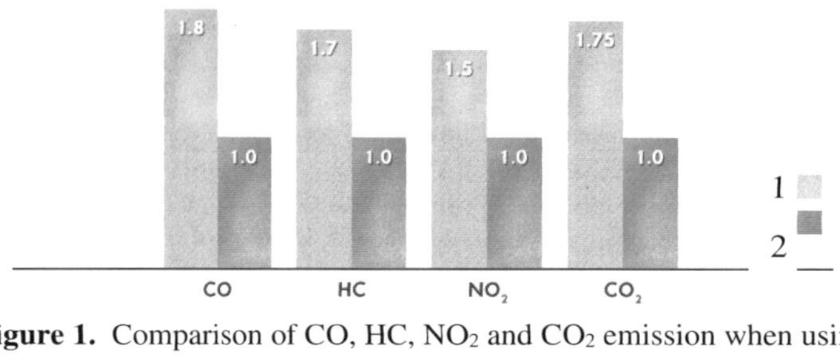

Figure 1. Comparison of CO, HC, NO_2 and CO_2 emission when using petrol (1) and gas (2)

Compressed natural gas is several times safer in operation than petrol and explosive liquefied gas (propane-butane). It is lighter than air and in case of a leak it vanishes; besides maximum allowable concentration in the air for fulminating is from 5 to 18 per cent. The same number for petrol is 70 per cent that proves the fact that an explosion of huge gas masses in case of a leak inside a vehicle is impossible.

Content from this work may be used under the terms of the Creative Commons Attribution 3.0 licence. Any further distribution of this work must maintain attribution to the author(s) and the title of the work, journal citation and DOI.
Published under licence by IOP Publishing Ltd

On the background of highly developing usage of hybrid and electro mobiles we should consider that compressed and liquefied natural gas is a fuel of the future. Nowadays comfortable conditions for spreading this type of fuel are created everywhere. This tendency exists in Europe which is worried about two main problems: how ecological and cost efficient the fuel is. There are several good conditions for development of natural gas motor fuel market in Russia: gas-transport system, low cost of natural gas and finally high reserves of gas.

We can claim that nowadays The Russian Federation has vehicle production which engines work using alternative fuels (Lada Vesta CNG и Lada Largus CNG). Among such fuels the most interesting one is natural gas (methane), which has low prime cost and its cost is more attractive in comparison with traditional fuels. Moreover its energetic efficiency in comparison with petrol AI 95 is equal to numbers of AI 140 (for example one liter of petrol with octane number AI 92 is equal to approximately 0,8 $м^3$ of compressed gas with octane number AI 140). So, using of natural compressed gas as a fuel will solve several tasks: it will increase ecological characteristics of Russian vehicles, from 3 to 10 times decrease the consumers' expenses on fuel and increase engine lifetime in 1,5-1,8 times.

3. Requirements to technology of production of high-pressure cylinders

Balloons for natural compressed gas which is used as a fuel for vehicles must be extremely light and fulfill all safety requirements for vessels which work under pressure. For efficient work of high-pressure cylinders construction it is necessary to choose right material and technology and its production. When choosing it is necessary to follow the rules of optimality and take into consideration not only existing materials but also the possibility of its modernization and possibility of producing new materials which will be designed for certain working conditions.

For producing of high-pressure cylinders it is necessary to solve several tasks:
- researching of strain-stress state of cylinders;
- selection of joining and reinforcing filler for composite shell of high-pressure cylinder;
- optimization of cylinder construction in shape and mass with changes in production technology;
- selection of technology of components modification in order to improve the qualities of adhesive contact;
- selection of technological parameters of production of adapter for spooling pressure shell of a cylinder;
- selection of technological parameters of spooling process, dipping and polymerization of composite layer of high-pressure cylinder;
- analysis of main technological and economical markers of production of high-pressure cylinder and also its prime cost;
- analysis of ecological safety of technological process of cylinders' production [1].

On the basis of working conditions there are several requirements to the construction of pressure cylinders made of composite materials:
- safe operation management;
- contentment of strength characteristics;
- providing optimality at which performance function is mass minimization of pressure cylinder;
- processability of construction which involves interrelation and interdependence of a product, technology of production, storage and usage requirements.

It is well-known that polymeric composites are nonhermetic, that is why constructions made of them, depending on the level of basis of design requirements, include additional pressure-sealed layers, shells etc. Pressure-sealed shell (liner) is usually used as an adapter for spooling layers of a pressure shell [4].

Original constructive and technological decisions which are the basis of creation of hermetic polymeric shell (liner), strengthened by glass fiber (or carbon fiber), let us to lower the mass of cylinders in 5 times in comparison with metal ones, to increase corrosion resistance, increase resistance to aggressive environment, fire and explosive prevention, guarantee destruction without shatters. Fatigue crack life is more than 15000 stress cycles, assurance coefficient after performing cyclic tests not less than 2,6; temperature operation range from 40 degrees below zero to 80 degrees above zero; service life not less than 10 years.

4. Design engineering of gas cylinder

Projecting design of gas cylinder comes down to definition of design parameters of the construction, which allow the fulfillment of the following conditions:
- minimal mass of a designed construction;
- preserving required strength and ruggedness during design load;
- technologic realizability;

- supporting minimal cost of a cylinder.

Main parameters of a cylinder which are defined during engineering process are:

- contour of a shell bottom;
- spreading of reinforcing angles with fiber;
- number of layers of reinforcement material which form the shell.

Main loads which appear in the cylinder under internal pressure are pictured in figure 2, and dependence of contour shape of the cylinder under different parameter λ values, defining the shape of the meridian of optimized shell is in figure 3.

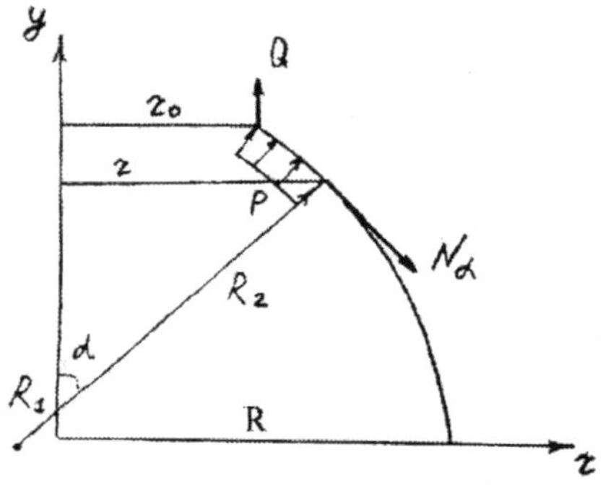

Figure 2. Spreading the loads in a pressure shell depending on coordinates: P- internal pressure; R_1, R_2 – principal radii of a shell curvature; Q – axial force; R-radius of a cylindrical part of the cylinder

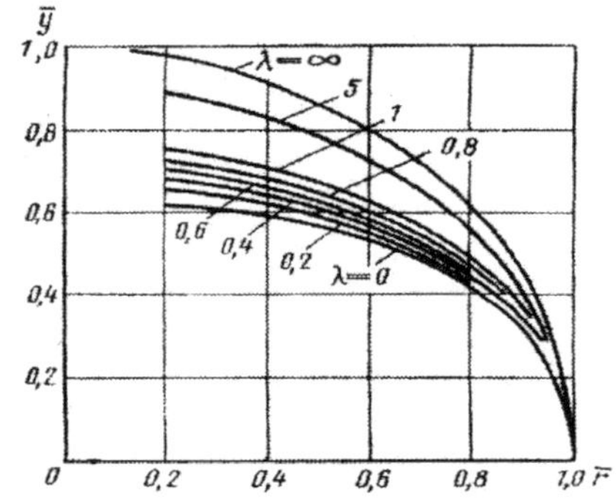

Figure 3. Diagram of dependence of contour shape of the cylinder under different parameters of λ for laying of fiber angles on the centerline φ = 7°49': $\bar{y}$ and $\bar{r}$ - nondimensional coordinates referred to radius value of centerline R

Enumerated dependencies are used in the starting stage of the engineering at defining geometrical characteristics of optimal cylinder, which let to set up strength and ruggedness characteristics in size of maximum operating pressure P, which is 2,4 MPa, on the next stage.

Producing composite cylinder of high-pressure was done by method of "wet" winding of reinforcement fiber: glass (fiberglass roving PBMH 19-1480-80), carbon (gasket "Tornel T700"), basaltic (roving RVMN-19-1200-KB42), combined with epoxide joining EDU to metal or plastic liner. Spooling machine WM2.800 was used for winding (figure 4). Hardening of the products was done by standard method using an automated system of heating process control [7].

Figure 4. Technological process of spiral winding of pressure cylinder

On figure 5 there are images of metalcomposite cylinder BK-7 before testing and after destruction due to triple overpressure (80 MPa).

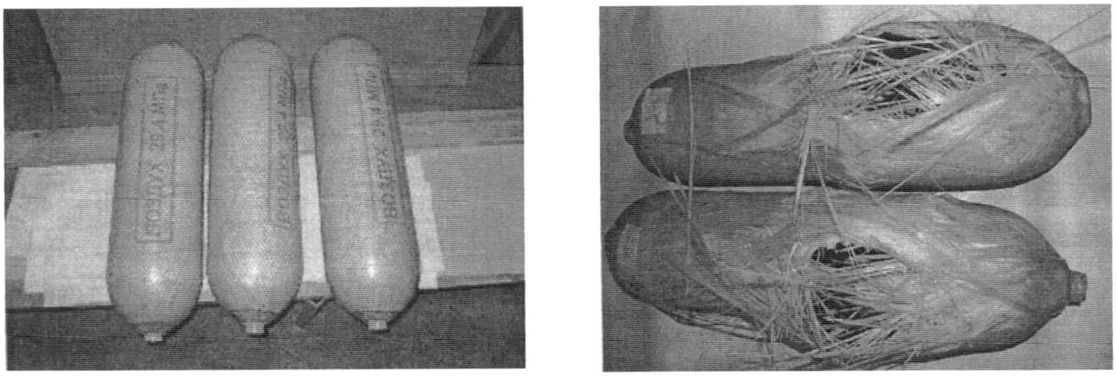

(a) (b)

Figure 5. Metalcomposite high-pressure cylinder before destruction tests (a) and after the tests (b) under pressure of 80 MPa

Simultaneousness of creation material and construction from polymeric composites defines a necessity of fundamental material research, usage of the most modern methods of modification of materials on the level of components, and also structuring research which is directed to optimization of features of material in construction.

In table 1 there is data about composite high-pressure cylinders in which metal and plastic liners were used, and strength shells were made of fiberglass, carbon fiber composite and basalt fiber reinforced polymer. The cylinders are designed for working pressure which is 29 MPa.

Table 1. The results of the tests of high-pressure cylinders made of composite materials

№ п.п.	Liner		Strength shell	Fracture pressure, MPa
	Material	Weight, kg		
1	Steel	1,5	Carbon fiber composite	81,0
2	Steel	1,5	Fiberglass	78,0
3	Steel	1,5	Basalt fiber reinforced polymer	80,0
4	Polypropylene	0,4	Fiberglass	65,0

5. Conclusion

The weight of composite cylinders with metal liner is not more than 3 kg. When using carbon fiber composite as strength shell the weight goes down to 1,8 kg under the same working pressure 29 MPa.

Metalcomposite cylinder of high pressure have almost triple safety factor which allows increasing its lifecycle to 10 years or more with obligatory control and service.

Fracture behavior of the cylinders at reaching critical pressure is with no shatters.

Economic efficiency of using composite cylinders for gas-fuel automobile systems is defined with two main criteria: sales volume of vehicles and fuel costs. Transition to bifuel system has several suppositions for gas-engine market:

-huge natural gas (metane) reserves in Russia;

-unstable situation at the world oil market;

-natural gas is less volatile in comparison with liquefied gas;

-usage of natural gas decreases rate of wear of an engine;

-serious decrease of cylinder weight and increase of its working pressure.

Therefore polymeric composite materials are becoming more and more popular in modern automotive industry, they tend to decrease specific amount of metal and specific amount of material in products and constructions. It can be considered that these materials are the materials of the XXI century, following the fifth level of technological wave.

References

[1] Vorobey V V and Markin V B 2014 *The basics of engineering and technology of superlight composite of high-pressure cylinder: monograph* (Barnaul: Altai State Technical University Press) p 166
[2] Vorobey V V and Markin V B 2003 *The basics of engineering and technology of rocket motor cases* (Novosibirsk: Science Press) p 164
[3] Myagkova N A 2016 *Central-Asian Material Science Journal* **7** 17-9
[4] Markin V B, Ananieva E S and Novikovskiy E A 2015 *Central-Asian Material Science Journal* **1** 45-50
[5] Markin V B and Myagkova N A 2015 *Central-Asian Material Science Journal* **1** 51-5
[6] Blaznov A N, Markin V B and others. 2018 *Technologies and equipment of chemical, biotechnological and food industries. Materials of XI Russian science-practical students' conference, post-graduate students and young scientists with international participation.* (Biisk) pp 80-88
[7] Markin V B and Saginganova I K 2019 *Izvestia of national academy of science of the Republic of Kazakhstan* **1** 63-7

Influence of dimension chains on product assembly sequences

A N Bozhko

CAD Department, Bauman Moscow State University, 5/1, 2-ya Baumanskaya, Moscow, 105005, Russia

E-mail: bozhkoan@bmstu.ru

Abstract. The paper considers the influence of design dimension chains on the assembly sequence of a complex product. When assembling products, the rule of connectivity of the dimension chain must be followed. For this reason, the dimension chain induces a binary partial order relation on the set of product parts. The assembly sequences for which the connectivity rule is applied are linear extensions of this order relation. The number of the linear extensions for the orders generated by the main methods of dimensioning is calculated. It is shown that the chain method of dimensioning is the most rigid, and the coordinate method is the most "tolerant". The combined method occupies an intermediate position. The principle of matching the dimension and mechanical structures of a complex product is proposed. The results obtained can be used to generate rational dimension schemes in integrated CAD / CAM systems.

1. Introduction

The design dimension system is an important part of the construction of any complex product. The set of dimensions and associated structural elements (parts, surfaces, axes, etc.) forms a dimension structure (DS) of the product. The dimension structure has a significant impact on the assembly sequence and the content of assembly operations. For example, long dimension chains make it difficult to assemble using the cost-effective methods of interchangeability. Dimension chains with common links can cause conflicts when the implementation of one chain entails the loss of accuracy of the other. The system of dimension chains creates two binary relations on the set of product parts: order and equivalence. These relationships should not contradict the ordering of parts in the process of assembling the product and dividing the product into assembly units. A simple example shown in figure 1 illustrates the contradiction between the mechanical and dimension structures of the product. Dimension A is a component dimension. It is easy to see that after assembling the fragment consisting of the wheel (18), the bearing (17) and the shaft (8), the closed (unsolvable) dimension chain (DC) will arise, which is formed by dimension A and the overall dimensions of these three parts.

In this article, a mathematical model of the dimension structure is developed and the principle of its matching with the mechanical structure of the product is proposed. The results obtained can be used in integrated CAD/CAM/CAE systems to implement modern design paradigms: Design for manufacturing, Design for assembly, Concurrent engineering [1].

The fifth chapter of the fundamental book [2] contains basic information from the theory of dimension chains. The articles [3] – [15] discuss various aspects of the synthesis of the dimension structure of a complex product in CAD systems. Methods of generating the DS from the 3D model of a product are proposed in [3] – [9]. Optimal distribution of the closing dimension tolerance over the links of the dimension chain is considered in [10]. Article [11] is devoted to the synthesis of 3D model of a product, taking into account a given system of dimensions. A method for modeling dimensional chains is proposed in [12]. The work uses the apparatus of graph theory. Article [13] discusses the synthesis of the dimensional structure, which can be implemented using methods of complete and incomplete interchangeability. Assignment of asymmetric tolerances in complex multiply connected dimension chains is described in [14]. The problem of finding assembly dimension chains is considered in [15]. The work proposes a multilevel model, which at each level of the hierarchy allows obtaining a rational dimension chain.

The considered works on the computer-aided design of DS have a number of disadvantages:

Content from this work may be used under the terms of the Creative Commons Attribution 3.0 licence. Any further distribution of this work must maintain attribution to the author(s) and the title of the work, journal citation and DOI.
Published under licence by IOP Publishing Ltd

IOP Conf. Series: Materials Science and Engineering 1100(2021) 012012 doi:10.1088/1757-899X/1100/1/012012

1. The main attention is paid to the synthesis of the dimension structure and the optimization of the tolerances for the constituting dimensions. The impact of DS on the assembly of a product is not considered.

2. Two criteria are used to select the optimal dimension structure: the minimum length of the dimension chain and the minimum number of common links. The problem of choosing a rational DS according to the criterion of maximum matching with the mechanical structure of a product is not discussed.

2. Basic rules for the implementation of the dimensional chains

The dimension chains that are created during the design of a product are called design dimension chains. Let's divide them into two groups. The first group of DC includes the dimensions that determine dimension relationships between geometric elements of parts. The second group of DC combines the dimensions that determine the coordination of different parts and precision parameters of the product. Such DCs are called assembly chains [2]. The component dimensions of the assembly DCs can be own dimensions of a part (type 1) and dimensions between different parts (type 2). The article discusses only the assembly design dimension chains.

In the assembly DCs dimensions and its tolerances are determined according to the working and assembly conditions of a product. The closing dimension is obtained automatically after mounting of the parts that carry the component dimensions of the DC. This means that the dimension chain induces an ordinal relation on the set of parts. This relation should not contradict the assembly sequence and the division of the product into assembly units.

When assembling, the following rules must be followed, which describe the relationship between the mechanical and dimension structures of a product.

1. The rule of connectivity of «a growing dimension chain». In the assembly operation, it is possible to implement such a component dimension of DC, which is incident to the assembled piece of the product.

2. The rule of dimension chain closure. The parts that are carriers of the component dimension of DC must belong to one assembly unit.

3. Formalization of dimensional constraints on the ordering of parts during assembly

Let all the dimensions of a dimension chain belong to type 1. We represent the dimension chain in the form of a graph, in which the edges correspond to the dimensions, and the vertices describe their incidence (contacts of surfaces, coincidence of axes, etc.). We consider two dimension graphs corresponding to the main methods of dimensioning: chain dimensioning (figure 2 a) and coordinate dimensioning (figure 2 b).

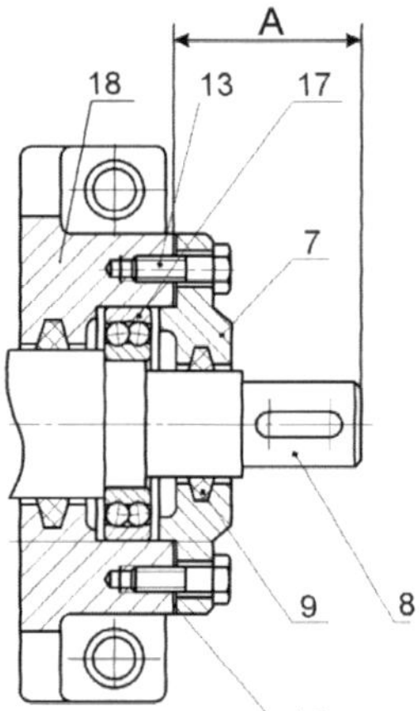

Figure 1. An example of a product with inconsistent dimension and mechanical structures.

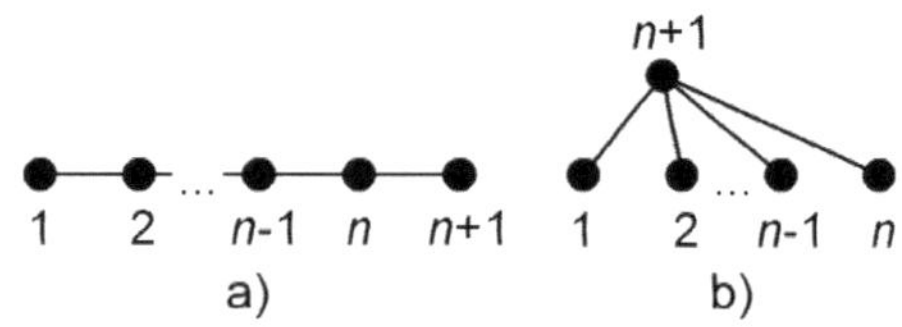

Figure 2. Dimension graphs.

We denote CR_n^1 – the number of all sequences of realization of the (n+1)-vertex chain (figure 2 a) that satisfy the rule of connectivity (RC), and TR_n^1 – the number of all such sequences for the (n+1)-vertex tree. It is easy to see that any sequence of tree size realization preserves the chain connectivity, therefore $TR_n^1 = n!$.

Let's count the number CR_n^1. The number of valid realizations of the (n+1)-vertex chain depends on the choice of the first edge of the chain. If the implementation of the chain starts from the left (right) vertex, then only one sequence is allowed. It is described by the order of vertices 1,2,3, .., n (n-2, n-3, ..., 1). Let the first be a dimension corresponding to any inner vertex of the graph (figure 2 a). Then the chain will split into two subchains, the implementation of which can be continued in one way. This turns the subchains into ordered sets. Let $r_i = (i, i+1)$ be the edge that connects the vertices with numbers i and i + 1. Figure 3 shows the orderings of dimensions generated by the choice of the first dimension.

We denote as $L(r_i)$ the number of admissible realizations of the dimension links of the chain, provided that the dimension r_i is executed first. It's obvious that $CR_n^1 = \sum_{i=1}^{n} L(r_i)$. For the calculation CR_n^1, we use the method of enumerating the number of linear extensions of a discrete sum of ordered sets [16].

Let $(P_i, \le_i)$, $i = \overline{1,r}$ be a family of ordered sets.

Definition. An ordered set $(P, \le)$ is called a discrete sum of ordered subsets $(P_i, \le_i)$, $i = \overline{1,r}$ if $P = \bigcup_{i=1}^{r} P_i$ and $\forall a,b \in P$, $a \le b$ if and only if $\exists i \in \{1,...,r\} \mid a,b \in P_i$, $a \le_i b$ [16].

The number of linear extensions of the ordered set $(P, \le)$ can be found by the formula

$$N = \frac{m!}{m_1! \times m_2! \times ... \times m_r!} N_1 \times N_2 \times ... \times N_r, \tag{1}$$

where $m_i = |P_i|$, $m = |P| = \sum_{i=1}^{r} m_i$, and N_i is the number of linear extensions of the ordered set $(P_i, \le_i)$, $i = \overline{1,r}$ [16].

Using formula (1), we find the number of linear extensions of the ordered set, the general view of which is shown in figure 4. In terms of graph theory, it is a rooted tree, each subtree of which is a chain.

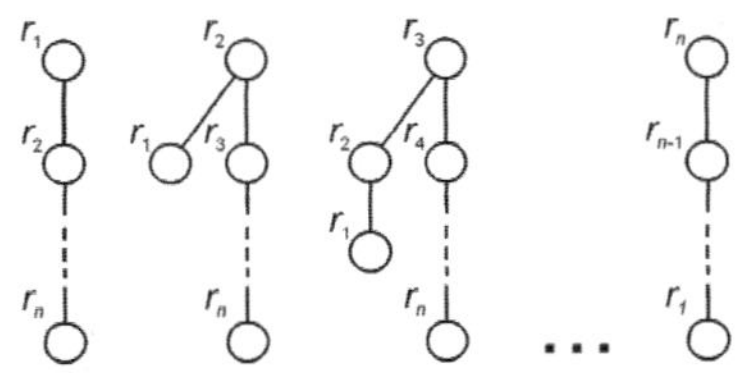

Figure 3. The orderings that are generated by the choice of the first dimension.

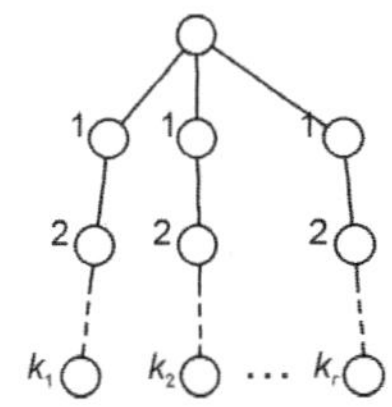

Figure 4. A hierarchical ordering.

All linear extensions of this order start from the root vertex. If we remove this vertex, we get the discrete sum of r ordered subsets, each of which is a chain. Any chain has only one linear extension, so $N_1 = N_2 = ... = N_r = 1$ and the formula (1) will take the form

$$N = \frac{(k_1 + k_2 + ... + k_r)!}{k_1! \times k_2! \times ... \times k_r!}, \tag{2}$$

where k_i, $i = \overline{1,r}$ is the number of the vertexes in the i-th branch growing from the root of the tree.

Each ordered set in figure 3, except for the first and the last, is a discrete sum of two chains consisting of $i-1$ and $n-i$ vertices. Using formula (2), we get $L(r_i) = \frac{(n-1)!}{(i-1)!(n-i)!}$. Hence it follows that $CR_n^1 = \sum_{i=2}^{n-1} \frac{(n-1)!}{(i-1)!(n-i)!} + 2$. We perform elementary transformations of this expression and get

$$CR_n^1 = \sum_{j=0}^{n-1} \frac{(n-1)!}{j!(n-j-1)!} = 2^{n-1}. \tag{3}$$

Consider a dimensional chain in which all dimensions are of the second type (type 2). In other words, they connect the geometric elements of different parts. If from the parts that form the assembly DC, select the base part (that is, the part that is mounted first), then this chain induces a hierarchical order relation. It is easy to see that the parts involved in the formation of the dimension chain can only be mounted in a sequence that is a linear continuation of the hierarchical order.

We express this statement in precise terms. Let $A \subseteq X = \{x_i\}_{i=1}^{n}$ be the set of parts of a product *X*, which, with their geometric elements, participate in the formation of a DC. We denote by $x_k R x_l$ the situation when two parts x_k, x_l, $x_k, x_l \in A$, are connected by a dimension. Let $x.y.z \in A$, and *z* be the part that is installed first. The parts *x* and *y* are in relation of hierarchical order if and only if there is a sequence $x_{i_1}, x_{i_2}, ..., x_{i_r} \in A$ such that $x_{i_1} = z$, $x_{i_r} = y$ and $\forall k = \overline{1, r-1}\ x_{i_k} R x_{i_{k+1}}$, $\exists t \in \{2, r-1\} \mid x = x_{i_t}$.

Ordered sets of this type are represented in the form of rooted trees, in which the base part is the root, and the arcs represent the precedence relation induced by dimension relationships. Figure 5 shows examples of two hierarchical orders that are generated by chain (a) and coordinate (b) dimensioning.

We denote $T(r)$ the hierarchical order generated on the set *A* by choosing the root vertex with number *r*. $N(T(r))$ is the number of linear extensions of order $T(r)$. CR_n^2 is the number of realizations of an *n*-vertex chain, and TR_n^2 is the number of realizations of an *n*-vertex tree that satisfy the connectivity rule. All the links of the dimension chains belong to the second type (type 2). It is easy to see that the formulas are valid; $CR_n^2 = \sum_{r=1}^{n} N(T(r))$ and $TR_n^2 = \sum_{r=1}^{n} N(T(r))$.

Each hierarchical order $T(r)$ can be associated with one and only one rooted tree of the family shown in figure 3. This implies the equalities

$$CR_n^2 = \sum_{r=1}^{n} N(T(r)) = \sum_{i=1}^{n} L(r_i) = 2^{n-1}. \tag{4}$$

Consider the hierarchical ordering *T* (figure 5 b), which generates the coordinate dimensioning method. It is easy to see that the given tree rooted at vertex *k* has $(n-1)!$ linear extensions. In addition, the implementation of the dimension chain can be started with any of the parts with numbers from 1 to $n - 1$. It gives $(n-1)(n-2)!$ linear orders. The total number of is

$$TR_n^2 = (n-1)! + (n-1) \times (n-2)! = 2(n-1)!. \tag{5}$$

The numbers CR_n^1, CR_n^2, TR_n^1, TR_n^2 are measures of the constraints that the dimension structure imposes on the assembly sequences of the product. The relations $CR_n^1 = CR_n^2 = CR_n < TR_n^2$, $TR_n^2 < TR_n^1, (n = 4, 5, ...,)$ are fulfilled. If $CR_2^1 = CR_2^2 = TR_2^1 = TR_2^2$, then $\lim_{n\to\infty} \frac{CR_n}{TR_n^2} = 0$, and $\lim_{n\to\infty} \frac{TR_n^1}{TR_n^2} = 0$. Simple calculations show that $\lim_{n\to\infty} \frac{CR_n}{TR_n^1} = 0$.

In engineering practice, the so-called combined dimensioning method is most often used. The dimension chains that the combined method generates are described by general trees. Consider a dimension scheme in which all constituent dimensions belong to the second type.

As before, the choice of the base part, with which the implementation of the dimension chain begins, turns the dimension tree into a hierarchically ordered set. We denote $T(r)$ the hierarchical order that generates the choice of the root *r* on the tree, and PR_n^2 – the number of realizations of the dimension chain that satisfy the RC. This number, as before, is equal to the sum of linear extensions of hierarchical orders $T(r)$ taken over all root vertices *r*, that is $PR_n^2 = \sum_{r=1}^{n} N(T(r))$.

The number of linear the extensions of hierarchical order $N(T)$ can be found by the formula

$$N(T) = \frac{|T|!}{|T| \times |T_1| \times |T_2| \times ... \times |T_r|}, \tag{6}$$

where $|T|$ – the number of vertices of the tree T; $|T_i|$, $i=\overline{1,r}$ – the number of vertices in all subtrees containing more than one vertex [16].

Let $T_1 = T_2 = T_3$ be n-vertex hierarchical orders of different types, that is, $|T_1|=|T_2|=|T_3|$. T_1 is the chain obtained by choosing a root at any internal vertex (figure 6 a), T_2 is a tree of arbitrary structure (figure 6 b) and T_3 is a one-level tree (figure 6 c).

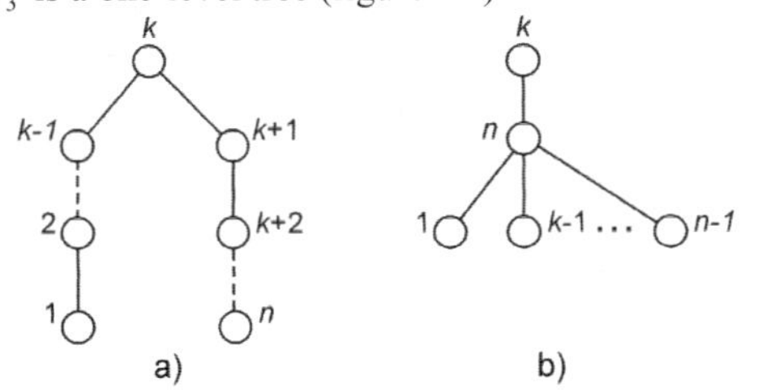

Figure 5. The hierarchical orders for chain (a) and coordinate (b) methods.

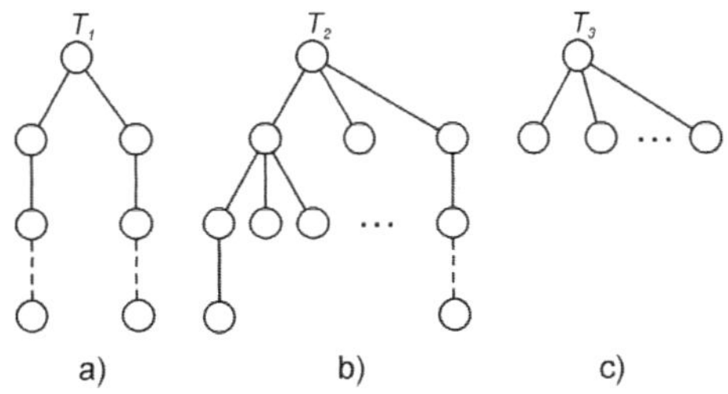

Figure 6. Hierarchical orders T_1, T_2, T_3.

From formula (6), It is easy to obtain inequalities $N(T_1) \le N(T_2) \le N(T_3)$ connecting the numbers of the linear extensions of hierarchical orders of different types. Term-by-term summation of these inequalities over all n root vertices gives the following relations

$$CR_n^2 \le PR_n^2 \le TR_n^2 . \qquad (7)$$

Using the given calculation method, it can be proved that the relationships $CR_n \le PR_n \le TR_n$ are fulfilled for any chains of equal power, in which the dimensions of the first and second types are used. The values CR_n, PR_n, TR_n and relationships between them (1) – (7) give numerical estimates of the constraints imposed by the dimension structure on the assembly sequences of a product.

The numbers CR_n, PR_n, TR_n are big enough. From (1) - (7) it follows that their lower bound is $CR_n = 2^{n-1}$, where n is the number of vertices of the dimension graph. In most cases, a dimension chain connects a limited number of parts, that is $n < |X|$, where X is the set of the product parts. Therefore, the mechanical structure of the product imposes much stronger restrictions on the assembly sequences. It is easy to give examples of products for which there is only one assembly sequence [17]. This allows us to formulate the following principle of matching the mechanical and dimension structures of the product.

The dimensional and mechanical structures of the product are consistent if for each assembly sequence of the product the rule of connectivity of «a growing dimension chain» is fulfilled.

4. Conclusion

The influence of the dimension structure on the assembly sequences of a complex product is considered. It is shown that when assembling, the rules of connectivity and closure of dimensional chains must be fulfilled. Dimensional chains induce ordinal relations on the set of the parts of a product. The formalization of the connectivity rule is proposed. It is performed for such sequences of product assembly, which are linear extensions of the hierarchical orders generated by dimensional chains. The number of linear extensions (CR_n, PR_n, TR_n) has been calculated for various methods of dimensioning. It is shown that the most severe limitations on the assembly sequences are imposed by the chain method of dimensioning, the softest - by the coordinate method. The combined method takes an intermediate position. The principle of matching the mechanical and dimension structures of a complex technical system is proposed.

References

[1] Anderson D M 2020 *Design for Manufacturability: How to Use Concurrent Engineering to Rapidly Develop Low-Cost, High-Quality Products for Lean Production* (New York: Productivity Press)
[2] Whitney D 2004 *Mechanical Assemblies. Their Design, Manufacture, and Role in Product Development* (New York: Oxford University Press)
[3] Zhenbo G Jing W Yanlong C and Jiangxin Y 2016 Automatic generation of 3D assembly dimension chain based on feature model *Procedia CIRP* **43** 70-75
[4] Chengtong W T 2005 Automatic generation of 3d assembly dimension chains *Chinese Journal of Mechanical Engineering* **6**

[5] Wang Z Z and Zhang W Q 2002 Automatic Generation of dimension chain based on the assembly database *Machine Design and Manufacturing Engineering* **5**
[6] Wang H N R and Yan Y 2006 Simulated toleranced CAD geometrical model and automatic generation of 3D dimension chains *The International Journal of Advanced Manufacturing Technology* **29(9-10)** 1019-1025
[7] Louhichi B Tlija M Benamara A and Tahan A 2015 An algorithm for CAD tolerancing integration: generation of assembly configurations according to dimensional and geometrical tolerances *Computer-Aided Design* **62** 259-274
[8] Desrochers A and Clément A 1994 A dimensioning and tolerancing assistance model for CAD/CAM systems *The International Journal of Advanced Manufacturing Technology* **9(6)** 352-361
[9] Kandikjan T Shah J J and Davidson J K 2001 A mechanism for validating dimensioning and tolerancing schemes in CAD systems *Computer-Aided Design* **33(10)** 721-737
[10] Haghighi P Mohan P Shah J J and Davidson J K 2014 Automatic detection and extraction of tolerance stacks in mechanical assemblies 2014 *International Design Engineering Technical Conferences and Computers and Information in Engineering Conference* **46322** V02BT03A018
[11] Singh P K Jain S C and Jain P K 2005 Advanced optimal tolerance design of mechanical assemblies with interrelated dimension chains and process precision limits *Computers in Industry* **56(2)** 179-194
[12] Zhang Y and Zhang L 2017 Theory solution method to solve the complex assembly dimension chain *Proceedings of the 2017 7th International Conference on Mechatronics, Computer and Education Informationization* (*MCEI 2017*) 661-664
[13] Gao Z Wang Z Wu Z and Cao Y 2015 Study on generation of 3D assembly dimension chain *Procedia CIRP* **27** 163-168
[14] Liu J D Chang Z Y Mo R Zhang D and Wei J F 2008 Computer aided selective assembly model for multi-dimension chains *Computer Integrated Manufacturing Systems* **5**
[15] Liu Z Huang M Tang Z and Liu T 2020 Selection and evaluation of assembly dimension chain based on analytical hierarchy process *Proceedings of the Seventh Asia International Symposium on Mechatronics* 870-878
[16] Schröder B 2016 *Ordered sets* (Basel: Birkhäuser)
[17] Bozhko A N 2019 Hypergraph model for assembly sequence problem *IOP Conference Series: Materials Science and Engineering* **560(1)** 012010

Influence of the exhaust gas recirculation on formation of NO_x in the hydrogen engine working on the leaked mixture (Experiment and 3D modeling)

R Z Kavtaradze[1,2,3], T M Natriashvili[1], M G Glonti[1], Baigang Sun[3], Yichun Wang[3], Cheng Rongrong[2,3] and Chang Citian[2,3]

[1]Dvali Institute of Machine Mechanics, 10 Mindeli Str., Tbilisi, 0186, Georgia;
[2]Bauman Moscow State Technical University, 5 2-aia Baumanskaia Str., Moscow, 105005, Russia;
[3]Beijing Institute of Technology, 5 South Zhongguancun Str., Haidian District, Beijing, 100081 China.

E-mail: kavtaradzerz@mail.ru

Summary. Nitric oxides are the only components from the harmful ones limited by the legislation found in the exhaust gases of the hydrogen engine. For the purpose of their minimization and prevention of the anomaly burning processes as well (detonation combustion, early ignition, back fire), the use of the leaked hydrogen-air mixture is offered. The influence of the exhaust gas recirculation (EGR) on the intra-cylindric processes, namely on formation of nitric oxides in the hydrogen engine is studied for the first time experimentally and with the help of the mathematical model based on the three-dimensional non-stationary equations of type of Navier-Stokes. The results obtained for the wide ranges of variation of the advance angle of ignition, excess air ratio and rate of EGR confirm that at successful combination of these controllable parameters the problem of minimization of nitric oxides exhausted by the hydrogen engine is removed.

Keywords: hydrogen engine, experiment, 3D modeling, nitric oxides.

1. Introduction

The reciprocating internal combustion engines (RICE) are widespread in the transport and stationary energetics due to, first of all, high efficiency, ensuring minimum (in comparison with other heat engines) specific fuel consumption. It is evident that the prospects for further perfection of RICE are directly related to improvement of their ecological indices. The most reliable way in this direction is the use of alternative fuels, among which hydrogen occupies a special place [1, 2, 3]. Besides the fact that hydrogen reserves are almost inexhaustible, as motor fuel it has remarkable properties that far exceed those of traditional fuels of petroleum origin: the lowest burning heat (approximately 3 times higher), wide range of variation of ignition limits (under atmospheric conditions ignition occurs at an excess air ratio α_{air}=0.13-10), high flame propagation rate (for laminar flame under atmospheric conditions in air at α_{air}=1, its value is 2.3 m/s, which is about 5-6 times higher compared with traditional fuels), minimum energy for ignition (at α_{air}=1 it is equal to 0.017 mJ in air, which is about 14 times less than for traditional fuels) [3]. From the environmental point of view the advantage of hydrogen is that from four harmful substances (nitric oxides NO_x, soot, carbon monoxide CO and unburned hydrocarbons CH), emission of which is limited by the legislation, the carbon-containing substances and also "greenhouse" gas CO_2 are absent in the exhaust gases of hydrogen RICE. It is clear that improvement of ecological indices of the hydrogen RICE with both, self-ignition [4] and forced ignition [5] implies minimization of the concentration of NO_x in the combustion products.

In spite of the fact that hydrogen has rapidly spread as a motor fuel in recent years, the combustion process of which leads to significant changes in intra-cylinder processes in comparison with the engines operating on traditional fuels, a number of important problems, in particular, influence of EGR on formation of nitric oxides have not been

Content from this work may be used under the terms of the Creative Commons Attribution 3.0 licence. Any further distribution of this work must maintain attribution to the author(s) and the title of the work, journal citation and DOI.
Published under licence by IOP Publishing Ltd

practically studied yet. It is obvious that solution of the problem is of practical importance, especially at conversion of serial traditional engines on hydrogen.

The results of the experimental study and 3D mathematical modeling of the working process and formation of nitric oxides in the combustion chamber of hydrogen RICE with external mixing and spark ignition are presented below. With the purpose of minimization of NO_x emission, as well as prevention of abnormal combustion processes (detonation combustion, early ignition, back fire) [6, 7], the experimental hydrogen engine operated in the mode $\alpha_{air}>1{,}5$. We note that high heat of hydrogen combustion provides acceptable power indices of the engine at operating on the leaked hydrogen-air mixture. The hydrogen engine under study was also equipped with the EGR system for study its influence on the environmental and effective indices.

2. Brief description of the experimental stand

Experimental researches were carried out on the test bench at the research laboratory of the Beijing Institute of Technology (BIT) [5, 6]. The engine under study is a hydrogen version of the 4-cylinder Chinese made atmospheric gasoline RICE CA20 with electronic, multipoint converted on hydrogen in BIT, technical data of which are given in Table 1. We note that the experimental hydrogen engine retains the basic parameters of the base engine CA20 without significant changes. In contrast to base gasoline RICE, where independent ignitions have the pairs of cylinders (1-4 and 2-3), in order to prevent misfires that can occur at operating on hydrogen-air super-leaked mixture, an independent ignition is used on the hydrogen engine for each cylinder. Besides, the system of hydrogen gas supply into the inlet system is changed. Considering the low hydrogen density and its high volumetric flow rate required for providing the necessary power of the engine, two hydrogen injectors are installed on each cylinder, i.e. an injector is located in each inlet. The hydrogen cyclic supply and excess air ratio are monitored, controlled and registered for each mode of operation during the experiment. The basic technical data of the experimental hydrogen RICE: Number of cylinders – 4; Diameter of the cylinder D=86 mm; Piston stroke S=86 mm; Degree of compression $\varepsilon=10$; Cooling system – Water; Rated power $N_e=60$ kW at frequency of rotation $n=5000$ min^{-1}; Maximum torque $M_\kappa=111$ Nm at frequency of rotation $n=4000$ min^{-1}.

The experimental bench is equipped with a measuring system ensuring measurement of all the necessary parameters (power, torque, fuel consumption, frequency of rotation of the crankshaft) [5, 6]. The devices for special measurements are also installed on the bench, primarily for indicating the engine and determining emission of the nitric oxides. The measured frequencies of rotation of the engine and pressure in the cylinder are introduced into the combustion analyzer for determining the heat release rate.

3. 3D model of intra-cylinder processes of the hydrogen RICE

The mathematical model of non-stationary turbulent transfer processes is based on the fundamental equations of momentum (Navier-Stokes), energy (Fourier-Kirchhoff), concentration (Fick) and continuity [4, 8, 9], averaged according to the method of Favre and using density ρ as a weight function [9, 10]. These equations of transfer, as a result of averaging, take the form of Reynolds:

$$\begin{cases} \bar{\rho}\dfrac{D\overline{W}_i}{D\tau} = \overline{G}_i - \dfrac{\partial \bar{p}}{\partial x_i} + \dfrac{\partial}{\partial x}\left[\mu\left(\dfrac{\partial \overline{W}_i}{\partial x_j} + \dfrac{\partial \overline{W}_j}{\partial x_i} - \dfrac{2}{3}\delta_{ij}\dfrac{\partial \overline{W}_k}{\partial x_k}\right) - \bar{\rho}\overline{W_i'W_j'}\right] ; \\ \bar{\rho}\dfrac{D\overline{H}}{D\tau} = \overline{G}_j\overline{W}_j + \dfrac{\partial \bar{p}}{\partial \tau} + \dfrac{\partial}{\partial x_j}\left(\bar{\tau}_{ij}\overline{W}_j\right) + \dfrac{\partial}{\partial x_j}\left(\lambda\dfrac{\partial \overline{T}}{\partial x_j} - c_p\bar{\rho}\overline{T'W_j'}\right) + \overline{w_r}\overline{Q_r} + \dfrac{\partial \bar{q}_{R_j}}{\partial x_j} ; \\ \dfrac{D\overline{C}}{D\tau} = \dfrac{\partial}{\partial x_j}\left(D_c\dfrac{\partial \overline{C}}{\partial x_j} - \overline{C'W_j'}\right) + \overline{\dot{m}} ; \\ \dfrac{\partial \bar{\rho}}{\partial \tau} + \dfrac{\partial}{\partial x_j}\left(\bar{\rho}\overline{W}_j\right) = 0 , \end{cases} \qquad (1)$$

where $D/D\tau$ is a substance derivative; ρ – density of the working body (WB), kg/m^3; p – pressure of WB, Pa; G_i – projection of density vector of volumetric forces, N/m^3 on the axes Ox_i of the rectangular Cartesian coordinate system;

C – concentration, kg/m^3; H – total specific energy, J/kg; μ – dynamic viscosity, kg/(m·S); c_p – thermal capacity at constant pressure, J/(kg·K); w_r – rate of the chemical reaction per unit volume, kg/(m^3·S); Q_r – amount of released heat per unit mass, J/kg; λ – thermal conductivity, W/(m·K); δ_{ij} – Kronecker symbol; D – diffusion coefficient, m^2/S; $\dot{m}$ - intensity of the mass source (rate of change of the chemical component mass in unit volume), kg/(m^3·S); W – velocity vector. The dashed parameters denote turbulent pulsations and the averaged parameters are denoted with a dash on top.

The system of the equations of Reynolds is ended by the k-ζ-f model of turbulence specially developed for modeling of processes in the piston engines [11]. For describing the processes in the near-wall layers are used so-called near-wall functions [9, 11].

The combustion process is taken into account by intensity of the internal heat source q_v, W/m3 and mass expenditure $\dot{m}$, kg/(m^3·S). The values of these parameters can be calculated using rate w_r of the chemical combustion reaction [3.9]: $q_v = Q_r w_r$; $\dot{m} = -w_r$, where Q_r is amount of the heat per unit mass released as a result of the chemical reaction, J/kg. The rate of the combustion process is determined on the base of the coherent flame model (CFM) described in [9]. The formation of nitric oxides [NO_x] is modeled on the base of the extended mechanism of Y. B. Zeldovich [9]. We note that the concentration is donated here and further by square brackets. The described 3D-mathematical model of intra-cylinder processes is realized with the help of the CRFD program FIRE, developed by AVL List GmbH (Austria) [9]. The FIRE kernel is based on the numerical method of control volumes using the advanced algorithm SIMPLE. Sampling of the transfer equations (1) is carried out using interpolation circuit of the second order of accuracy, which is optimal, both in accuracy and in time of counting.

4. The results of calculation and experimental studies

Verification of the mathematical model of intra-cylinder processes of hydrogen RICE, based on equations (1), was carried out by comparing the results of modelling with experimental data. In particular, were compared the experimental and calculated indicator diagrams, as well as changes in temperature of the working body and heat release rates obtained on their basis. Despite the good congruence of these parameters (the difference between the experimental and calculated values of instantaneous parameters did not exceed: by pressure and temperature 1 - 3%, by heat release rate 10 - 12%), for evaluation of reliability of the developed 3D mathematical model, comparisons were made of the concentrations of the final product of the working process - nitric oxides obtained by direct measurement in the outlet system of the hydrogen RICE and modelling.

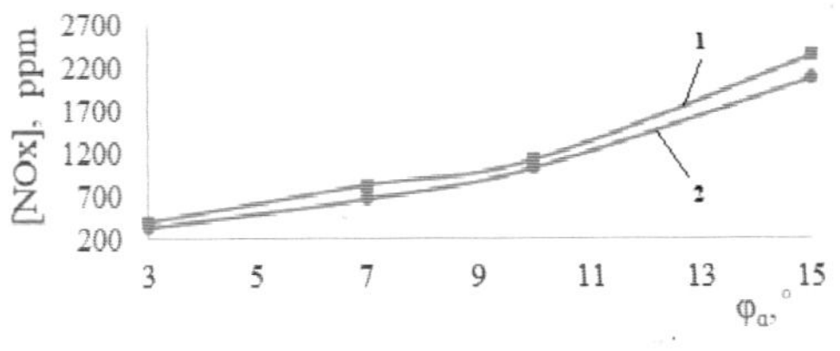

Figure 1. Influence of the advance angle of ignition on the emission of nitric oxides (1. Modeling, %. 2. experiment). The mode of operation of the hydrogen RICE, n=3000 min^{-1}.

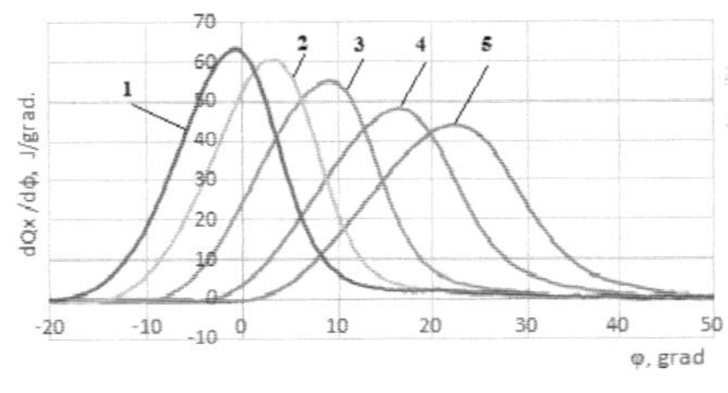

Figure 2. Change of heat release rate $dQ_x/d\varphi$ in the hydrogen RICE depending on the advance angle of ignition at operation mode of the hydrogen RICE n = 3000 min^{-1} (1. φ_a =25^0; 2. φ_a = 20^0; 3. φ_a =15^0; 4. φ_a = 10^0; 5. φ_a =5^0), obtained as a result of processing of the experimental indicator diagrams.

We note that the verification of the 3D model was carried out in the wide range of changes in frequency of rotation of the engine, EGR degree, excess air ratio and advance angle of ignition. When converting the traditional RICE (in this case, gasoline) to hydrogen, it becomes necessary to ascertain the optimal (from the point of view of the ecological compatibility and fuel economy of the engine) advance angle of ignition φ_a. The results of the modeling and experimental study of influence of φ_a on emission of NO_x are given in Figure 1. The maximum deviation of the results of modeling from the experimental data, as shown, does not exceed 11%. Analysis of the experimental indicator diagrams and obtained on their basis heat release rates $dQ_x/d\varphi=f(\varphi)$ (Figure 2) indicates that for this mode of operation of the engine the most optimal is the advance angle of ignition $\varphi_a = 15^0$.

The positions of these peak values relative to the upper dead point at $\varphi_a = 15^0$ are the most favorable that indicates the efficiency of the working cycle of the hydrogen engine (indicator efficiency at this mode according to the experimental data η_i=40.8 %). It is obvious that this is the result of intensification of the combustion process, leading simultaneously to the rise in p_z and increase in both, the maximum cycle temperature (in this case T_z=2375 K) and local temperatures at the flame front. The increase in the temperature level of the cycle is, of course, accompanied by the increase of emission of nitric oxides. This explains the increase of [NOx] with increase of φ_a (Figure 1).

For the first time on the hydrogen RICE was studied the influence of the EGR on the working process of the hydrogen RICE. The experimental results given in Figure 3 show that the use of EGR (degree of recycling z=15%) at n = 1000 min^{-1} results in the significant reduction in emission of NOx (almost 3 times). It can also be seen that with increase of α_B the maximum emission compared to the combustion process without EGR, moves away relatively further from the stoichiometric value α_B=1. Besides, the changes of [NOx] depending on α_B occur more smoothly, than for the case without EGR (Figure 3). Figure 4 shows the changes in concentrations of nitric oxides in the exhaust gases of the hydrogen engine under study, depending on the excess air ratio at frequency of rotation of the crankshaft n = 3000 min^{-1} and degree of EGR z=15 %, obtained experimentally and by modeling, that confirm the reliability and trustworthiness of the developed 3D model of the intra-cylinder processes. It is noticeable that the maximum amount of nitric oxides during hydrogen combustion is formed at slightly enriched hydrogen-air mixture (α_{air}=1,1…1,2) and its value relatively little depends on the frequency of rotation of the crankshaft. However, as already noted, the operation of the hydrogen engine at α_{air}=1,1…1,2 or close to these values, is not desirable due to the danger of detonation combustion, back fires and early ignition [6,7].

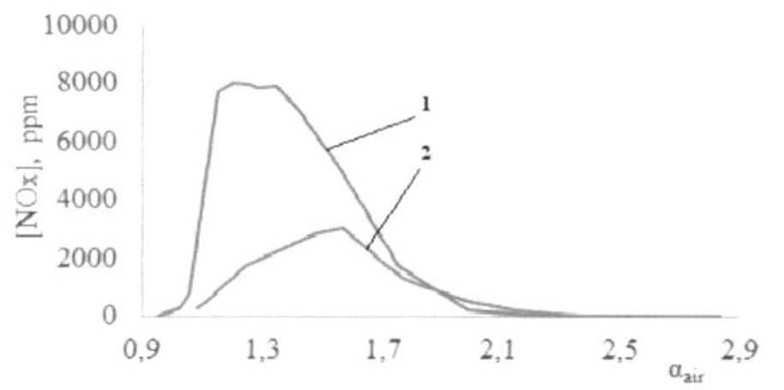

Figure 3. The results of the experimental study of the nitric oxides emission depending on degree of EGR (1. z=0, 2. z=15%). Operation mode of the hydrogen RICE, n = 1000 min-1.

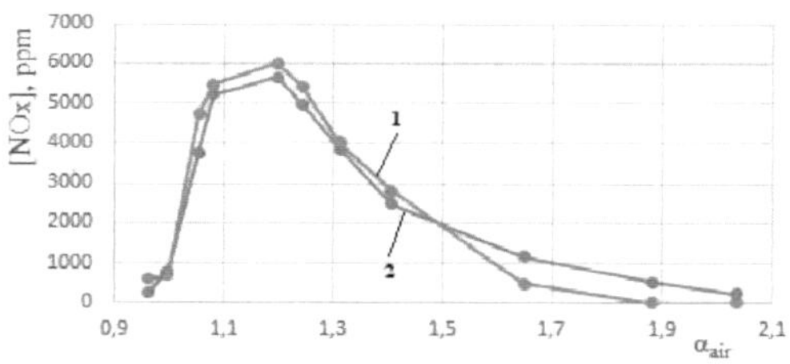

Figure 4. Influence of the excess air ratio on the emission of nitric oxides at constant frequency of rotation of the crankshaft n = 3000 min^{-1} and degree of EGR, z=15 % (1. modeling, 2. Experiment.)

The results of the study showed that the effect of the excess air ratio is more significant than that of the frequency of rotation. At low frequency of rotation of the engine n = 1000 min^{-1} (when α_{air} decreases from 2.0 to 1.2), the concentration of nitric oxides in the exhaust gases increases from 200 ppm to its maximum value (approximately 6000 ppm) respectively. In general, judging by the nature of the change in the dependence $[NO_x]=f(\alpha_{air})$, the formation of nitric oxides in the hydrogen engine occurs similarly to the engines operating on traditional oil fuels in the whole speed range [8]. Absence of the free oxygen at combustion of enriched mixture ($\alpha_{air}<1,0$) and temperature decrease at combustion of leaked mixture ($\alpha_{air}>1,0$) prevent combination of the nitrogen and oxygen and limit formation of nitric oxides. Obviously, increasing the deviation of α_{air} in any direction from the stoichiometric value (α_{air}=1,0) exacerbates this effect. The two characteristic areas of sharp reduction of NO_x depending on α_{air} are clearly seen in Fig. 4:

1. Area of the air excess ratio reduction from α_{air} =1.1...1.2 to α_{air} =0.9;
2. Area of the air excess ratio rise from α_{air}=1.1...1.2 to α_{air}=2.0.

At the same time, the second area is about 2 times higher than the first one in the range of change of α_{air} and the change of [NO_x] occurs more slowly in it than in the first area. The reduction of [NO_x] by increasing α_{air} (super-leaked hydrogen-air mixtures) or by decreasing α_{air} (super-enriched hydrogen-air mixtures) beyond the mentioned limits of change of α_{air} occurs much slower. The presence of areas of sharp reduction (or sharp increase) of [NO_x] is explained by the presence of a high temperature – the result of the hydrogen combustion having a high combustion heat and by the peculiarity of changing this temperature in these areas. In this case, in the first zone (in the zone of rich mixtures) there is a high concentration of hydrogen H_2 in the cylinder, which is chemically more active than nitrogen N_2. Therefore, it reacts faster than N_2 with oxygen O_2, the amount of which in the rich mixture is small that leads to the sharp decrease in emissions of NO_x (Figure 4). In the second zone, with increase of α_{air} towards super-leaked mixtures, with increase of the air amount, the temperature of the working body decreases in the cylinder, however, the rate of the temperature decrease is lower here than in the first one, and the rate of change of NO_x (Fig.4) correlates with the rate of the temperature change.

Both, the experimental studies and simulations showed that even with a substantially leaked mixture (α_{air}=1.5) and degree of EGR, z=15 %, the emission of nitric oxides in the exhaust gases of the hydrogen RICE under study is quite high ([NO_x]=2000 ppm) (Fig. 4). The further increase in the degree of EGR, $z \geq 20\%$ can lead to the noticeable decrease in effective indices of the engine. It should be noted that the capabilities of the EGR in hydrogen engines have not been practically studied to date. Obviously, the exhaust gas in this case is water vapor H_2O, which, when entering the inlet system and stirring with the cooler hydrogen-air mixture, can partially condense, and then evaporate again with increase of the temperature during the compression process. This can lead to the drop of temperature in the cylinder and decrease of nitric oxides similarly to the engines with water injection into the cylinder [12]. Obviously, the temperature of the exhaust gases bypassed into the hydrogen engine inlet system can play a role here. Indeed, according to the results of measurement at the mode n = 1000 min^{-1}, we have t_{vap}=431^{0}C, which is approximately 1.6 times less than the temperature t_{vap}=679^{0}C at n = 4500 min^{-1} with the almost unchanged composition (α_{air}=1.61 – 1.64) of the hydrogen-air mixture. The exhaust gases presenting water vapour, with such temperatures enter the inlet system having atmospheric parameters p_0 and t_0, where the steam condensation temperature is equal to $t_{\kappa H20}$=100 ^{0}C. Under such conditions, due to the lack of time, the steam can presumably only partially condense before entering the cylinder. During the compression stroke, the steam condensation process continues until the temperature in the cylinder reaches the boiling point (condensation) of water (about 180^{0}C at 10 bars). The water vapor is not a reactive gas and is not involved in the combustion process, but its presence in the cylinder, both as gas and condensed water, leads to decrease in the temperature of the working body and consequently, to decrease in the emission of nitric oxides. Thus, the EGR process in the hydrogen engine is accompanied by the phenomena similar to the process in the engines in combustion chambers of which water is injected for reduction of NO_x emissions.

5. Conclusion

The results of the experimental study, as well as the mathematical modeling of the working process of the hydrogen engine with injection of hydrogen gas into the inlet system and ignition of the hydrogen-air mixture using an electric spark, allow us to conclude that the choice of the excess air ratio α_{air} is the most important factor for preventing abnormal phenomena (back fire, early ignition, detonation combustion) and reduction of the nitric oxides content in exhaust gases, as well as ensuring normal, stable engine operation. It is confirmed that there is a relationship between α_{air} and emission of nitric oxides NO_x, which is relatively little dependent on the engine speed mode. It is established that in the limits of change of air excess ratio α_{air}=1.6-2.4, within the range of change of frequency of rotation of the crankshaft speed (n = 1000-5000 min^{-1}) the stable operation of hydrogen engine is provided without abnormal phenomena and misfires. The speed characteristic of the hydrogen engine, built with a fully open air throttle, when at all modes the condition of constancy of the excess air ratio ($\alpha_{air}\approx$1.6-1.7) was observed, showed that the minimum specific effective hydrogen flow rate g_{emin}=87.24 g/(kWh) was reached at n = 2500 min-1. In idling mode, the specific flow rate of hydrogen reaches 96 g/(kWh). In mode n = 5000 min^{-1}, we have g_e=92.2 g/(kWh) and N_e=51.0 kW, which is 15% less than the declared maximum power of the base gasoline engine N_e=60 kW. Achieving such a maximum power on the experimental hydrogen engine is possible at α_B=1.35 in mode n = 5500 min^{-1}, but in this case the signs of abnormal combustion appear. They are expressed primarily in appearance of back fires due to the high temperature of the residual combustion products and in high speeds of movement of the flame front. The latters lead to increase in the rates of heat release and pressure, observed during the analysis of indicator diagrams. A study of influence of EGR on the working process, conducted for the first time for hydrogen RICE, shows that its use during the engine operation on the leaked hydrogen-air mixture (α_{air}>1.6) practically eliminates the problem of NO_x emission. A verification of the 3D model of the working process of the hydrogen engine was carried out based on the results of experimental studies of the hydrogen engine and fundamental equations of three-dimensional non-stationary turbulent transfer, in a wide range of changes in the controlled and mode parameters of the engine. This model, successfully tested earlier for a hydrogen diesel [3, 4, 10], was first used for the hydrogen engine with spark ignition. Good coincidence of the results of modeling and experimental data, especially on indicator diagrams and emissions of nitric oxides, obtained by changes in such engine parameters as advance angle of ignition; excess air ratio and degree of EGR, makes it possible to conclude that in the form of a verified model a tool has been created for conducting numerical experiments on study of intra-cylinder processes, environmental and efficient characteristics of promising hydrogen engines with forced ignition. The use of this model can significantly reduce the time and cost of creation of promising ICE of this type.

On the whole, the effectiveness of operation of the engine with spark ignition on the leaked hydrogen-air mixtures and expediency of its creation have been confirmed. An instrument in the form of a verified 3D mathematical model has been developed for further research and improvement of the working process of this engine.

The work is carried out by the financial support of Rustaveli National Science Foundation of Georgia (Project FR-18-122).

References

[1] da Rosa A V 2005 *Fundamentals of Renewable Energy Processes*. London, New York, Oxford, elsevier Press

[2] Smil V 2010 *Energy Myths and Realities: Bringing Science to the Energy Policy Debate*. Enterprise Institute for Public Policy Research, Washington

[3] Kavtaradze R Z 2011 *Thermophysical Processes in Diesel Engines Converted to Natural Gas and Hydrogen*, Publishing house of Bauman Moscow State Technical University (BMSTU) p 238 (in Russian)

[4] Kavtaradze R, Natriashvili T and Gladyshev S 2019 *Hydrogen-Diesel Engine: Problems and Prospects of Improving the Working Process,* SAE Technical Paper 2019-01-0541 p 15

[5] Sun Baigang, Duan Junfa and Liu Fushui 2014 *NOx Emission Characteristics of Hydrogen Internal Combustion Engine* vol 23, Journal of Beijing Institute of Technology 3 pp 339-44

[6] Duan Junfa, Liu Fushui and Sun Baigang 2014 *Backfire Control and Power Enhancement a Hydrogen Internal Combustion engine* International Journal of Hydrogen Energy 39 pp 4581-89

[7] Kavtaradze R Z, Zelentsov A A and Krasnov V M 2018 *Local Heat Transfer in the Combustion Chamber of a Diesel Engine Converted to Natural Gas and Hydrogen* vol 56, High Temp 6, pp 900–09

[8] Merker G, Schwarz Ch and Stiesch G 2016 Verbrennungsmotoren. *Simulation der Verbrennung und Schadstoffbildung.*Teubner-Verlag: Stuttgart, Leipzig, Wiesbaden p 820

[9] FIRE 2019 *User's Manual Version* AVL List GmbH Graz (Austria)
[10] Kavtaradze RZ, Natriashvili TM, Glonti MG, Wang Yichun and Sakhvadze G Zh 2019 *Local Heat Transfer in the Combustion Chamber of a Hydrogen Diesel* Vol 39, 10, Russian Engineering Research, Allerton Press. pp 831–36
[11] Hanjalić K, Popovać M and Hadziabdić M 2004 *A Robust Near-Wall Elliptic-Relaxation Eddy-Viscosity Turbulence Model for CFD* // Int. J. Heat Fluid Flow, 5. pp 897-01
[12] Böhm M, Mährle W, Bartelt H and Rubbert St 2016 *Funktionale Integration einer Wassereinspritzung in den Ottomotor*. MTZ 1 pp 38-43

Tool for assessing the quality of electricity from renewable energy sources

I V Belitsyn, S O Khomutov and A V Shendrik

Altai State Technical University, 46 Lenin av, Barnaul, 656038, Russian Federation

E-mail: b_i_w@mail.ru

Abstract. The article deals with the problem of quality control of electrical energy. A method for increasing the accuracy of measurements at the software level is considered in detail, using the example of a Hall sensor. Determination of the Hall sensor hardware function is based on the assumption of a stationary current density distribution. It is shown that the hardware function of the sensor has a characteristic maximum, which must be taken into account when determining the quality indicators of electrical energy in electrical networks. The results of modeling the potential distribution in the Hall sensor are presented. The influence of the geometric linear dimensions of the sensor on the instrumental function is analyzed.

1. Introduction

In the Altai Territory, there are two problems: a shortage of its own electricity and, as a result, a high tariff for electricity [1]. Altai Territory has a high potential for the construction of power plants operating on renewable energy sources (sun, wind). This requires investment and unoccupied land.

Within the framework of the Agreement between Altayenergosbyt OJSC, Altai Territory Administration, IDGC of Siberia OJSC and Vent Rus LLC dated 04.05.2010 on cooperation in the development of wind energy in the Altai Territory, work is underway to implement investment projects for the construction of wind power plants.

The hydropower potential of the rivers of the Altai Territory is able to significantly reduce the power supply deficit in rural areas and villages remote from the existing power system, as well as areas with single-circuit and radial physically worn out power lines. LLC EC Energia plans to implement the project "Construction of small and micro-hydroelectric power plants in the Altai Territory with a total installed capacity of 31.6 MW", which includes the following small hydroelectric power plants: Gilevskaya Micro-HPP with a capacity of 2.4 MW at the Gilevsky reservoir in the Loktevsky district; Soloneshenskaya SHPP with a capacity of 1.2 MW on the Anuy River in the Soloneshensky District.

The generated electrical energy must comply with the regulations [2, 3]. In particular, indicators of the quality of electrical energy are determined by the state standard GOST 32144–2013 "Electrical Energy. Electromagnetic compatibility of technical means. Electricity quality standards in general-purpose power supply systems.

Electricity quality indicators require continuous monitoring [4]. To process the measurement results, modern algorithms are used, including those based on neural networks [5]. Evaluation of the parameters of the quality of electrical energy is also necessary for the management of urban electrical networks within the Smart Grid [6].

2. Mathematical model

Evaluation of indicators of the quality of electrical energy is possible based on the analysis of the parameters of the electromagnetic field. To determine the characteristics of the magnetic field, it is proposed to use a Hall sensor [7].

If the measured magnetic field is small enough, as well as the sensor current, the Hall sensor can be considered linear. In this case, the Hall potential difference and the measured value of the magnetic field have a linear relationship, which can be represented as a convolution integral:

Content from this work may be used under the terms of the Creative Commons Attribution 3.0 licence. Any further distribution of this work must maintain attribution to the author(s) and the title of the work, journal citation and DOI.
Published under licence by IOP Publishing Ltd

$$u_{\mathrm{H}}(x,y) = \frac{IK_{\mathrm{H}}}{S} \iint_S r(x',y')B(x-x',y-y')dx'dy' \tag{1}$$

where I is the current flowing in the Hall sensor through its contacts; S is the cross-sectional area of the Hall sensor; K_{H} is the coefficient (constant) of the Hall sensor; $r(x,y)$ is the hardware $r(x,y)$; $B(x,y)$ is the magnetic field distribution inside the sensor volume.

Relation (1) shows that the potential difference at the potential contacts of the Hall sensor is determined by the hardware function $r(x,y)$, which is invariant to the field distribution and the current flowing through the sensor.

Obviously, when $r(x,y) = 1$, the minimum change in the measured value that the sensor can detect is primarily determined by its dimensions, ideally $r(x,y) = S_{\delta}(x,y)$. If the sensor is placed in a uniform magnetic field, it is necessary to enter the normalization

$$\iint_S r(x,y)\,dxdy = S \tag{2}$$

Let us assume that the general view of the apparatus function $r(x,y)$ depends on the geometric dimensions of the sensor and its contacts and the current density in the sensor.

To substantiate (1) and determine the hardware function of the Hall sensor, we will assume that a direct current I flows through its contacts. The x axis passes through the current contacts located on the left and right edges of the crystal. The y axis passes through potential contacts located on the upper and lower faces of the crystal. In the absence of a magnetic field, a stationary current density distribution $j_0(x,y)$is established inside the sensor, which is potential [8]

$$rotj_0(x,y) = 0 \tag{3}$$

$$divj_0(x,y) = 0 \tag{4}$$

According to [9], in this case there should be a scalar potential function φ_0, which determines the distribution of the electric field inside the sensor

$$E = -grad(\varphi_0) \tag{5}$$

The current density is calculated according to Ohm's law, written in differential form

$$J_0 = \sigma E = -\sigma grad\varphi_0 \tag{6}$$

where σ is the conductivity of the Hall sensor.

Substituting (6) into equation (4), we obtain the Laplace equation for the potential distribution $\varphi_0(x,y)$,

$$\Delta\varphi_0(x,y) = 0 \tag{7}$$

3. Results and Discussion

Let us assume that the conductivity of the material of the sensor contacts, both current and potential, is much higher than the conductivity of the Hall sensor crystal, in this case only the normal component of the electric field strength will be present at the border of the sensor and the contact.

In the absence of a magnetic field inside the sensor, there is no current through the upper potential faces, therefore, from the side of these contacts $j_{0y}\left(y = \pm\frac{b}{2}\right) = 0$, then, taking into account (7), we obtain the first boundary condition

$$\frac{\partial\varphi_0}{\partial y}(x, y = \pm\frac{b}{2}) = 0 \tag{8}$$

From the side of the current contact $(x = \pm\frac{a}{2}, |y| \leq \frac{\varepsilon}{2})$, taking into account the constancy of the current and its uniform distribution over the entire current contact, only the x-component of the electric field strength will exist, therefore the current density

$$j_{0x} = \left(x = \pm\frac{a}{2}, |y| \leq \frac{\varepsilon}{2}\right) = \frac{I}{d\varepsilon} \tag{9}$$

where I is the total current flowing through the contact; d is the Hall sensor thickness; ε is the current contacts width.

From the side of the Hall sensor at the boundary $(x = \pm\frac{a}{2}, |y| \leq \frac{\varepsilon}{2})$, the potential φ_0 and the current density $j_{0x}$are related by relation (7), therefore the second boundary condition for the potential φ_0

$$\frac{\partial\varphi_0}{\partial x}\left(x = \pm\frac{a}{2}, |y| \leq \frac{\varepsilon}{2}\right) = -\frac{I}{d\varepsilon\sigma} \tag{10}$$

where $\sigma$is the conductivity of the Hall sensor.

For current density outside the current contacts

$$divj_0(x,y) = 0\frac{\partial\varphi_0}{\partial x}\left(x=\pm\frac{a}{2},|y|\leq\frac{\varepsilon}{2}\right) = -\frac{I}{d\varepsilon\sigma}=0 \quad (11)$$

Expressions (9) - (11) represent the well-known Neumann boundary value problem [10] for the potential distribution φ_0.

If the width of the current and potential contacts of the sensor is much less than the linear dimensions of the sensor, the conductivity of the contact material is multiples of the conductivity of the crystal, the sensor is placed in a uniform magnetic field directed along the z axis (there is no tangential component of the magnetic induction vector), then the potential distribution φ_0 along the z axis can be considered uniform and one can pass from the three-dimensional model of the Hall sensor to its two-dimensional approximation. In the case of a rectangular sensor with dimensions $a \times b$, the potential distribution is found by separation of variables.

The potential φ_0(x, y)must satisfy the symmetry conditions. In particular, the potential along the y-axis must be even, and odd along the x-axis. Then, taking these requirements into account, we obtain the potential distribution

$$\varphi_0(x,y) = -\frac{I}{bd\sigma}x - \frac{I}{\varepsilon d\sigma}b\sum_{m=1}^{\infty}\frac{\sin(\frac{2\pi m\varepsilon}{b})}{(\pi m)^2 ch(\frac{\pi ma}{b})}sh\frac{2\pi mx}{b}\cos\frac{2\pi my}{b} \quad (12)$$

This distribution can be obtained alternatively by numerically integrating the Laplace equation (7), taking into account the boundary conditions (9-11).

Let us transform the integral in expression (1) into a multiple integral. To do this, replace the variables of integration x, y with x′, y′ taking into account that the center of the Hall sensor is at the point with coordinates x, y, the parity of the functions $v(y)$, $j_0(x,y)$:

$$u_H(x,y) = \frac{1}{ane}\int_{-\frac{b}{2}}^{\frac{b}{2}}\int_{-\frac{a}{2}}^{\frac{a}{2}} v(y')j_{0x}(x',y')B(x-x',y-y')dx'dy' \quad (13)$$

Comparison of (13) with (1) gives an expression for the apparatus function of the Hall sensor

$$r(x,y) = \frac{v(y)S}{anelK_H}j_{0x}(x,y) \quad (14)$$

Integrating (14) over the entire surface of the sensor and taking into account the normalization condition (2), we obtain

$$K_H = \frac{1}{anel}\int_{-\frac{b}{2}}^{\frac{b}{2}} v(y)\left[\int_{-\frac{a}{2}}^{\frac{a}{2}} j_{0x}(x,y)dx\right]dy \quad (15)$$

The distribution of x component of the current density $j_0(x,y)$ which is included in (15), can be determined from the distribution (12) of the potential $\varphi_0(x,y)$:

$$j_{0x}(x,y) = -\sigma\frac{\partial\varphi_0}{\partial x} = \frac{I}{bd} + \frac{I}{\varepsilon d}\sum_{m=1}^{\infty}\frac{2\sin(\frac{2\pi m\varepsilon}{b})}{\pi m ch(\frac{\pi ma}{b})}ch\frac{2\pi mx}{b}\cos\frac{2\pi my}{b} \quad (16)$$

To find the current density distribution $j_E(x,y) = -\sigma grad\,\varphi_H$, we first determine the potential distribution $\varphi_H(x,y)$ caused by the hollowing of the potential difference u_H on the potential contacts of the sensor.

The solution to this problem, similarly to (12)

Respectively

$$\varphi_H(x,y) = -\frac{u_H}{ad\sigma R}y - \frac{u_H}{\varepsilon d\sigma R}a\sum_{m=1}^{\infty}\frac{\sin(\frac{2\pi m\varepsilon}{a})}{(\pi m)^2 ch(\frac{\pi mb}{a})}sh\frac{2\pi my}{a}\cos\frac{2\pi mx}{a} \quad (17)$$

where R is the output resistance of the sensor.
Respectively

$$j_E(x,y) = -\sigma\frac{\partial\varphi_H}{\partial y} = \frac{u_H}{adR} + \frac{u_H}{\varepsilon dR}\sum_{m=1}^{\infty}\frac{\sin(\frac{2\pi m\varepsilon}{a})}{\pi m ch(\frac{\pi mb}{a})}sh\frac{2\pi my}{a}\cos\frac{2\pi mx}{a} \quad (18)$$

After making the transformations, we get the expression for the auxiliary function $v(y)$

$$v(y) = \left[\frac{b}{2} + \frac{a^2}{\varepsilon}\sum_{m=1}^{\infty}\frac{\sin(\frac{2\pi m\varepsilon}{a})sh(\frac{\pi mb}{a})}{(\pi m)^2 ch(\frac{\pi mb}{a})}\right]\int_{-\frac{a}{2}}^{\frac{a}{2}}\frac{\frac{1}{a}+\frac{2}{\varepsilon a}\sum_{m=1}^{\infty}\frac{\sin(\frac{2\pi m\varepsilon}{a})ch(\frac{\pi my}{a})}{\pi m ch(\frac{\pi mb}{a})}\cos\frac{2\pi mx}{a}}{\frac{b}{2}+\frac{a^2}{\varepsilon}\sum_{m=1}^{\infty}\frac{\sin(\frac{2\pi m\varepsilon}{a})sh(\pi mb/a)}{(\pi m)^2 ch(\frac{\pi mb}{a})}\cos\frac{2\pi mx}{a}}dx \quad (19)$$

Integral (19) in the general case is found numerically, but for the particular case ($a = b$)) it is greatly simplified and gives $v(y) = 1$. Then

$$K_H = \frac{1}{anel}\int_{-\frac{a}{2}}^{\frac{a}{2}}\left[\int_{-\frac{b}{2}}^{\frac{b}{2}} j_{0x}(x,y)dy\right]dx = \frac{1}{ned}, \quad (20)$$

$$r(x,y) = 1 + \frac{2a}{\varepsilon}\sum_{m=1}^{\infty}\frac{1}{\pi m ch(\pi m)}\sin\frac{2\pi m\varepsilon}{a}ch\frac{2\pi mx}{a}\cos\frac{2\pi my}{a}.$$

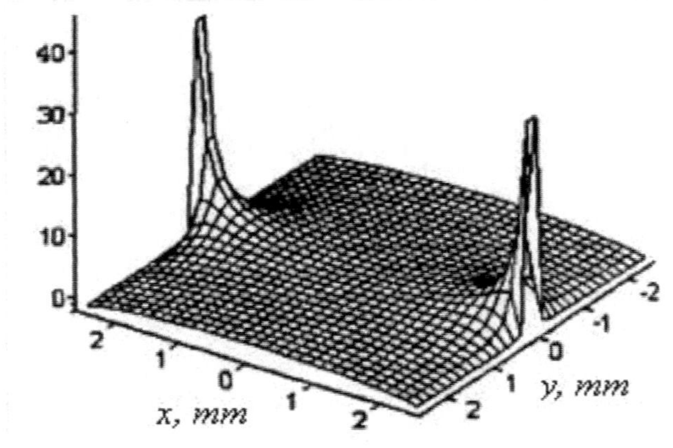

Figure 1. Hall sensor hardware function surface.

Let us present the results of mathematical modeling of the apparatus function of the Hall sensor of the SS495A type with a crystal 5 × 5 mm in size, 1 mm thick and with a width of potential and current contacts mm. For modeling and calculating the instrumental function, as well as for modeling the potential distribution, the first thirty terms of the series were used, while the calculation showed that the error does not exceed 0.1%. The total number of calculated grid points was 34x35. Figure 1 shows the surface of the hardware function; it has pronounced extreme values at the location of the current contacts.

4. Conclusion

For an ideal point sensor, there is a single and narrow extremum of the apparatus function r (x, y) of the form of a delta function. Physically, this is possible if the linear dimensions of one of the current contacts are multiples of the linear dimensions of the other.

Thus, the use of Hall sensors is possible and advisable for continuous monitoring of indicators of the quality of electrical energy. Hall sensors of the following types SS49E can be used for these purposes. SS494B SS495A1, SS495A2, SS496A, Their Hall potential difference and the measured value of the magnetic field have a linear relationship. Since these types of sensors have the same crystal size, the results of mathematical modeling of the instrumental function can be used for them.

References

[1] Kurenkova A and Lyubimova N 2018 *Bulletin of the University.* **7** 97-103
[2] Belitsyn I V, Kotyrlo T V and Makarov A V 2008 *Bulletin of the Tomsk Polytechnic University.* **4** 61-5
[3] Belitsyn I V and Khomutov S O 2017 *International Journal of Applied Engineering Research.* **12(3)** 376-81
[4] Deri A 1981 *EEE Power Engineering Review.* **4** 31–2
[5] Kostin N S 2013 *Intellectual potential of the XXI century: stages of knowledge* . **5** 91–5
[6] Davronov Sh 2018 *Bulletin of science and practice*. **4(2)** 213-6
[7] Goryachkin Y, Odnolko A, Pavlyuk M, Svistunov A and Demidov D 2017 *ARPN Jour. Of Enginiering and Appl.Science.* **12(19)** 5597-606
[8] Korolev M, Pavlyuk M and Davlekanovo S 2017 *Izvestiya vuzov. Electronics.* **2** 166-70
[9] Malykh A, Leonov A, Mordkovich V and Pavlyuk M 2016 *Experimental devices and techniques* **5** 104-8
[10] Eleev V A and Zhemukhova Z Kh 2002 *Vladikavkaz math. magazine.* **4(4)** 8-18

IOP Conf. Series: Materials Science and Engineering 1100 (2021) 012015 doi:10.1088/1757-899X/1100/1/012015

Methods for reducing electricity losses in main electrical networks

I V Belitsyn[1], S O Khomutov[1] and A I Belitsyn[1]

[1] Altai State Technical University, 46 Lenin av, Barnaul, 656038, Russian Federation

E-mail: b_i_w@mail.ru

Abstract. The article deals with the problem of reducing the loss of electrical energy when transmitting it over long distances in electrical networks. In particular, technical measures are considered that can compensate for the inequality in the capacity of overhead power lines. A mathematical model is presented that makes it possible to determine the parameters of the electromagnetic field of a power line taking into account its real design. Effective measures have been proposed for equalizing electric charges in phases. Specific recommendations are given for a 500 kV line with horizontal phases. Also, safety issues are considered, namely, the dependences of the electric field strength near power lines are obtained.

1. Introduction

High voltage electrical installations, which include overhead power lines (PTL) of high and extra-high voltage (over 220 kV), are a source of an electromagnetic field (EMF) [1]. The study of the electromagnetic field created by high-voltage electrical installations, in particular, created by overhead power lines of ultra- and ultra-high voltage, is of great importance, both for technical and medical and social research [2].

As shown by energy surveys of a number of electrical network enterprises [3], measures to reduce electricity losses can be divided into six groups:

- Measures to optimize the modes of electrical networks and improve their operation.
- Measures for the construction, reconstruction, technical re-equipment and development of electrical networks, putting into operation energy-saving equipment.
- Measures to improve the settlement and technical accounting, metrological support for electricity measurements.
- Measures to clarify the calculations of loss standards, balances of electricity for feeders, power centers and the electrical network as a whole.
- Measures to identify, prevent and reduce theft of electricity.
- Measures to improve the organization of work, stimulate the reduction of losses, improve the qualifications of personnel, control the effectiveness of its activities.

2. Mathematical model

Thus, the problem associated with the transmission of electrical energy over long distances is the non-symmetry of currents and voltages in the transmission line, which increases with an increase in its length due to different values of inductance and capacitance of the phases. This problem is traditionally solved with the help of wire transposition; one of the disadvantages of transposition is the installation of expensive transposition supports.

As shown in [4], to equalize the charges in phases, you can use:

- an increase in the suspension height of the middle phase wire;
- an increase in the equivalent radius of the conductors of the extreme phases.

To analyze these measures and determine linear charges, we will use the first group of Maxwell's formulas in a complex notation [5] whose matrix form:

$$[\alpha][\dot{\tau}]=[\dot{\varphi}] \qquad (1)$$

Content from this work may be used under the terms of the Creative Commons Attribution 3.0 licence. Any further distribution of this work must maintain attribution to the author(s) and the title of the work, journal citation and DOI.
Published under licence by IOP Publishing Ltd

where $[\alpha]$ is a matrix of potential coefficients that depend only on the geometric dimensions of the conductors, their mutual arrangement and on the properties of the medium; $[\dot{\varphi}]$ is the matrix of conductor potentials, expressions for their definition are given in [6]. However, they do not take into account the influence of a massive metal structure, which is represented by the support of an overhead line of a voltage class of 220 kV or more, as well as sagging of wires.

It is known that structurally a high-voltage support consists of corners and strips of metal, which for the proposed technique can be replaced by cylindrical conductors of circular cross-section with such a radius that the capacitance of round and non-round conductors per unit length is the same, such a radius is called equivalent. Equivalent radii can be determined, for example, by the formulas given in [7], so the equivalent radius for a strip of small thickness and width a

$$r_э = \frac{a}{4} \tag{2}$$

and for an equilateral corner with side a

$$r_э = \frac{a}{2{,}5} \tag{3}$$

So, the metal support can be replaced by a set of N cylindrical conductors with length L_k and equivalent radii $r_{эk}$, where $k \in N$. The charge induced on the conductors due to electric induction per unit length of the kth conductor is denoted by τ_k. It is not constant along the entire length of the conductor and depends on its specific point. The total number of wires and lightning protection cables suspended on a support will be denoted by N_{np}. The linear charge on the wire is not constant along the entire length and is denoted by τ_{np}, and the radius of the wire is r_{0np}

When choosing the coordinate system xyz so that $z=0$ corresponds to the earth's surface, the potential of an arbitrary observation point $M(x_j, z_j, y_j)$, will be determined from the principle of superposition and is equal to the algebraic sum of potentials from all N conductors and from mirror images, as well as from all N_{np} of wires and their images

$$\varphi_M = \frac{1}{4\pi\varepsilon\varepsilon_0}\left[\sum_{k=1}^{N}\int_{L_k}\left(\frac{\tau_k(l_k)dl_k}{R_{kM}} + \int_{L_k}\frac{\tau'_k(l_k)dl_k}{R'_{kM}}\right) + \sum_{v=1}^{N_{np}}\int_{l_{np}}\left(\frac{\tau_{npv}(l_v)dl_v}{r_{vM}} + \int_{L_k}\frac{\tau'_{npv}(l_v)dl_v}{r'_{vM}}\right)\right] \tag{4}$$

where $\tau_k(l_k)$ is the linear charge of the kth conductor on its elementary section dl_k; R_{kM} is the distance from the elementary section dl of the kth conductor to point M; R'_{kM} is the distance from the elementary section dl of the mirror image of the kth conductor to point M; r_{vM} is the distance from the vth elementary section dl of the wire to the point M r'_{vM} is the distance from the elementary section dl of the mirror image of the vth wire to the point M, l_{np} is the length of one span.

Functional equation (4) includes an integral transformation over the unknown function $\tau_k(l_k)$ and $\tau_{np}(l_{np})$ is an integral equation, namely the Fredholm equation of the first kind, taking into account that $\tau'_k = -\tau_k$ and $\tau'_{np} = -\tau_{np}$, we will rewrite it in the form

$$\varphi_M = \frac{1}{4\pi\varepsilon\varepsilon_0}\left[\sum_{k=1}^{N}\left(\int_{L_k}\frac{\tau_k(l_k)dl_k}{R_{kM}} - \int_{L_k}\frac{\tau_k(l_k)dl_k}{R'_{kM}}\right) + \sum_{v=1}^{N_{np}}\int_{l_{np}}\left(\frac{\tau_{npv}(l_v)dl_v}{r_{vM}} - \int_{L_k}\frac{\tau_{npv}(l_v)dl_v}{r'_{vM}}\right)\right]. \tag{5}$$

We place the observation point one by one on the surface of all cylindrical conductors and, taking into account that the potential of all conductors is zero, and then, sequentially on the surface of all wires, we obtain a system of integral equations

$$\begin{cases} 0=\dfrac{1}{4\pi\varepsilon\varepsilon_0}\left[\sum_{k=1}^{N}\left(\int_{L_k}\dfrac{\tau_k(l_k)dl_k}{R_{k1}}-\int_{L_k}\dfrac{\tau_k(l_k)dl_k}{R'_{k1}}\right)+\sum_{v=1}^{N_{np}}\int_{l_{np}}\left(\dfrac{\tau_{npv}(l_v)dl_v}{r_{v1}}-\int_{L_k}\dfrac{\tau_{npv}(l_v)dl_v}{r'_{v1}}\right)\right] \\ \dots \\ \varphi_{Nnp}=\dfrac{1}{4\pi\varepsilon\varepsilon_0}\left[\sum_{k=1}^{N}\left(\int_{L_k}\dfrac{\tau_k(l_k)dl_k}{R_{kN.np}}-\int_{L_k}\dfrac{\tau_k(l_k)dl_k}{R'_{kN.np}}\right)+\sum_{v=1}^{N_{np}}\int_{l_{np}}\left(\dfrac{\tau_{npv}(l_v)dl_v}{r_{vNnp}}-\int_{L_k}\dfrac{\tau_{npv}(l_v)dl_v}{r'_{vNnp}}\right)\right]. \end{cases} \quad (6)$$

One of the most common methods for solving integral equations and their systems is the method of quadrature formulas (finite sums) and consists in replacing the integral equation with an approximating system of algebraic (finite) equations for the discrete values of the desired function and its solution [8, 9].

To compose a system of linear algebraic equations, we divide each conductor into g equal sections, length h_k, and each wire and cable into g_{np} of equal sections, length $l_{эл}$. We will assume that the linear charge density at each section of the conductor, wire and cable τ_i remains constant. In this case, the total charge of each electrode with a smooth change in linear density and with a stepwise change is the same. In order not to increase the number of equations in the system, taking into account expression (3), we write down the resulting potential from the section g, k[th] conductor and its image, as well as from the section g_{np}, v^{th} wire (cable) and its image

$$\tau_i\left(\alpha_{ij\Pi}-\alpha_{ij3}\right)=\tau_i\alpha_{ij} \text{ or } \alpha_{ij}=\alpha_{ij\Pi}-\alpha_{ij3} \quad (7)$$

where $\alpha_{ij\Pi}$ is the coefficient due to the charge of the section of the conductor of the wire or cable; α_{ij3} is the coefficient due to the charge of the mirror image of the same section of the conductor, wire or cable.

Thus, the system of integral equations (6), taking into account (7), can be replaced by a system of linear algebraic equations, which in matrix form has the form

$$\begin{cases} \left[\alpha_{ij}\right]\left[\tau_i\right]=[0] \\ \left[\alpha_{ij}\right]\left[\tau_i\right]=\left[\varphi'_i\right] \end{cases} \quad (8)$$

3. Results and Discussion

We will consider the identification of the effectiveness of measures to equalize electric charges in phases using the example of a 500 kV power transmission line, made by the supports shown in Figure 1. The equivalent height of the suspension of the wires is assumed to be 15 m. The equivalent phase radius is 0.16 m, for the AC 400 wire and the number of conductors in the split phase 3. Distance between the phases is 12 m.

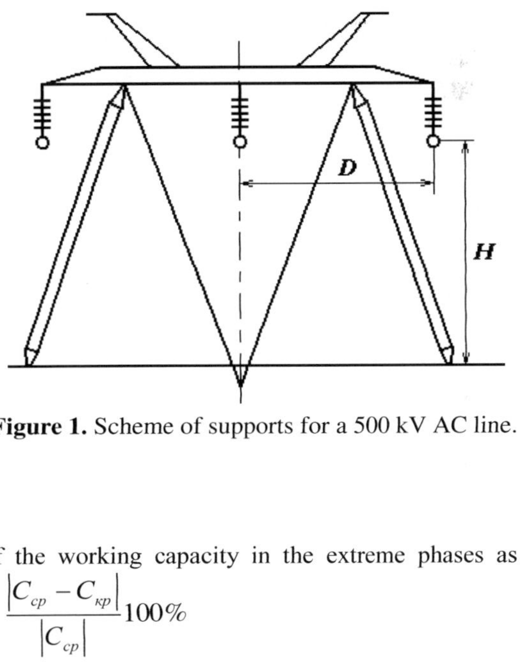

Figure 1. Scheme of supports for a 500 kV AC line.

Let us introduce the coefficient ***K*** showing the ratio of the equivalent radius of the extreme phase to the equivalent radius of the middle phase. The dependences of the electrical parameters, given as a percentage of the parameters of the middle phase, on the coefficient ***K***, are shown in Figures 2 and 3. They show 1 - the deviation of the charge at the extreme phases as a percentage of the charge at the middle phase $\dfrac{\left|\tau_{cp}-\tau_{\kappa p}\right|}{\left|\tau_{cp}\right|}100\%$; 2 - deviation of the working capacity in the extreme phases as a percentage of the working capacity in the middle phase $\dfrac{\left|C_{cp}-C_{\kappa p}\right|}{\left|C_{cp}\right|}100\%$

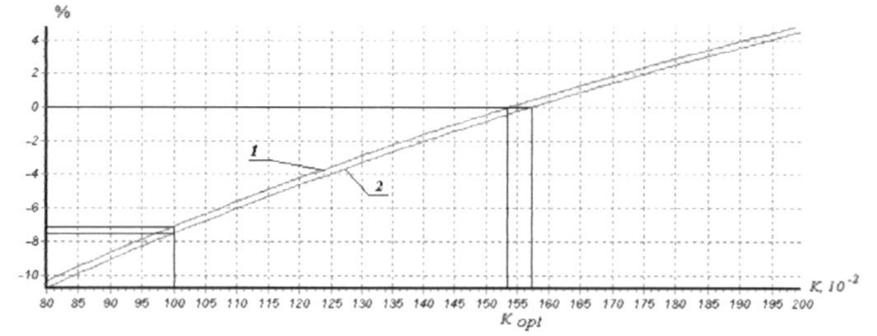

Figure 2. Dependence of electrical parameters on the ratio of the equivalent radius of the extreme phase to the equivalent radius of the middle phase.

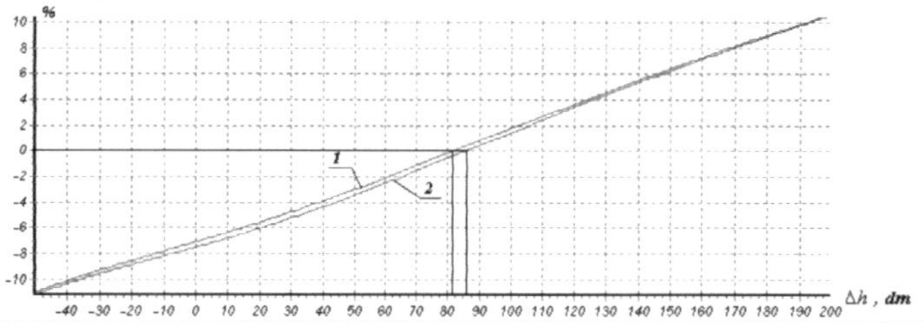

Figure 3. Dependence of electrical parameters on the change in the height of the suspension of the middle phase Δh.

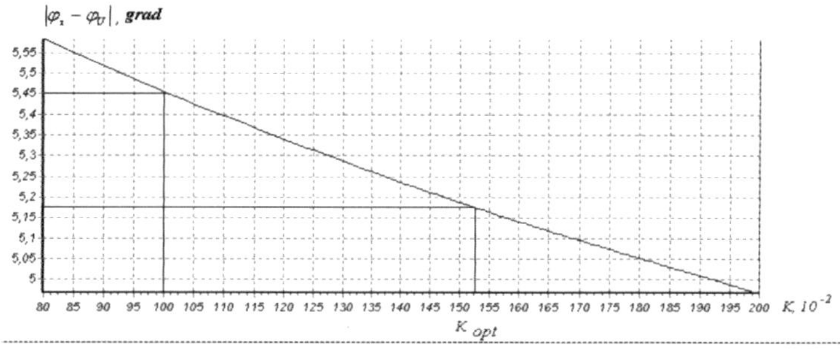

Figure 4. Dependence of the difference between the modulus of the arguments of the charge and the potential of the extreme phases $|\varphi_\tau - \varphi_U|$ on the ratio of the equivalent radius of the extreme phase to the equivalent radius of the middle phase.

It follows from the graph that charge modules do not differ from each other at all phases, if the equivalent radius of the outer phases is 1.53 times the equivalent radius of the middle phase, or if the height of the middle phase is increased by 8.2 m. phases are the same provided that the ratio of the equivalent radius of the extreme phase to the equivalent radius of the middle phase is 1.57, or with an increase in the height of the middle phase by 8.6 m.

It is known [10] that due to geometric symmetry, the charge and potential at the middle phase coincide with each other, i.e. the phase difference between them is zero. The dependence of the difference between the modulus of the arguments of the charge and the potential of the extreme phases $|\varphi_\tau - \varphi_U|$ on ***K***, and the change in the height of the suspension of the middle phase is shown in Figures 4 and 5.

It follows from the graphs that with the optimal ratio of the equivalent radii of the extreme and middle phases, the difference between the modulus of the arguments of the charge and the potential of the extreme phases decreases, compared with the initial value equal to 5.45, to 5.17 degrees, or by 5%.

4. Conclusion

Based on the above, the following conclusions can be drawn:

The sequence of the phase arrangement does not affect the dependence of the electrical parameters on the ratio of the equivalent radius of the extreme phase to the equivalent radius of the middle phase.

To equalize the capacities of each phase, first of all, it is necessary to focus on the alignment of the charge modules, and not their arguments.

The working capacitances of all phases are the same, provided that the ratio of the equivalent radius of the extreme phase to the equivalent radius of the middle phase is 1.57, or with an increase in the height of the

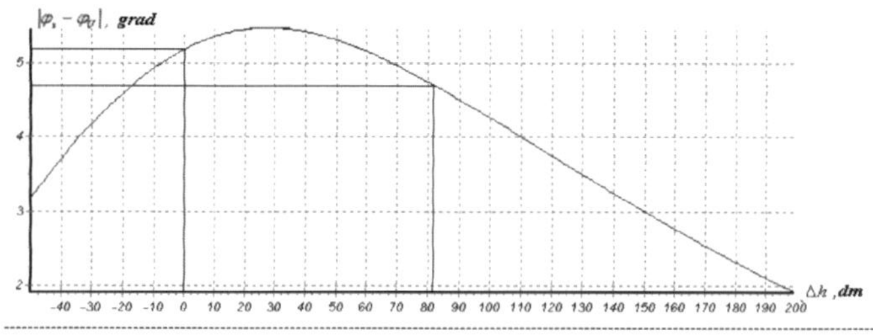

Figure 5. Dependence of the difference between the modulus of the arguments of the charge and the potential of the extreme phases $|\varphi_\tau - \varphi_U|$ on the change in the height of the suspension of the middle phase Δh.

middle phase by 8.6 m.

For all considered options for suspending the wires of power lines with a voltage of 500 kV, the discrepancy between the effective value of the electric field strength and the effective value of the electric field strength along the major semiaxis of the polarization ellipse is insignificant, almost completely disappears outside the power line limited by the extreme wires. Therefore, to control the level of the electric field created by traditional transmission lines, it is possible to use the effective value of the electric field strength with an error not exceeding 5.5%.

References

[1] Belitsyn I V and Khomutov S O 2017 *Modeling of the electric field of overhead power lines taking into account the influence of the earth and edge effects* (Novosibirsk: Nauka) p 144
[2] Eskov E K and Karev V A 2009 *Proceedings of the Samara scientific center of the Russian Academy of Sciences* **11(1)** 127-32
[3] Belitsyn I V, Kotyrlo T V and Makarov A V 2008 *Bulletin of the Tomsk Polytechnic University.* **4** 61-5
[4] Belitsyn I V and Khomutov S O 2017 *International Journal of Applied Engineering Research.* **12(3)** 376-81
[5] Postolaty V M 2019 *Problems of the regional energetics.* **1(45)** 68-85
[6] Shevchenko N Yu, Lebedev Y V, Ugarov G G and Solinov A G 2010 *Electrical Equipment equipment: operation and repair.* **12** 8-12
[7] Goryachkin Y, Odnolko A, Pavlyuk M, Svistunov A. and Demidov D. 2017 *ARPN Jour. Of Enginiering and Appl.Science.* **12(19)** 5597-606
[8] Guseva N V, Shevchenko N Yu and Soshinov A G *Bulletin of the Lobachevsky chevskyUniversity of Nizhny Novgorod.* **5(1)** 277-80
[9] Shevchenko N Yu 2010 *Scientific and technical Bulletin of the Saint Petersburgstate Polytechnic University.* **4** 66-72
[10] Shevchenko N Yu, Lebedev Y V and Solinov A G 2010 *Modern problems of science and education.* **6** 102-5.

Effects of structural formation in the implementation of the technology for obtaining asphalt concrete mixtures with phosphogypsum and other additives

A A Ignat'yev[1], V M Gotovtsev[1] and P B Razgovorov[1]

[1]Yaroslavl State Technical University, Moskovsky prospect, 88, Yaroslavl, 155023, Russia

E-Mail: ignatyevaa@ystu.ru, gotovtsevvm@ystu.ru, razgovorovpb@ystu.ru

Abstract. The paper considers the effects of structural formation in dispersed systems and composite materials obtained using the technology of asphalt concrete with up to 30% phosphogypsum included in the mixture. The study of physicochemical and consumer properties of such systems and materials involved using the method of study of surface effects (measurement of a contact angle). One of such materials is bitumen with a non-traditional additive polyethylene terephthalate. The presented microphotographs of modified asphalt concrete mixtures allow to study the changes in the state of bitumen surface. Differential-scanning calorimetry is used to register thermal effects arising from the implementation of technology to modify the properties of materials. It is revealed that obtaining asphalt concrete mixture by pelletizing provides the necessary indicators of the hardened material's compressive strength. It is determined that the water resistance of asphalt concrete increases at simultaneous introduction of polyethylene terephthalate and industrial waste powder — phosphogypsum — into dispersed system. It is also determined that the specified additives synergistically affect the consumer properties of bitumen and achieve the positive effect of structural formation in asphalt-binding materials, which reveals new prospects of their use in road building.

Keywords: composite materials, disperse systems, asphalt concrete mixtures, modification, phosphogypsum, polyethylene terephthalate additive, pelletizing, structural formation, consumer properties

1. Introduction

The term "structure" in composite (dispersed) materials means a spatial framework formed by the bonds of atoms, ions, molecules, micelles, crystalline terminations, and particles of colloidal sizes [1].

Coagulation (coagulation-tixotropic) and condensation (condensation-crystallization) dispersed systems (DS) are distinguished by the type of contacts. In coagulation DS, the particles interact through a layer of liquid dispersed medium and are bound by the van der Waals attractive force. Condensation DSs are formed in the extraction of a new phase from oversaturated vapors, melts, and solutions [2].

The structure formation effect and detailed study of colloidal-chemical properties [3] are the basis for the creation of most modern composite materials. P.A. Rebinder formalized the main provisions for obtaining ideal composites. He argued: "The easiest way to increase the strength of any solid to almost the ideal ceiling is to grind it to particles that are in order of magnitude corresponding to the distances between the dangerous weak points. If such particles are densely packed or glued with the finest interlayers, the resulting material will be dense, impermeable to liquids and gases, high-strength and durable" [4].

Asphalt concrete is a composite material in which crushed stone, river sandб and mineral powder are used as fillers, and bitumen acts as a binding agent. The main structure-forming component of the mixture is a mineral powder, a finely dispersed fraction with a highly developed specific surface. According to [5], up to 95% of the bitumen introduced into the asphalt concrete mixture - the most expensive component of the mixture - is used for wetting the surface of the mineral powder.

Compliance with P.A. Rebinder's requirements is possible due to pelletizing mineral powder (with bitumen as a binder) [6]. Previously [6] we obtained granulated asphalt binder, which has high strength,

Content from this work may be used under the terms of the Creative Commons Attribution 3.0 licence. Any further distribution of this work must maintain attribution to the author(s) and the title of the work, journal citation and DOI.
Published under licence by IOP Publishing Ltd

minimum water saturation, lack of tendency to clumping. Further research has allowed us to develop the technology framework aimed at solving a global environmental problem — phosphogypsum utilization (waste phosphoric acid and fertilizer production). The accumulated volumes of this product in Russia are estimated at hundreds of millions of tons and are increasing annually. Large useful areas are wasted on them, and they impose a real threat to the environment.

The first attempts to use phosphogypsum as a mineral powder for asphalt concrete mixture, which is obtained by pelletizing technology, allowed the researchers to obtain a material that meets the GOST requirements for most physicochemical and consumer properties, except for water resistance. As a result of long searches, it was possible to find a modifier for bitumen that provides the necessary level of water resistance of asphalt concrete. As such, it is now possible to use household waste polyethylene terephthalate (PET), from which a large share of plastic containers (plastic bottles in particular) is made. The found technical solution is protected by the RF patent [7]. Taking into account the results above, the industrial implementation of the developed technology will allow to lay not only quality road pavements, but also contribute to the solution of the environmental problem — PET waste utilization (estimated in millions of tons annually).

It should be noted that the developed technology is based on the effects of structural formation, the implementation mechanisms of which have not yet been sufficiently studied. In this regard, the new information resulting from our studies appears very relevant.

2. Work objective

The aim of the work is to scientifically justify the effect of the increase of water resistance of asphalt concrete described earlier [6, 7] using phosphogypsum as a mineral powder when modifying bitumen with polyethylene terephthalate. The use of polymer-bitumen binders is among the actively implemented world technologies for construction and repair of road surfaces. In Russia, copolymers of butadiene and styrene as well as solutions of synthetic rubbers are used to modify bitumen [8]. The purpose of introducing polymers into bitumen is to increase their deformability at low ambient temperatures (about 0°C), heat resistance, and elasticity.

Improved performance properties of bitumen can be achieved when the polymer dissolves or swells in the dispersion medium of bitumen, forming a structural mesh. In contrast to thermoelastic polymers, the molecular structure of polyethylene terephthalate appears to be linear, and therefore the mechanism of polymer exposure to bitumen should be different from the one described previously. In addition, in the process of pelletizing mineral powder particles form an ordered structure with interlayers of a binder of minimum thickness between them (meets the requirements of P.A. Rebinder).

The nature of the interaction between bitumen and polymer in this case is unknown. This calls for the study of a mechanism to increase the water resistance of asphalt concrete with added phosphogypsum. The importance of such works is related to the introduction of technology of manufacture of new asphalt concrete mixtures and achieving high ecological and economic effect. Specificity of the developed technology is presented in [9].

3. Experiment

To achieve this goal, modern methods are used to study the structure of obtained materials, including electronic microscopy and differential scanning calorimetry to determine phase transitions in the source systems and their modified states. At the final stage, we evaluate the stability at storage of operational (consumer) properties of the received composite materials according to GOST 12801-98. Conducting theoretical research in this case is problematic. The process of asphalt concrete mixture formation involves taking into account the mutual influence of many factors on each other, the degree of influence of which on the final result is not quite obvious.

4. Results discussion

The water resistance of asphalt concrete is the most important factor determining the durability of the road surface. This factor has the greatest impact during the off-season, when freezing alternates with thaws, and moisture penetrates the coating and expands when there is the next freezing, thus tearing it. However, studies of water resistance of asphalt concrete based on phosphogypsum were conducted at room temperature, which excludes such an effect. The strength of samples based on traditionally used mineral powder and similar with the added phosphogypsum are comparable, but the water resistance of the latter is significantly lower. Ensuring the required strength of the material when using phosphogypsum is evidence of its sufficient adhesion to bitumen. The decrease in water resistance is probably due to specific chemical problems.

The increase in water resistance of the material when introducing PET into bitumen can be explained in two ways. The first explanation assumes the formation of a true solution of bitumen with polyethylene terephthalate, which is thermodynamically stable at all temperatures and has high water resistance. The second theory assumes that the mixture of bitumen and plastic during cooling is divided into initial phases. It is possible that a new material structure is formed which corresponds to an increase in internal pressure in the system which prevents water from diffusing into the composite structure.

A simple method to study surface effects is to measure the contact angle θ_0 [3]. Its value allows to determine the ratio of cohesion forces in the liquid phase and adhesion forces in the contact area of liquid and solid phases. Drops of bitumen were applied on the glass slides with PET additives (5,10, 15, 20 wt. %), kept at 100 °C for 20 minutes, then we surveyed the surface and measured the contact angle (Fig. 1).

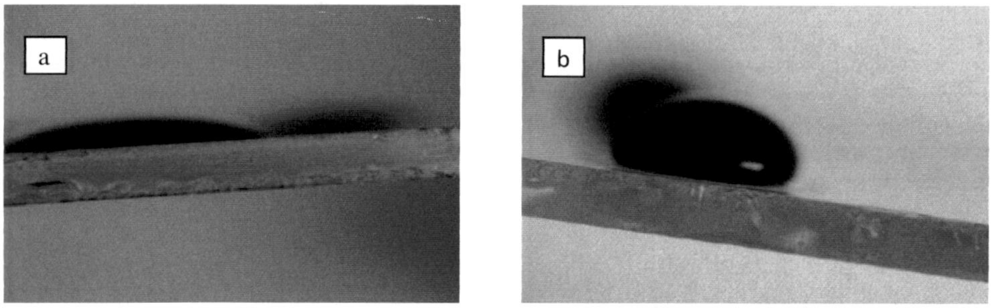

Figure 1. Photographs of bitumen drops with 5 (*a*) and 20 (*b*) wt. % of PET content.

As seen in Table 1, increasing PET content in bitumen leads to a sharp increase in the equilibrium contact angle of the surface.

Table 1. Change of the contact of the surface as PET content in the droplet increases.

Polymer concentration, %	0	5	10	15	20
Equilibrium contact angle, deg.	6	28	30	30	120

There are several possible explanations to that:

a) increased plastics content reduces the adhesion of bitumen to the hard surface;

b) increase in plastic content enhances cohesion of liquid;

c) both of the above factors apply simultaneously.

The single study we conducted is not enough to give an unambiguous answer to the nature of the change in water resistance of the material when introducing PET in bitumen. Note that the melting point of PET is 260 °C, and softening of bitumen happens earlier. When PET is introduced into bitumen heated to its melting point, a visually homogeneous mass is formed during mixing. This leads to the initial theory of a true solution of plastic in bitumen. The question is: what state will PET be in when the mixture is cooled to a temperature below its melting point?

It is reasonable to assume that when cooled, PET can remain in bitumen, forming a homogeneous mixture, as well as to form dispersion in the bituminous matrix. Since it is impossible to identify plastic particles in the bitumen mass, we conducted an additional experiment in the cyclohexanone medium. It is a good solvent for this polymer. When heated, we obtained a homogeneous solution, and when cooled, we observed its dispersion. A picture of dispersion of PET in cyclohexanone is shown in Fig. 2. The maximum particle size of the dispersed PET is 5-8 microns.

Coagulation of small particles with the formation of enlarged aggregates is a factor that negatively affects the homogeneity of the product.

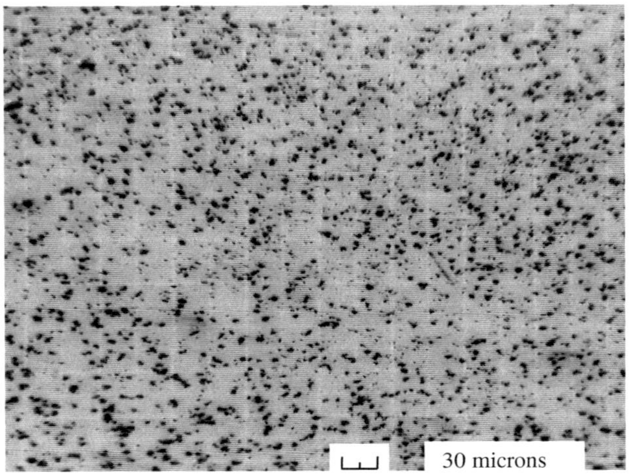

Figure 2. General view of the dispersion of PET in cyclohexanone (30 microns).

At first glance, using bitumen as a solvent makes coagulation more difficult, i.e. conditions are created to distribute small particles of PET uniformly in the binder volume. However, the theory was not confirmed after observing the behavior of the modified bitumen during its cooling. This indicates that intermolecular interaction between PET particles is more intensive than their interaction with bitumen, despite the direct phase contact in the latter case. It becomes clear why the equilibrium contact angle increases when the content of PET in bitumen increases as well.

Consequently, the provision about the cohesive force strengthening in the binder is confirmed when PET is added. However, the problem of specifying the form of plastic presence in the binder (dispersion or solution) remains open. To solve it, we studied samples of bitumen, PET, and modified bitumen (PET 10% added) using DSC 200 Phox by NETZSCH for differential scanning calorimetry. The method consisted in recording thermal effects in the studied sample of PET when heated at a constant rate compared to a standard sample (in this case — an empty crucible).

As seen in the thermograms (Fig. 3), the sample of PET (curve *2*) has three transitions. The first transition is observed in the temperature range corresponding to the beginning of glass transition (76.5 °C). This process consumes energy due to the increase in mobility of polymer molecules during its transition from amorphous to glassy state. The second transition (155 °C) corresponds to the beginning of polymer crystallization. In this case, energy is released due to the arrangement of molecules in the lattice, limiting their mobility. The third transition (with absorption of thermal energy) starts at 200 °C and ends at about 260 °C. It corresponds to the melting interval of this polymer, accompanied by the removal of restrictions on the mobility of molecules. An important thing to note is that peak areas during crystallization of the material are almost completely indistinguishable from the peak areas during its melting, which indicates the amorphous state of the modifier (PET) and disorder of macromolecules.

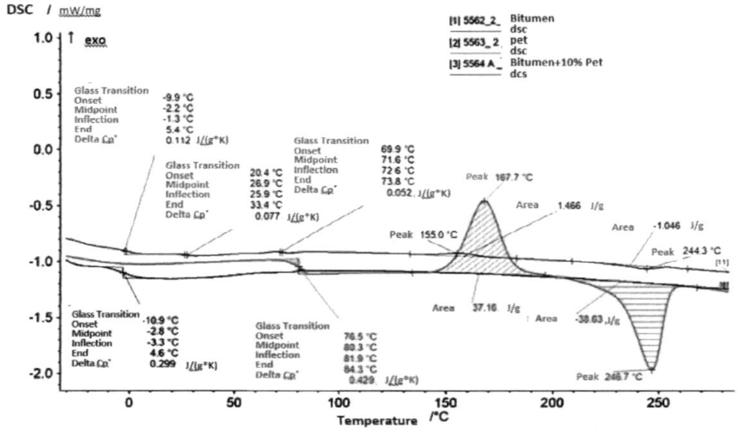

Figure 3. Thermograms taken for samples of bitumen (*1*), PET modifier (*2*), and bitumen-PET modifier system (*3*).

The curve *1* shows the transition that determines the temperature of glassing of bitumen in the range of temperatures (from -10.9 to 4.6 °C). The thermogram of the bitumen sample with the PET additive (curve *3*) is quite diverse. It has transitions corresponding to the temperatures of bitumen glass transition and PET, as well as the crystallization and melting temperatures of the modifier. The latter are not expressed explicitly, which is explained by the low content of the additive in the composite material (1 wt. %). One specific aspect of this thermogram is a new transition in the temperature range (20.4 - 33.4 °C) corresponding to the glass transition.

Analysis of Fig. 3 allows us to draw some important conclusions. First, the presence of characteristic transitions, which are inherent in the original components, on the curve *3* indicates the absence of their qualitative mixing. There is no true solution formation; the modifier is distributed in bitumen in the dispersion form [10]. Secondly, the decrease of energy to support a given mode of temperature change indicates the emergence of additional bonds that limit the internal mobility of the system. Third, the glass transition temperature of the mixture (20.4 - 33.4 °C) determines the moment of coagulation of dispersed particles of PET with the formation of sediment in bitumen [11].

This process can be presented as follows. When PET is dispersed, the work is done in bitumen to create new interfacial areas and increase the internal energy of the system. When the mixture cools down, coagulation of PET particles generates energy, determining the transition to glass transition [12]. The disperse phase of the modifier, which is distributed in the bitumen volume, acts as power centers, in a way. These centers create additional internal pressure in the system to prevent fluid from entering the binder. In the theory of surface phenomena, this is called the structuring effect, which most nanoeffects are based on [13].

The research made it possible to clarify the mechanism of PET influence on bitumen, which provides an increase of water resistance of asphalt concrete. Microphotographs of asphalt binder samples with the phosphogypsum powder additive obtained from PET-free bitumen (Fig. 4, *a*) and the modifier additive (Fig. 4, *b*) complete the picture and serve in favor of the theory of dispersion of the modifying additive of polyethylene terephthalate.

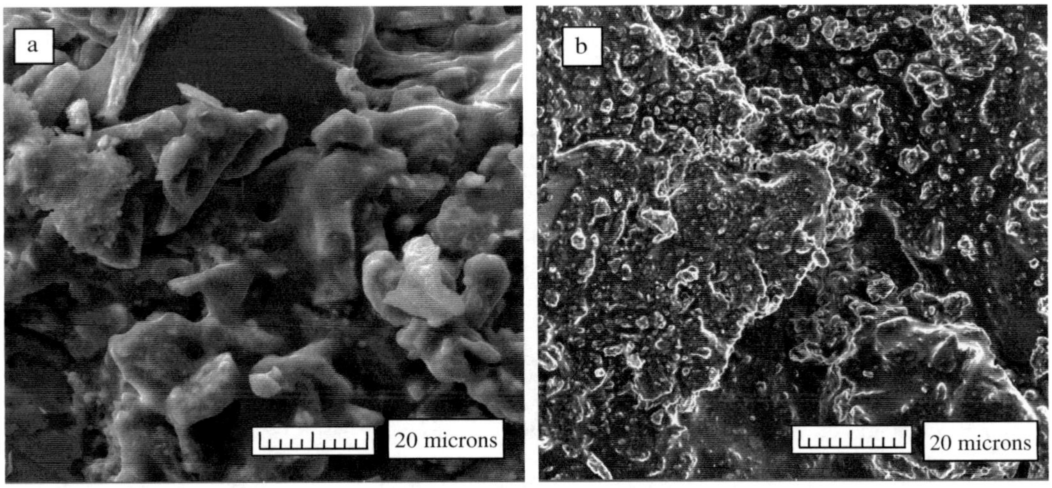

Figure 4. Microphotographs of a bitumen-phosphogypsum system (*a*) and bitumen with a PET-phosphogypsum additive (*b*).

Phosphogypsum crystals shown in Fig. 4, *a* are evenly wetted with bitumen, and their surface is smooth. The introduction of PET into the binder leads to a large number of small inclusions of the modifying additive on the crystallized surface (Fig. 4, *b*). Thus, the specified modifier is distributed in the asphalt binder as dispersed formations, which interact with each other at a distance of an additionally formed field of intermolecular forces. The presence of such a field prevents diffusion of moisture into the depth of the created composite material and provides an answer to a number of questions raised earlier in [14].

5. Conclusion

Recycling of industrial and domestic waste in modern conditions is an urgent problem. The results of numerous studies in this area tend to allow recycling in small quantities, which is not a global solution. The data we have obtained is of great interest in terms of updating and developing road construction technologies. In particular, it is the first time polyethylene terephthalate is offered to be used as a modifier of road bitumen, and we studied the problems of physicochemical mechanics of the pelletized systems with the specified additive with phosphogypsum being in system at the same time. It was found out that the obtained composite materials (asphalt concrete mixtures) can be stored in a cooled state for a long time without losing their operational (consumer) properties. This makes it possible to manufacture them for future use without limiting the production volume.

References

[1] Prokofyev V Yu, Razgovorov P B, Iliyn A A Osnovy fiziko-khimicheskoi mekhaniki ekstrudirovannykh katalizatorov i sorbentov. [Basics of physical and chemical mechanics of extruded catalysts and sorbents.] Moscow: KRASAND. 2012. 314 p. (in Russian)

[2] Tadmor Z, Gogos C G *Teoreticheskie osnovy pererabotki polimerov.* [Principles of Polymer Processing.] Moscow: Khimiya. 1984. 632 p. (in Russian)

[3] Ignat'yev A A, Gotovtsev V M, Gerasimov D V, Razgovorov P B The Modeling of Interfacial Contacts in Composites Using the Sitting Drop - Solid Body System as an Example. *New Polymer Composite Materials II Selected peer-reviewed full text papers from XVI International Scientific and Practical Conference "New Polymer Composite Materials".* Key Engineering Materials Submitted: 2020-04-12 ISSN: 1662-9795, Vol. **869**, pp 400-407 Accepted: 2020-05-07 doi:10.4028/www.scientific.net/KEM.869.400 Online: 2020-10-27 © 2020 Trans Tech Publications Ltd, Switzerland.

[4] Rebinder P A *Izbrannye trudy. Poverkhnostnye yavleniya v dispersnykh sistemakh.* Fiziko-khimicheskaya mekhanika. [Selected Works. Surface Phenomena in Disperse Systems. Physical and Chemical Mechanics.] Moscow: Nauka. 1979. 469 p. (in Russian)

[5] Rumyantsev A N, Nanenkov A A, Lomov A A, Gotovtsev V M, Sukhov V D Strukturirovannyy asfal'tobeton — novoe dorozhnoe pokrytie [Structured Asphalt Concrete — the New Road Surface]. //

Aktual'nye napravleniya nauchnykh issledovaniy XXI veka: teoriya i praktika. [Recent research trends of the XXI century: Theory and Practice]. Voronezh: Voronezh State University of Forestry and Technologies. 2013. Pp. 23-25. (in Russian)

[6] Patent RU2182136C2. Method of asphalt concrete mixture producing. 10.05.2002

[7] Patent RU2701007C1. Method of producing granulated asphalt-binding based on phosphogypsum 24.09.2019

[8] Kalinin V V, Masyuk A F, Khudyakova T S Osobennosti struktury i svoistv bitumov, modifitsirovannykh polimerami [Structure and properties of bitumen modified with polymers] // *Dorozhnaya tekhnika i tekhnologii* [Road building machinery]. 2003. No. **4**. Pp.. 174-181. (in Russian)

[9] Gerasimov D V, Ignat'yev A A, Gotovtsev V M, Golikov I V Perspektivy ispol'zovaniya fosfogipsa v proizvodstve asfal'tobetona [Prospects for using phosphogypsum in asphalt concrete production] // *Dorogi i mosty* [Roads and Bridges]. 2018. No. **40**. Pp. 304-315. (in Russian)

[10] Tager A A Fiziko-khimiya polimerov [Physics and chemistry of polymers]. Moscow: *Nauchniy mir*. 2007. 576 p. (in Russian)

[11] Kochnev A M Fizikokhimiya polimerov [Physics and chemistry of polymers]. Kazan: "FEN" Publ. House. 2003. 512 p. (in Russian)

[12] Gotovtsev V M, Ignat'yev A A The effect of structuring composite building materials. IOP *Conference Series: Mat. Sci and Eng*., V. **666** (1) (2019), doi:10.1088/1757-899X/666/1/012079 Retrieved from www.scopus.com. 2019.

[13] Ignat'yev A A, Gerasimov D V, Golikov I V, Gotovtsev V M and Dispersed-filled composites with a structured nanoscale. IOP *Conference Series: Materials Science and Engineering*, V. **365** (2018). Article ID 032064 - doi:10.1088/1757-899X/365/3/032064; http://iopscience.iop.org/article/10.1088/1757-899X/365/3/032064.

[14] Golikov I V, Ignat'yev A A, Gotovtsev V M, Kalashyan A R Osobennosti modifikatsii dorozhnogo bituma polietilentereftalatom [Specific aspects of road bitumen modification with polyethylene terephthalate] // *Dorogi i mosty* [Road and Bridges]. 2020. No. **42**. Pp. 192-202.

Lightweight concrete based on foam glass gravel: composition, structure and properties

G V Sopegin

Perm National Research Polytechnic University, 29, Komsomolsky Ave., Perm, 614990, Russia

E-mail: sp.georg@yahoo.com

Abstract. The article discusses the research results, the purpose of which is to provide the possibility of increasing the physical and mechanical as well as heat-shielding properties of lightweight cement concrete based on foam-glass gravel. The authors also investigated the possibility of reducing the alkali-silica reaction occurring in such concretes. To evaluate the course of the alkali-silica reaction, a method is used based on the study of the microstructure of lightweight concrete samples. The article also assesses the physical and mechanical as well as thermal engineering properties of lightweight concrete based on foam-glass gravel. The determination of properties is carried out according to standard methods. Analysis of the microstructure lightweight concrete samples allowed to fix flowing alkali-silica reaction. To reduce the alkali-silicate interaction, the authors propose introducing into the concrete composition a complex additive consisting of a siliceous modifier and a plasticizing component. The developed compositions of lightweight concrete, belonging to the class of structural and thermal insulation, have a compressive strength of 6.84-9.05 MPa, a density mark grade of D500-D1500 and a thermal conductivity of 0.218-0.282 W/(m·°C).
Keywords: foam glass gravel, lightweight concrete, high-active metakaolin, porous aggregate, compressive strength, thermal conductivity, alkali-silica reaction.

1. Introduction

One of the main criteria in assessing the energy efficiency of building materials is their ability to have required physical and mechanical as well as heat-engineering properties. Such materials must have high thermal resistance, sufficient mechanical strength, resistance to aggressive environments, and be non-toxic and non-combustible. The market and research data show that lightweight concrete with porous aggregates has the most effective properties for the construction of buildings and structures [1–5]. A promising direction in improving the reliability and durability of lightweight concrete is the use of foam-glass gravel in their composition, which allows for improving their heat-insulating and strength properties [3, 6].

The research results cited in [7] indicate that foam glass produced in the form of granules and gravels has a bulk density of 150-350 kg/m^3, thermal conductivity of 0.045-0.1 W/(m·K) and a strength of at least 0.5 MPa. The use of foam glass as an aggregate makes it possible to obtain heat-insulating and structural-heat-insulating lightweight concretes with an average density of up to 500 kg/m^3 and 500-1500 kg/m^3, respectively.

However, when using foamed glass gravels in cement binder composites, one should take into account the fact that the aggregates contain a large amount of amorphous silica. As a result of the interaction of amorphous silica with cement alkalis during long-term operation an alkali silica hydrogel is formed in the structure of FGG-concrete. The consequence of this interaction is the expansion of concrete as well as a decrease in its strength and durability. The described mechanism relates to the occurrence of alkali-silica reaction (ASR), which is a known problem that has been dealt with in many studies [8-16].

One of the effective ways to reduce the likelihood of ASR is adding high-volatile reactive additives to the concrete composition, which includes metakaolin. Metakaolin reacts with calcium hydroxide formed during hardening of cement. As a result, the number of new growths in the concrete body increases helping to reduce the formation of an alkali silica hydrogel. It is also known that metakaolin improves the structure of concrete, increases its physical and mechanical properties as well as thermal engineering properties [17, 18].

Content from this work may be used under the terms of the Creative Commons Attribution 3.0 licence. Any further distribution of this work must maintain attribution to the author(s) and the title of the work, journal citation and DOI.
Published under licence by IOP Publishing Ltd

The aims of this work are to study the properties and possibilities of using foam-glass gravel in lightweight concrete, determination of alkali silica interaction in lightweight concrete based on foam-glass gravel in the study of the microstructure of samples as well as study of the possibility of improving the physical and mechanical as well as heat-engineering properties of FGG-concrete by introducing a complex siliceous additive into its composition.

2. Materials and Methods

The study used the following materials: foam-glass gravel fractions 5-10 and 10-20 mm, having a bulk density of 200-250 kg/m^3 and compressive strength in the cylinder of 0.5-1.0 MPa; silica sand with a bulk density of 1485 kg/m^3 and particle size modulus M_k = 1.96; portland cement with a compressive strength of 28 days at least 32.5 MPa; silica modifier – high-active metakaolin (HAM) with a specific surface area of 12630 cm^2/g; water-reducing component - S-3 plasticizer (S-3).

To select the composition of the concrete mixture, the apparatus of mathematical planning of the experiment was used. Since the foam-glass gravel has a porous structure, the preparation of the concrete mixture was carried out in the following sequence: initially, the foam-glass gravel was mixed with 2/3 of the volume of water, kept for 5 minutes, and then mixed with a previously prepared dry mixture of sand and cement. At the last stage, the remainder of the calculated amount of water was introduced into the resulting mixture.

Determination of compressive strength and average density of FGG-concrete samples was carried out according to standard methods at the age of 28 days.

The microstructure of FFG-concrete samples was studied using a HITACHI S-3400 N high-resolution scanning electron microscope.

3. Results and Discussion.

From the analysis of published data [3], it follows that when glassy porous aggregates are used in lightweight concrete, cement consumption for concrete grades in density mark D500–D1500 is 280–300 kg/m^3. In this case, the optimal water-cement (W/C) ratio is in the range of 0.4-0.6.

Another important characteristic when selecting the composition of FGG-concrete is the ratio of small (5–10 mm) and large (10–20 mm) aggregate fractions. It was shown in [3] that for various ratios of coarse and fine fractions of the foam-glass gravel and depending on the value of the W/C ratio, the content of cement paste in the concrete mix can be from 14 to 25%. Moreover, a decrease in the content of the coarse aggregate fraction entails an increase in the volumetric content of the cement paste. Thus, to ensure a more favorable consumption of cement and aggregate in the study, a mixture of two fractions (5-10 and 10-20 mm) of foam glass gravel in a ratio of 60/40 respectively, was used.

The content of quartz sand and foamed glass gravels (S/FGG) was taken in the following ratios: 80/20, 60/40, 40/60

After testing, regression equations were obtained for the average density response function (y_ρ) and compressive strength (y_R) respectively:

$$y_\rho=1222.94 - 238.88x_1 - 13.67x_2 - 11.75x_1^2 - 115.25x_2^2 - 193.25x_1x_2 \quad (1)$$

$$y_R=7.06 - 2.57x_1 - 0.71x_2 - 0.17x_1^2 - 0.33x_2^2 + 0.75x_1x_2 \quad (2)$$

The response surfaces describing the changes of FGG-concrete properties depending on the W/C and S/FGG are shown in Figure 1.

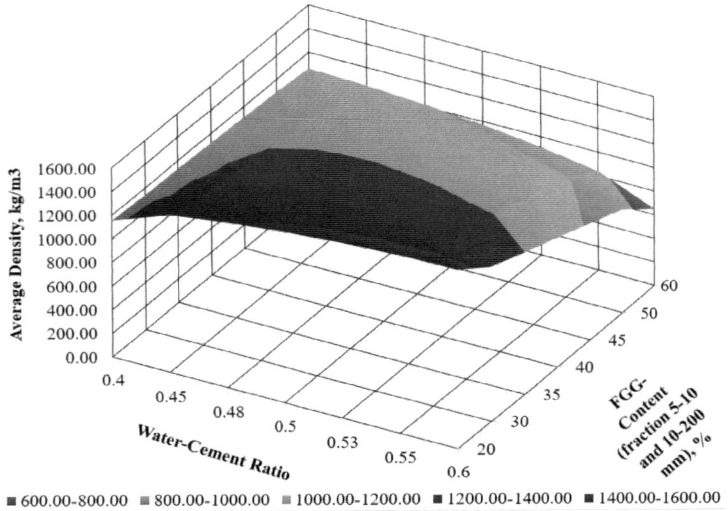

(a)

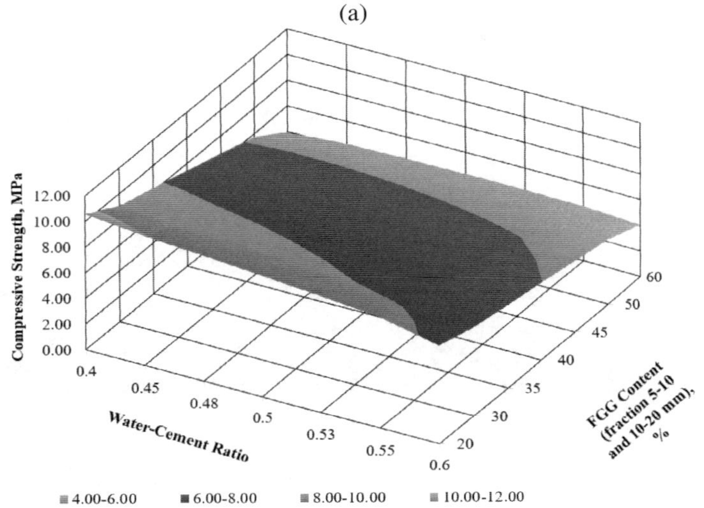

(b)

Figure. 1. The response functions of the average density (a) and compressive strength (b) of FGG-concrete from the water-cement ratio and the content of foamed glass gravel in the ratio S / FGG.

When analyzing the data in Figure1, it was found that the greatest compressive strength is achieved with a W/C of 0.4-0.5 and a content of FGG from 20 to 40%, while the average density of FGG concrete does not exceed 1500 kg/m^3. The compressive strength varies inversely with the W/C value. It can be explained by the fact that the increased W/C results in the increased porosity of the cement matrix, leading to a decrease in strength of the composite.

The results of the physical and mechanical as well as heat-engineering properties of FGG-concrete are shown in Table 1

Table 1. Physical and mechanical as well as heat-engineering properties of FGG-concrete.

W/C	S/FGG	Compressive Strength, MPa	Thermal Conductivity, W/(m·°K)	Average Density, kg/m^3
0.4	80/20	10.93	0.272	1230
0.4	60/40	6.53	0.254	992
0.4	40/60	4.52	0.263	1120
0.5	80/20	9.66	0.375	1480
0.5	60/40	7.08	0.267	1190
0.5	40/60	4.04	0.258	1040
0.6	80/20	7.14	0.324	1411
0.6	60/40	6.84	0.282	1321
0.6	40/60	3.74	0.207	528

The Table 1 and Figure 1 show that – at the same values of the water-cement ratio – the compressive strength of FGG-concrete decreases with an increase in the content of FGG in its composition.

The thermal conductivity of FGG-concrete according to Table 1 varies from 0.207 to 0.375 W/m·K. The lowest thermal conductivity is achieved at W/C = 0.6 and a FGG content of 60%. The amount of porous aggregate in the structure of the material significantly affects the value of thermal conductivity: the higher the content of FGG in the composition of concrete, the lower the value of thermal conductivity is, which is fully confirmed by the test results.

Thus, the composition of FGG-concrete with a foam-glass gravel content of 20-40% and a water-cement ratio of 0.4 to 0.5 is optimal and most effective. This composition provides the best combination of physical and mechanical as well as heat-engineering properties. Microstructures of FGG-concrete of this composition are shown in Figure 2.

(a) (b) (c)

Figure 2. Microstructure of FGG-concrete: (a) – same, with FGG fractions of 10–20 mm, magnification ×500; (b) – the boundary of the cement-sand mortar with FGG fractions of 5-10 mm, magnification ×500; (c) – same, with FGG fractions of 5-10 mm, magnification x3000.

Figure 2 shows that the structure of FGG-concrete is loose, there are amorphous inclusions of calcium hydrosilicates in it (Figure 2a, pos. 1, 2), which indicates the completeness of cement hydration reaction. Flat or prismatic crystalline formations of calcium hydroxide (Figure 2b, pos. 4, 5) are observed only in cement-sand mortar. The absence of such formations at the "aggregate-cement stone" phase boundary is explained by the interaction of calcium hydroxide with amorphous aggregate silica. The result of this reaction is needle-shaped neoplasms (Figure 2a, b, c, pos. 3, 6-9), which indicates the occurrence of alkaline-silica interaction in FGG-concrete [8, 19].

In order to exclude the occurrence of ASR, the composition of the FGG-concrete was modified with a complex silica admixture consisting of highly active metakaolin (HAM) and water-reducing plasticizer S-3 (S-3). The content of the HAM was 5, 10 and 15%, replacing the corresponding amount of cement. The plasticizer dosage for all samples with a complex additive was 0.5% by weight of cement. The microstructure of the modified FGG-concrete with an indication of the interface between the aggregate and cement stone is shown in Figure 3.

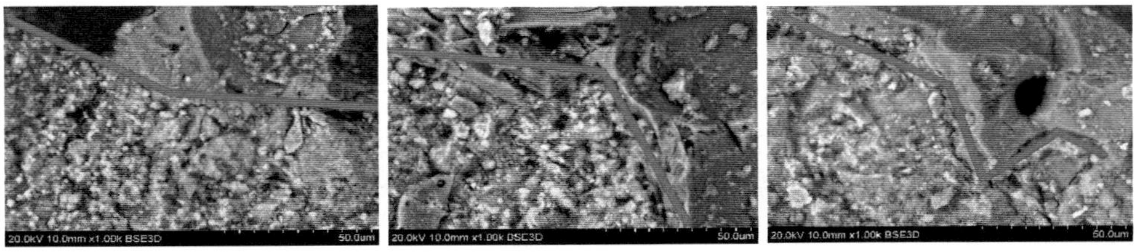

(a) (b) (c)

Figure 3. The microstructure of modified FGG-concrete at x1000 magnification, containing metakaolin in an amount of 5% (a), 10% (b), 15% (c).

Figure 3 shows that after introducing a complex additive into the composition of FGG-concrete, a significant decrease in acicular neoplasms at the "aggregate-cement stone" phase boundary is observed due to the binding of calcium hydrosilicate plates to metakaolin, which results in a decrease of ASR. At this stage, there is also an improvement in the uniformity of the structure of FGG-concrete in comparison with the unmodified composition. The effect of the complex additive on the compressive strength, average density and thermal conductivity of FGG-concrete is shown in Figure 4.

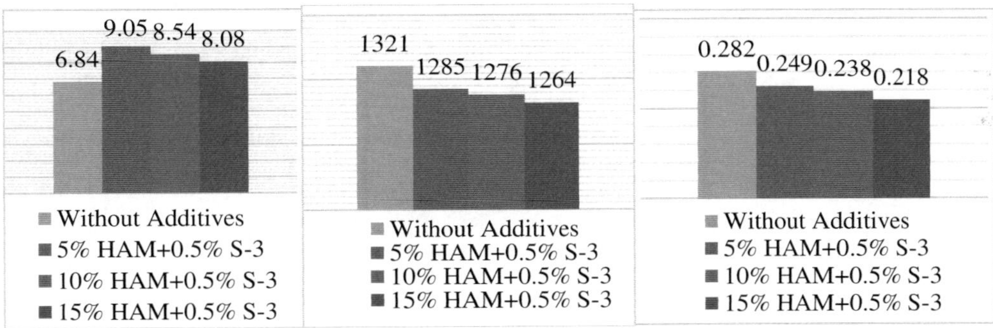

(a) (b) (c)

Figure 4. Change in the properties of FGG-concrete depending on the content of the complex additive: (a) – compressive strength, MPa; (b) – average density, kg/m^3; (c) – thermal conductivity, W/(m·°K).

The data analysis in Figure 4 showed that a complex admixture contributes to an increase in the compressive strength of FGG-concrete compared to the control composition. The best effect is achieved when the complex additive contains 5% metakaolin.

Figure 4 also shows that with a decrease in the average density of FGG-concrete, its thermal conductivity decreases from 0.282 to 0.218 W/(m·K), depending on the content of the complex additive.

4. Conclusions.

According to the study, the following conclusions:

1. It has been experimentally confirmed that on the basis of FGG it is possible to obtain lightweight concrete with a wide range of properties in terms of density, strength and thermal conductivity. The resulting compositions made it possible to obtain structural and heat-insulating lightweight concrete with a compressive strength of 3.74-10.95 MPa, an average density of 528-1480 kg/m3 and a thermal conductivity of 0.207-0.375 W/(m·K).
2. The analysis of the microstructure of FGG-concrete with an unmodified composition showed that needle-shaped neoplasms are observed at the interface between foam glass gravel and cement stone, which are the result of the interaction of calcium hydroxide with amorphous silica aggregate, resulting in an alkaline silica reaction.
3. The study of the microstructure of modified FGG-concrete revealed a significant decrease in acicular neoplasms at the aggregate-cement stone interface due to the binding of calcium hydroxide plates to metakaolin, which leads to a decrease in the course of the alkaline-silica reaction. At this stage, there is also

an improvement in the uniformity of the structure of FGG-concrete in comparison with the unmodified composition.

4. The joint use of metakaolin and S-3 plasticizer helps to improve the physical and mechanical as well as heat-engineering quality indicators of FGG-concrete, which can contribute to increasing its effectiveness when used in wall enclosures of buildings and structures for various purposes.

References

[1] Emon M A B, Manzur T and Yazdani N 2015 *Construction and Building Materials.* **106** 575-583
[2] Oreshkin D, Semenov V and Rosovskaya T 2016 *Procedia Engineering.* **153** 638-643
[3] Roberz F, Loonen R C G M, Hoes P and Hensen J L M 2017 *Energy and Buildings.* **138** 432-442
[4] Zhou H and Brooks A L 2019 *Construction and Building Materials.* **198** 512-526
[5] Gunawan P and Setiono. 2014 *Procedia Engineering.* **95** 433-441
[6] da Silva Fernandes F A, Arcaro S, Tochtrop Junior E F, Valdés Serra J C and Bergmann C P 2019 *Process Safety and Environmental Protection.* **128** 77-84
[7] Mugoni C, Montorsi M, Siligardi C, Andreola F, Lancellotti I, Bernardo E and Barbieri L 2015 *Ceramics International.* **41(3)** 3400-08
[8] Bumanis G, Bajare D, Locs J and Korjakins A 2013 *Construction and Building Materials.* **47** 274-281
[9] Hay R and Ostertag C P 2019 *Journal of Cleaner Production.* **212** 864-879
[10] Tora Bueno E, Paris M J, Clavier A K, Spreadbury C, Ferraro C and Townsend G 2020 *Journal of Cleaner Production.* **257** 120180
[11] Zhou J, Zheng K, Liu Z, Chen L and Lippiatt N 2019 *Cement and Concrete Research.* **124** 105817
[12] Nagrockienė D and Rutkauskas A 2019 *Construction and Building Materials.* **201** 599-609
[13] Shi Z and Lothenbach B 2019 *Cement and Concrete Research.* **126** 105898
[14] Kleib J, Aouad G, Louis G, Zakhour M, Boulos M, Rousselet A and Bulteel D 2018 *Construction and Building Materials.* **184** 295-303
[15] Kim T, Alnahhal M F, Nguyen Q D, Panchmatia P, Hajimohammadi A and Castel A 2019 *Cement and Concrete Composites.* **104** 103378
[16] Guo S, Dai Q and Si R 2019 *Cement and Concrete Research.* **115** 220-229
[17] Muduli R and Mukharjee B B 2019 *Journal of Cleaner Production.* **209** 398-414
[18] Shen P, Lu L, Chen W, Wang F and Hu S 2017 *Construction and Building Materials.* **152** 357-366
[19] Thomas M, Fournier B, Folliard K, Ideker J and Shehata M 2006 *Cement and Concrete Research.* **36(10)** 1842-56

Thread turn milling simulation

A A Malikov[1], O A Yamnikova[2] and A S Yamnikov[1]

[1]Tula State University, Lenin Pr., 92, Tula, 300012, Russia
[2] Ganichev SPLAV, AO, 33 Shcheglovskaya Zaseka, Tula, 300004, Russia

E-mail: yamnikovas@mail.ru

Abstract. The study shows that the high-performance thread turn-milling process with a synchronous rotation of the helical milling cutter and the workpiece is not widely adopted since there are no reliable methods to estimate the organic thread formation errors. The thread is machined as an envelope curve along the elementary surfaces of each cut shifted along the workpiece axis by a fraction of the thread pitch inversely proportional to the number of the cutter teeth. Domestic and international researchers tried to analytically estimate the thread formation errors at the part face. Such an approach failed to produce acceptable results since thread turn-milling is a three-dimensional process. We analyzed the thread formation errors in 3D using COMPAS 3D CAD system and developed a simulation model. We modelled a reference 3D thread machined on the workpiece; a 3D resulting thread; a single cutter tooth path in the plane perpendicular to the workpiece axis; 3D cut by a single cutter tooth. We also simulated the entire process and made a 3D model of the machining errors by subtracting the reference part geometry from the 3D thread model, and estimated the errors. The results were checked for consistency (max deviations) and acceptability for each dimension. The undercut dimensions were obtained by subtracting the 3D part model from the 3D reference model. Conversely, the overcut dimensions were obtained by subtracting the 3D reference model from the 3D part model.

1. Introduction

The thread turn-milling process was first presented in 1941 by S.I. Skukhtorov and S.I. Khlunov in their paper published in *Manufacturer* [1]. However, it is not widely adopted so far despite its theoretically high performance [2-5]. The cutting forces and tool life for thread turn-milling were investigated [6-11]. One of the reasons that inhibit the process adoption is the lack of an adequate simulation model of the thread formation by turn-milling. Paper [12] is an attempt to develop an analytical model of the thread turn-milling process. We propose to apply simulation for selecting the milling cutter as in [13-19].

2. Simulation

The input variables are the part and workpiece geometry, the output variables are the milling cutter properties (Figure 1). The independent variables, in this case, are the dimensional tolerances. They define the target function or the process criterion: minimizing the machined surface errors.

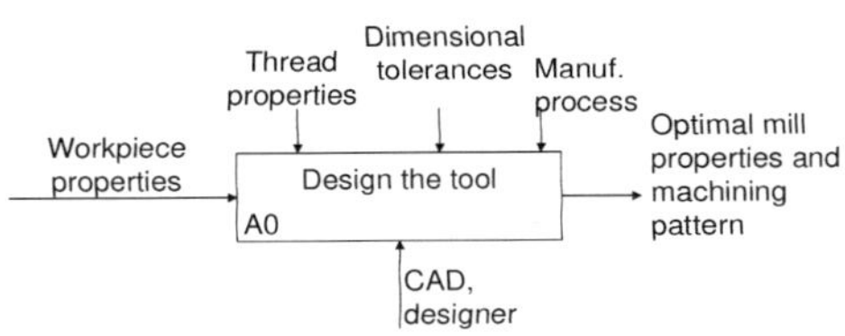

Figure 1. Structural and functional model of the machining strategy and milling cutter selection for thread turn-milling.

Content from this work may be used under the terms of the Creative Commons Attribution 3.0 licence. Any further distribution of this work must maintain attribution to the author(s) and the title of the work, journal citation and DOI.
Published under licence by IOP Publishing Ltd

For real-life applications, it is sufficient that the selected cutter properties meet the specified thread properties. The simulation tool was a 3D CAD such as COMPAS 3D.

The simulation steps were as follows:

1. 3D modeling of the reference thread on the workpiece.

2. 3D modeling of the part machined with turn-milling.

2.1. Drawing the cutting path for a cutter tooth in the plane perpendicular to the workpiece axis.

2.2. Creating a 3D model of the cut made with a cutter tooth.

2.3 Simulation of the entire machining process.

3. Creating a 3D model of the deviations by subtracting the reference model part from the 3D model of the resulting thread.

4. Generating projected views of the deviations.

5. Estimating the deviations quantitatively.

6. Verifying the results for consistency (max deviations) and meeting the max acceptable dimensional tolerances.

To make a reference thread model, the thread and workpiece geometry is required. To model the real tool marks on the workpiece, the machining strategy and the milling cutter properties are pre-selected. The selection should provide the least error possible. It is an iterative process since the optimal cutter parameters are re-adjusted until the deviations (undercuts and overcuts) stop decreasing.

One of the most significant stages of preparing input data for simulation is generating a cutter tooth path. For this, we should estimate the 3D point coordinates, and draw an enveloping contact curve. That is why special software tools are needed [14, 15].

We used a parametric CAD to rapidly rebuild the models as new cutter parameters are specified. One of such systems is COMPAS 3D.

For external threads, we simulated the outer contract as the milling cutter and the workpiece rotate in the same direction. It is the simplest machining strategy to implement (Figure 2).

We progressively changed the milling cutter diameter and the number of cutting edges.

Then we developed an equation for the cutter tooth tip path in a single cut.

Let us investigate the equation in detail. Figure 3 shows a diagram of the workpiece and milling cutter cross-sections. As the cutter tooth is at its max penetration the axis of the cutter and the workpiece are concentric at the line $O_\partial O_{u_1}$ while the tooth is at point B. As the workpiece rotates by the angle $\Delta\varphi$, the cutter tooth rotates around the milling cutter axis also by $\Delta\varphi$. As a result, the cutter axis would shift to O_{u_2}, and the tooth would move from B to B_1. Since the angles are equal, the line $B_1O_{u_2}$ is always parallel to the line $O_\partial O_{u_1}$ and to $\angle O_\partial AB_1 = 90°$. The coordinate y_{B_1} can be estimated from the triangle $O_\partial AB_1$:

$$y_{B_1} = dr_1 \cdot \sin\Delta\varphi .$$

Here the center-to-center distance is a constant value that does not depend on the rotation angle:

$$d_{r1} = O_\partial O_{u_2} = O_\partial O_{u_1} = R_u + R_3 - h = const ,$$

where R_u is the cutter radius, R_3 is the workpiece radius, h is the thread root-to-crest height.

The Ox coordinate is the difference between the second leg of the triangle and the cutter radius:

$$x_{B_1} = AO_{u_1} - B_1O_{u_1} = dr_1 \cdot \cos\Delta\varphi - R_u .$$

We used 3D modeling for building the reference geometry, the geometry of the thread machined by turn-milling, and the geometry of deviations.

A 3D model of the workpiece is used for this. It is modeled to the specified design. The model structure is shown in figure 3.

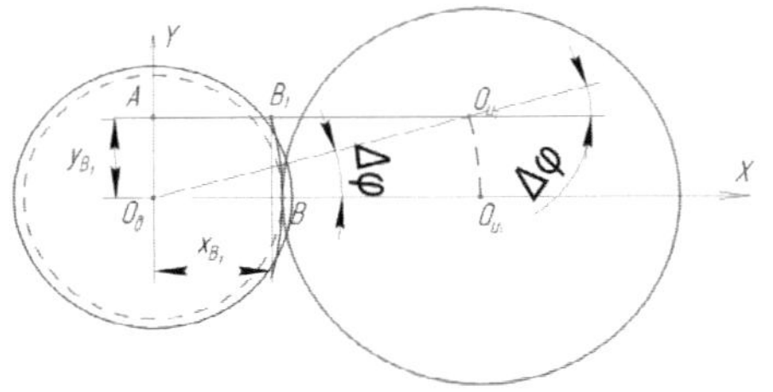

Figure 2. Cutter tooth coordinates estimation for external thread turn-milling with outer cutter-to-workpiece contact.

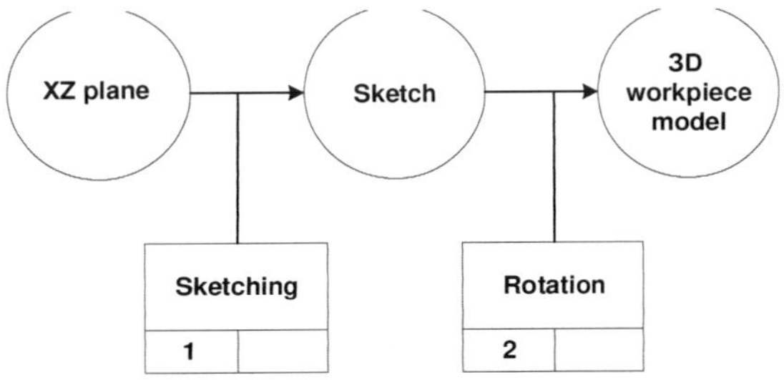

Figure 3. A 3D workpiece model building sequence.

The sequence is represented as a design tree in a CAD system such as COMPAS 3D.

The workpiece dimensions are constant, so the model was suitable for all the subsequent experiment campaigns.

Then a reference 3D model was built. This model does depend on the thread properties as specified. The reference model included the workpiece model and an extra operation: the tooth profile extrusion along a helix path (Figure 4).

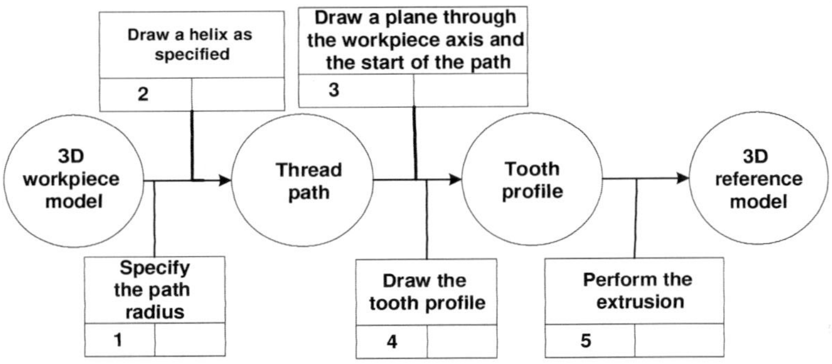

Figure 4. A 3D reference model building sequence.

The resulting model is shown in figure 5.

The core of the simulation is simulating the thread turn-milling process for a range of cutters. The simulation parameters are the milling cutter diameter and the number of teeth.

The tooth path depends on the machining strategy used. The number of teeth affects the thread profile formation. It is built with the Pattern Along Path command where the generatrix is the cutting profile (defined by the tooth path), and the directrix is a helix around the thread inner diameter.

Figures 6 and 7 show the sequence of building the 3D model of the resulting part machined by turn-milling.

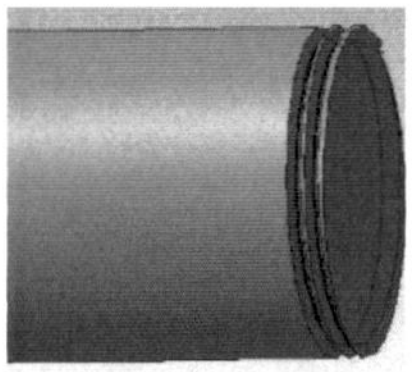

Figure 5. Example of a 3D reference model.

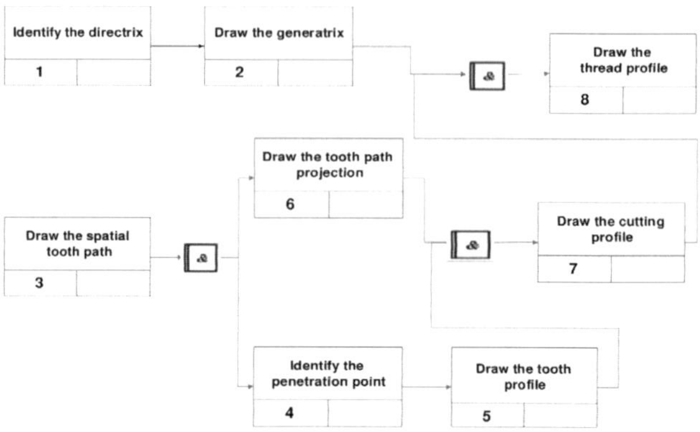

Figure 6. Building sequence for the 3D model of the machined thread.

Figure 6 shows the general sequence, and figure 7 shows how the model is progressively modified. The sequence is branched. One branch is building the basic directrix that accommodates the 3D models of the cuts. The one branch is building the cut profile (3D model).

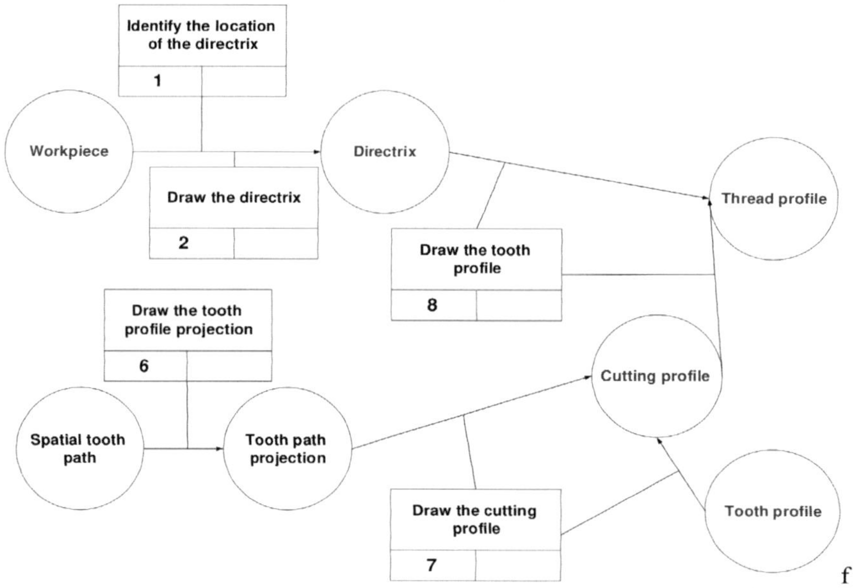

Figure 7. The simulation model state diagram and transformations.

First, the coordinates of a 3D curve representing the cutter marks on the workpiece are estimated. Then a spline is drawn that represents the spatial tooth path.

Then the tooth profile is drawn on a plane passing across the workpiece axis and the penetration point, i.e., the tooth path to the workpiece intersection point.

Then a profile of a single tooth cut is modeled by extrusion along a path, where the directrix is a 2D projection of the tooth path, and the generatrix is the tooth profile.

The last modeling stage is building the thread profile. The operation used is a pattern along a curve. The number of generatrix profiles along a revolution depends on the number of milling cutter teeth.

3. Simulation Results

As a result, we obtained a 3D model of the part with two oppositely directed threads machined by turn-milling (Figure 8). Figure 9 shows a simulated thread profile. The undercut dimensions were obtained by subtracting the 3D part model from the 3D reference model. Conversely, the overcut dimensions were obtained by subtracting the 3D reference model from the 3D part model (Figure 10).

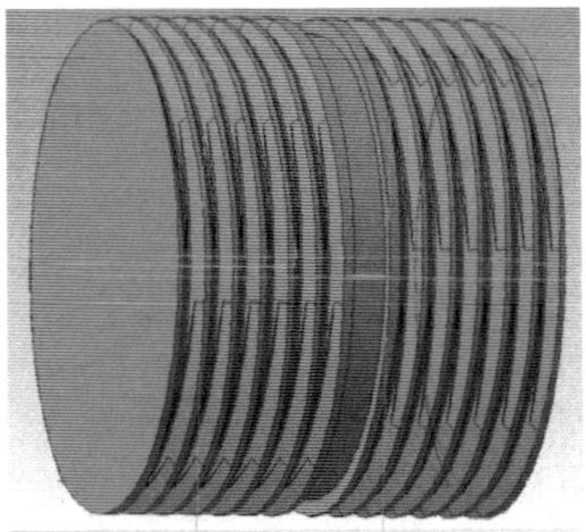

Figure 8. 3D model of a part machined by turn-milling.

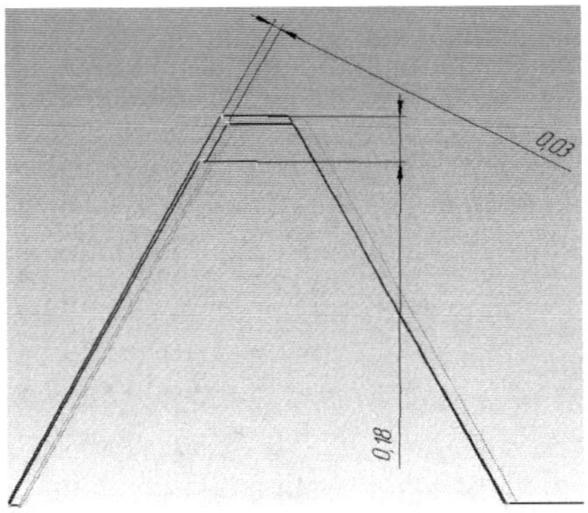

Figure 9. A typical simulated thread profile section.

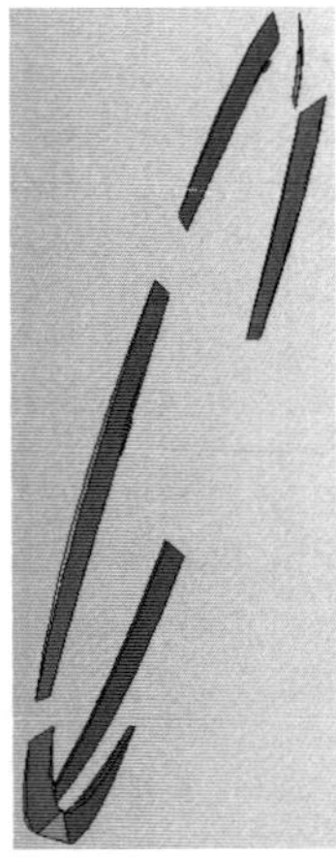

Figure 10. 3D model of the thread profile overcuts.

4. Results and Discussion

With the simulation, we estimated threading errors. By comparing the errors with the thread tolerances we can verify if the manufacturing process is correct.

References

[1] Skukhturov S I and Khlunov V N 1941 Thread Milling Generating Process *Mashinostroitel* N **2** pp 6-9
[2] Solyankin D Yu and Yamnikov A S 2010 Thread Milling with Radial Feed Helical Cutters *Izvestiya Tulskogo gosudarstvennogo universiteta. Tekhnicheskiye nauki* V **4** p **2** pp 79-5
[3] Voronov V N 1991 Threading with Helical Tools *Stanki i instrument* N **10** pp 14-6.
[4] Voronov V N and Serova E V 1993 Thread Geometry Generation with Radial Feed Helical Tool *Tekhnologiya izgotovleniya i sborki* (Tula State University) pp 87-5
[5] Solyankin D Yu and Yamnikov A S 2010 Relative Thread Turn Milling Productivity. *Fundamentalnye i prikladnye problemy tekhniki i tekhnologii* (Oryol State University) V **6 (284)** pp 109-4.
[6] Yamnikova O A 2011 Experimental Cutting Force Measurement in Thread Milling *Izvestiya YUgo - Zapadnogo gosudarstvennogo universiteta. Seriya «Tekhnicheskie nauki»* N **1(34)** pp 113-20
[7] Yamnikov A S Yamnikova O A 2015 Experimental Cutting Force Measurement in Thread Milling *Izvestiya Tulskogo gosudarstvennogo universiteta. Tekhnicheskiye nauki* V **1** pp 82-93
[8] Yamnikov S A Yamnikova O A and Troitsky D I 2015 Thread Turn Milling Simulation *International Journal of Innovative and Information Manufacturing Technologies* (SHEI Donetsk National Technical University & International Union of Machine-Builders) N **2** pp 73-7
[9] Yamnikov S A Yamnikova O A and Troitsky D I 2015 Physical Modeling of Cutter Lathe Thread Turn Milling Thread *Advanced Manufacturing Technologies and Equipment* N **1 (51)** pp 207-11
[10] Yamnikov S A and Yamnikova O A 2017 Thread Milling Physical Simulation *Izvestiya Tulskogo gosudarstvennogo universiteta. Tekhnicheskiye nauki* N **8** V **1** pp 48-58
[11] YamnikovA S Yamnikova O A Boriskin O I and Troitsky D I 2017 Physical Modeling of Cast Iron Radiator Nipple Oppositely Directed Thread Turn Milling *Iron and Steel Review* V **14** pp 40-44
[12] Lashnev S I and Serova E V 1993 Mill Broach Generatrix Parameters Evaluation for Round Thread Machining *Avtomatizirovannye stanochnye sistemy i robotizaciya proizvodstva* (Tula State University) pp 87-2

[13] Chuprikov A O Yamnikov A S and Troitsky D I 2019 Simulation of deformations in thin shells clamped in a gripper with wrap-around jaws *AIP Conference Proceedings* **2188**, **050028**
[14] Yamnikov S A Yamnikova O A and Solyankin D Yu 2011 Thread Milling Simulation *International Journal of Innovative and Information Manufacturing Technologies* (Donetsk National Technical University) N **5** pp 3-13
[15] Malkov O V and Stepanova M Yu 2015 The Analysis of Design Parameters of Thread Milling Cutters *Nauka i obrazovanie* (Internet edition of the Moscow State Technical University named after N. E. Bayman) N **07** pp 76-95
[16] Jun M B G and Araujo A C 2010 Modeling of the Thread Milling Operation in a Combined Thread/Drilling Operation: Thrilling *Journal of Manufacturing Science and Engineering* V **132** I **1** DOI: 10.1115/1.4000944
[17] Jun M B G and Araujo A C 2010 Modeling of the thread milling operation in a combined thread/drilling operation: Thrilling *Journal of Manufacturing Science and Engineering, Transactions of the ASME* V **132** N **1** Art N 014505
[18] Araujo A C Jun M B G Kapoor S G and DeVor R E 2007 Experimental Investigation of a Combined Drilling and Thread Milling Process: Thrilling *Transactions of NAMRI/SME* V **XXXV** pp 518-27
[19] Araujo A C Fromentin G and Poulachon G 2013 Analytical and Experimental Investigations on Thread Milling Forces in Titanium Alloy *International Journal of Machine Tools & Manufacture* V **67** pp 28-4

Influence of dynamic loads in joints on dimensional stability

A S Yamnikov[1], A A Malikov[1], O S Kashmin[1] and I D Troitsky[2]

[1]Tula State University, Lenin Pr., 92, Tula, 300012, Russia
[2] Ganichev SPLAV, AO, 33 Shcheglovskaya Zaseka, Tula, 300004, Russia
[3] Troitsky Translation Services Tula, 44 Mezentseva street, bldg. 2, Tula, 300026, Russia

E-mail: yamnikovas@mail.ru

Abstract. In small arms manufacturing the lock dimensional chain final link precision is secured with manual file fitting and removing the tolerance intentionally left on the recoil lugs. The manual process reduces the contact stiffness of the lock components, and the final link dimension (between the cylindrical gauge end that simulates a cartridge, and the breeching lock face) set during the assembly is distorted when the breech lock components are exposed to shot impact loads. For this reason, the cylindrical cartridge dummy used for the initial assembly is sized to allow for subsequent plastic deformation at the breech lock component interfaces. The deformation value is determined experimentally on an ad hoc basis. Polynomial expressions have been obtained experimentally for the interface plastic deformation vs. relative positioning errors and surface finish relation for parts made of typical military steel grades. To study the relation between the interface surface machining methods and contact stiffness flat specimens, 50RA steel quenched to HRC 40...45 (the roughness R_z is 10...15 μm) were machined as follows: peripheral pregrinding; flank milling; file crosswise (at 45°) movements. The results generated after processing the measurements made with a HOMMEL TESTER W55 instrument show that finish flank milling yields the best results.

1. Introduction. Problem Overview

Impacts between contact bodies commonly occur in modern machinery. For example, impacts present in rifle and automatic weapon locks. A part's contact stiffness depends on the surface layer properties [1...4]. The lock parts are exposed to high dynamic loads natural for small arms [5]. The loads cause contact deformations at the interface surfaces. For an initial service period as the lock components are exposed to a shot impact loads, plastic contact deformations develop. Finally, they result in the initial assembly accuracy degeneration. This fact shall be considered when selecting the manufacturing process for ensuring the closing dimension accuracy [6...10].

Figure 1 shows a typical assault rifle lock. The lock reliable operation a paramount for the rifle availability. The barrel (pos. 2) when assembled to the box (pos. 1) is located off the cylindrical surfaces for assembly. The barrel is fixed in the axial direction by fixing its end face with a predefined thread torque. The lock is an assembly of the bolt head (pos. 4) and the locking lug (pos. 5) with a pair of datum seating surfaces. The lock is located off these surfaces and absorbs most of the shot energy. Initial lock assembly shall ensure a $A_\Sigma = 0.05 \div 0{,}1$ mm clearance between the lock face (pos. 4) and the go-on gauge dummy (pos. 3) used for the lock assembly and adjustment. The dummy is installed in the barrel chamber (pos. 1.). Since the dimension A_Σ is inaccessible and cannot be inspected or measured directly, a set of go-on and o-go gauge dummies have to be used. When a go-on gauge is applied, the clearance shall be $A_\Sigma = 0.05 \div 0{,}1$ mm. To verify the $A_\Sigma < 0.1$ mm condition, the no-go gauge is applied. In this case, should the A_Σ dimension be incorrect, the lock would not close. As Fig. 1 shows, the clearance A_Σ between the gauge dummy end

Content from this work may be used under the terms of the Creative Commons Attribution 3.0 licence. Any further distribution of this work must maintain attribution to the author(s) and the title of the work, journal citation and DOI.
Published under licence by IOP Publishing Ltd

(pos. 3) and the lock (pos. 2) is a closing link of the dimensional chain consisting of six dimensions. Dimension A_6 itself depends on the combination of the barrel chamber and the gauge dummy dimensions: $A_\Sigma = A_3 - (A_1 + A_2 + A_4 + A_5 + A_6)$.

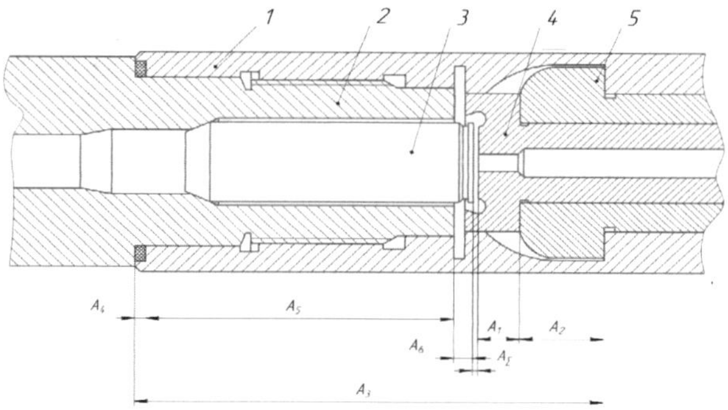

Figure 1. A typical assault rifle lock.

The required dimensional chain closing link accuracy is secured with manual file fitting and removing the tolerance intentionally left on the lock datum surfaces. The manual process reduces the contact stiffness of the lock components, and the final link dimension(between the cylindrical gauge end that simulates a cartridge, and the lock face) set during the assembly is distorted when the breech lock components are exposed to shot impact loads. For this reason, the cylindrical cartridge dummy used for the initial assembly is sized to allow for subsequent plastic deformation at the breech lock component interfaces. The deformation value is determined experimentally on an ad hoc basis. After acceptance tests, the gauge dummy is replaced with a longer one. The part interface surfaces are irregular (roughness, undulation, geometric shape Deviations). It significantly reduces the actual contact area compared to the rated one. For this reason, contact deformations prevail in the part motion balance [14]. Consequently, contact stiffness and surface geometry are closely related. Contact stiffness depends on the surface deviation height and shape. Roughness numerical properties can be analyzed in different ways. For example, in [15] a part surface layer load bearing capacity is estimated by analyzing the micro deviations: max micro deviation H_{max}, micro deviation smoothness height H_P; undulation W_a, W_P, W_Z, S_{mW}; roughness R_a, R_P, R_{max}, t_m, S_m, S, physical and mechanical surface properties described with microhardness parameters and residual strain $H\mu_0$, $\pm\sigma_0$, $h\mu_0$, $h\sigma_0$.

2. Problem Statement

Summarizing the above, we can assume that identifying the relation between the lock part interface surface plastic deformation value the material's physical and mechanical properties and the surface finish would help find an optimal fitting process. It will facilitate the designer to select appropriate surface layer parameters, while the manufacturing engineer will be able to select optimal machining modes to meet the design requirements.

3. Theory

As a result of machining, part contact surfaces have micro- and macro irregularities. Usually two interface surface initially contact at three points [1, 3, 4, 11]. The strain generated as a normal force is applied to such a part interface greatly exceeds the yield strength because the contact area is so small. Peaks that first come into contact are exposed to plastic deformations (Figure 2). The deformation value depends on the formation of a contact area capable of resisting the load applied to it [14]. It is expressed with an empirical relationship:

$$A_{ri} = \frac{P_i^{\upsilon}}{c[\sigma_T]}, \tag{1}$$

where A_{ri} is the actual *ith* contact area [mm^2], P_i is the load applied to this area [N],υ is the exponent Indicating the force vs. deformation relation (for initial loading υ =0.9), [σ_T] id the material's yield strength [N/mm^2], C is the surface layer yield strength enlargement factor.

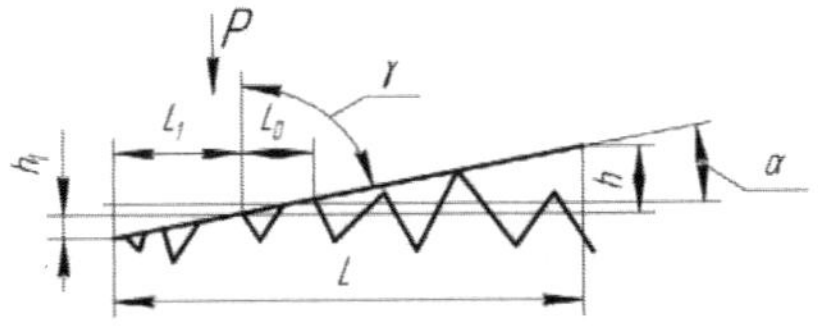

Figure 2. Surface profile after applying a normal force.

In terms of physics, the C factor is a relation between the average pressure p applied to the contact area and resulting in plastic yield and the σFS flow stress, i.e. C=p/σFS. Note it is in the range $C \approx 2.84... 4.78$ [14]. As a load is re-applied (not exceeding the initial load) without the surface shift, the peaks are elastically deformed.

Re-applying a higher load to immovable surfaces (exceeding the initial one) lead to the micro irregularity peak deformation transformation from elastic to plastic. As the load increases, structural changes occur. They change the surface layer's mechanical and physical properties. The yield strength is greatly enhanced. Besides, a higher load results in expanding the plastic deformation area, the actual surface contact area being the sum of the elementary contact areas

$$A = \sum_{i=1}^{n} A_{ri} \tag{2}$$

A further load increase leads to negligible plastic deformations [15]. Under a subsequent shift-free load the deformed peaks come into contact. Each peak has a certain contact area. If a contact area under initial loading varies directly with the pressure with a proportionality factor value near to 1 (υ =0.9) then under subsequent loadings the value belongs to the υ = 0.5...0.88 range [16].

Surface shift prior to each loading results in elastic deformations of the plastically deformed peaks and plastic deformations of the peaks that have not been contacted yet. Under numerous contact surface shifts and repeated loadings more and more peaks are elastically deformed, and less and less are plastically deformed. Multiple load applications cause contact destruction of the plastically deformed contact areas, and irregularity formation. A dedicated test bed was built to study the Y part approaches in an assembly model. The interface surfaces were filed as it is a common practice for rifle steel processing [1, 3, 11]. The factors were: Ra surface roughness [μm]; HRC is the Rockwell hardness; α° is the part interface expansion angle [degrees]. We hypothesize that the approach distance *Y* varies with the factors under consideration and can be expressed as the following regression equation:

$$Y = C Ra^{\beta} HRC^{\ \gamma} \alpha^{\phi}, \tag{3}$$

After logarithmation, it becomes

$$\ln Y = \ln C + \beta \ln Ra + \gamma \ln HRC + \phi \ln \alpha. \tag{4}$$

Finally, the experiment result is represented by a polynomial

$$U = b_0 + b_1x_1 + b_2x_2 + b_3x_3 + b_{12}x_1x_2 + b_{13}x_1x_3 + b_{23}x_2x_3 + b_{11}x_1^2 + b_{22}x_2^2 + b_{33}x_3^2 \tag{5}$$

In equation (5) $U = \ln Y$; x_1, x_2, x_3 are coded Ra, HRC, α° factor values. The equation (3) was verified by checking the polynomial (5) linear component validity.

For 30Cr grade steel the factor variation grades are:

$$x_1 = (Ra - 1.6)/1; x_2 = (HRC - 42.5)/5.5; x_3 = (\alpha^\circ - 0.13)/0.09.$$

The experimental model is:

$$U = 10.889 + 2.8333x_1 - 2.6667x_2 + 5.2778x_3 + 0.3333x_1x_3 + 0.94444x_3^2.$$

For 35HGSA grade steel the factor variation grades are:

$$x_1 = (Ra - 1.6)/1; x_2 = (HRC - 43)/5; x_3 = (\alpha^\circ - 0.13)/0.09.$$

The experimental model is:

$$U = 10.778 + 3x_1 + 2.0556x_2 + 5.5x_3 + 0.25x_1x_2 + 0.33333x_1x_3 + 0.25x_2x_3 + 0.94444x_3^2$$

For C50E grade steel the factor variation grades are:

$$x_1 = (Ra - 2.8)/2.2; x_2 = (HRC - 44.5)/4.5; x_3 = (\alpha^\circ - 0.085)/0.055.$$

The experimental model is:

$$U = 10.926 + 3.4444x_1 - 3.5556x_2 + 3.8889x_3 - 0.33333x_1x_2 +$$
$$+0.58333x_1x_3 - 1.9167x_2x_3 - 0.77778x_1^2 + 0.55556x_2^2 - 0.77778x_3^2$$

For 30HN2MFA grade steel the factor variation grades are:

$$x_1 = (Ra - 2.8)/2.2;\ x_2 = (HRC \quad - 43)/4;\ x_3 = (\alpha^\circ - 0.085)/0.055$$

The experimental model is:

$$U = 11 + 4x_1 - 2.7222x_2 + 2.72222x_3 - 0.66667x_1x_2 + 0.58333x_1x_3 - 0.75x_2x_3 - 1.4444x_1^2.$$

The roughness was mostly estimated with profile records within the static sample length l. The sample length is such that it is not affected by other irregularities (undulation, micro deviation) [17].

With rough surface profile records, it is possible to estimate a number of micro geometry properties important for the surface load bearing capability assessment. One of them is the bearing surface curve (Figure 3).

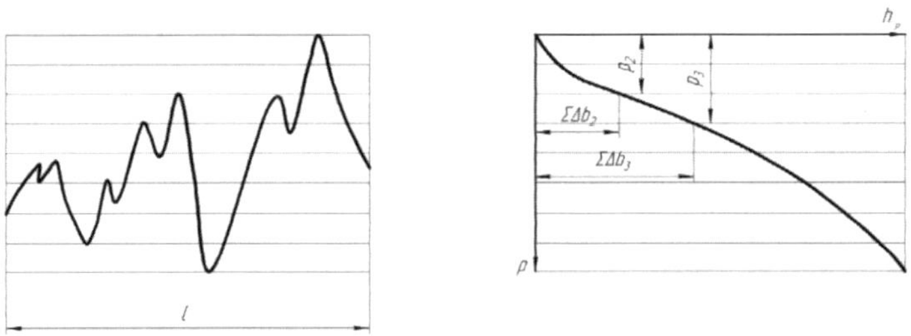

a b

Figure 3. A profile record (a) and the Abbott-Firestone curve (b).

Let us consider a profile section at a certain distance p from the profile peak line. Then the sum of the peak section lengths Δb_i at a certain level is called "reference profile length"

$$\eta_p = \Sigma_1^n \Delta b_i. \tag{6}$$

The reference profile length to the sample length is called "relative reference profile length t_p at the p level". Consequently, the following equation [13, 15] is true:

$$t_p = \frac{\eta_p}{l} = \frac{A_p}{A_c}, \tag{7}$$

where A_p is the rough layer section area(at the p level); A_c is the sample surface area.

It follows that the relative reference profile length is equal to the relative rough layer section area at a certain level. A curve expressing the A_p vs. p relation is called "bearing surface curve". It clearly shows the ratio of material in each rough material layer. Such curves were proposed by the US researchers Abbott and Firestone in 1933 [16]. In order for a profile record to represent both longwise and crosswise roughness fluctuations. The record was made at 45° to the machining marks.

The reference curve can be drawn in relative values. In this case the horizontal axis represents the ration of a cross section are at a given level A_p to a contour area A_c, while the vertical axis represents the ration of the approach distance a to the max peak height $Rmax$. The relative material section area at a certain level is η_s; the relative approach distance is ε. A contact area equal to the sum of contact patches between individual surface peak pairs is called "actual contact area" [16]. Actual contact patches are grouped at the deformed wave pattern tops creating individual zones that make up the contour contact area. The contour area concept is conventional to some extent. Still, the introduction of this concept enables solving the rough/wave-like surface contact problem. AS rough surfaces come into contact, in most cases only the highest peaks do contact. They make up the top of the bearing curve. For contact deformation analysis it is sufficient to consider a part of the bearing surface curve expressed by a simple equation:

$$\eta_s = b\varepsilon^{v}, \quad (8)$$

where η_s is the relative cross section area being the ratio of the profile section area sums at the y level to the sample length; y is the distance measured from a specified baseline; ε - is the relative approach of relative section depth expressed as $\varepsilon = y/R_{max}$; b and v are the bearing curve parameters that depend on the machining process; usually b=2÷4; v =1.5÷3.

To study the relation between machining process and part interface contact stiffness/stability in assembly [1, 3, 11] flat specimens made of 50RA steel quenched to HRC 40...45 were used. The surfaces were machined with processes common for assembly manufacturing (the specified roughness R_z = 10÷15 μm): peripheral pregrinding; flank milling; file crosswise (at 45°) movements.

The surface layer was tested with HOMMEL TESTER W55 from HOMMELWERKE. The profile records were processed, and specific roughness parameters t_m; Rp; Ra; $Rmax$ were obtained (refer to the table.) With these data, and (6) and (7) equations, b v bearing curve parameters were estimated. Their values are also listed.

Table. Specimen surface parameters for various machining processes

R_a	1.81	1.165	0.576
R_p	7.416	5.055	2.266
R_{max}	13.344	12.293	4.426
t_m,%	51.63	53.993	49.622
v	3.2	4.917	2.893
b	3.25	6.29	3.518

4. Conclusions:

1. Operation loads (first shots) change the lock part dimensions due to plastic deformations of the interface surface micro irregularities.
2. The plastic deformation value at the initial break-in is roughly proportional to the load. As the contact area increases, the plastic deformations are reduced, and the dimensions are stabilized.
3. Polynomial relations between the part interface plastic deformations, surface positioning errors and surface quality were identified for commonly used rifle steels.
4. It is proved that the micro irregularity plastic deformation at the interfaces is affected not only by the micro irregularity height, but the finishing process as well. The preferable process is finish flank milling.

References

[1] Gryazev V M and Yamnikov A S 2012 Methodical bases of the solution of dimension chains with norm of soot surface contact *European Science and Technology* October 30 th - 31 st Publishing office Vela Verlag Waldkraiburg (Munich – Germany) V **I** pp 161-8
[2] Kazakov K E Kurdina S P and Manzhirov A V 2017 Contact Interaction Between Surface Inhomogeneous Foundations and systems of Rigid Punches *Procedia IUTAM* V **23** pp 201-9
[3] Vasilyev A S and Gryazev V M 2012 Surface Layer Property Estimation for Various Parts Machining Processes *Fundamentalnye i prikladnye problemy tekhniki i tekhnologii* (Oryol State University) N **2-5 (292)** pp 3-6
[4] Vasilyev A S Gryazev V M and Yamnikov A S 2012 Functionally Dependent Assembly Dimensional Chains Ensuring the Required Surface Contact *Sborka v mashinostroenii priborostroenii* N **5** pp 3-6
[5] Bazrov B M Suslov A G Taratynov O V and Shoev A N 2013 Quality of Surface Modules in Machine Parts *Russian Engineering Research* V **33** N **11** pp 651-4
[6] Liu T Cao Y Wang J and Yang J 2016 Assembly Error Calculation with Consideration of Part Deformation *Procedia CIRP* V **43** pp 58-63

[7] Gavankar P and Bedworth D 1991 Stacked Tolerance Analysis and Allocation Using Assembly Models *Journal of Intelligent Manufacturing* V **2** I **6** pp 365–77
[8] Chatterjee B and Sahoo P 2012 Effect of Strain Hardening on Elastic-Plastic Contact of a Deformable Sphere against a Rigid Flat under Full Stick Contact Condition *Advances in Tribology* V **2012** Article ID 472794 8 p
[9] Söderberg R Wickman C and Lindkvist L 2008 Improving Decision Making by Simulating and Visualizing Geometrical Variation in Non-Rigid Assemblies *CIRP Annals - Manufacturing Technology* V **57** I **1** pp 175-8
[10] Samper S and Giordano M 1998 Taking into Account Elastic Displacements in 3D Tolerancing: Models and Application *Journal of Materials Processing Technology* **78(1)** pp 156-62
[11] Yamnikov A S Yamnikova O A and Kashmin O S 2015 Empirical Relations for Contact Part Approach Values under Impulse Loads *Sborka v mashinostroyenii. priborostroyenii* N **3** pp 42-8
[12] Neville K Lee S Grace Yu Joneja A and Ceglarek D 2006 The Modeling and Analysis of a Butting Assembly in the Presence of Workpiece Surface Roughness and Part Dimensional Error *The International Journal of Advanced Manufacturing Technology* V **31** I **5–6** pp 528–38
[13] Shi X and Polycarpou A A 2005 Measurement and Modeling of Normal Contact Stiffness and Contact Damping at the Meso Scale *ASME J Vib Acoust* **127** pp 52–60
[14] Tskhai E B Klopotov A A and Kozyreva R A 2015 *Surface Roughness for Various Part Applications: Course and Final Project Guidelines* (Izdatelstvo Tomskogo gosudarstvennogo arkhitekturnostritelnogo universiteta) 28 p
[15] Radchik V S Ben-Nissan B and Muller W H 2002 Semi-graphical Methods for the Calculation of Real Areas of Loaded Contact by Means of the Abbott-Firestone Bearing Curve *Journal of Tribology-transactions of The Asme* N **1** V **124** pp 223-6 DOI:10.1115/1.1398545
[16] Abbott E J and Firestone F A 1933 Specifying Surface Quality a Method Based on Accurate Measurement and Comparison *Mechanical Engineering ASME* V **55** pp 569-72
[17] Yamnikov A S MalikovA A Kashmin O S and Troitsky D I 2018 Ensuring Precision for Small Arms Lock Assembly With Consideration for Interface Contact Stiffness *CIS Iron and Steel Review* V **15** pp 39–2

IOP Conf. Series: Materials Science and Engineering 1100 (2021) 012020 doi:10.1088/1757-899X/1100/1/012020

Development of proposals of improving the risk-oriented approach in control and supervision activities on labor protection

A P Bordakovskaja, N V Rumyantseva, S O Kamenskaya and N G Hlebnikov

Peter the Great St. Petersburg polytechnic university, Polytechnicheskaya, 29, Saint-Petersburg, 195251, Russia

E-mail: bordakovskaya.ap@edu.spbstu.ru

Abstract. This work is devoted to the development of recommendations for improving the methodology for classifying the activities of employers to a certain risk category, which will make it possible to focus supervisory measures on enterprises in which professional risks to the life and health of workers are high.

1.Introduction

In 2017, in the field of inspections of compliance with labor law requirements, a number of regulations were adopted. Those provide important changes for both the employer and the supervisory authorities, namely the transition to a risk-based approach (RBA).

This approach is a method of organizing and implementation of control, according to which the number of scheduled inspections and measures to prevent violations is determined by classifying the employer's activities as a risk category.

The risk-based model is aimed at concentrating a limited amount of resources precisely in those zones where there is a chance of maximum risk, as well as at reducing the administrative load and preventing harm to legally protected values on conscientious employers [1,2]

The risk category, which makes it possible to estimate the activities of the enterprise, is calculated on the basis of a certain methodology, which includes several stages, and the calculations of the corresponding formulas. However, this methodology is based on a list of indicators that do not fully or do not always reflect the specifics of the activities of enterprises. For example, the same lists include enterprises whose nature of activity is fundamentally different from the point of view of the presence of occupational risks to the health and life of an employee. There is a dependence of checks on criteria that do not significantly affect the definition of the risk category.

The introduction of a risk-based approach to control and surveillance activities in the field of labor protection provides a change in the number, duration and frequency of scheduled inspections by dividing organizations into several categories – those which require more close attention and enterprises that can be characterized as relatively safe. [3,4,5]

This approach is based on decisions taken while taking into account the calculation of the degree of risk. Therefore, it is necessary to provide a justification for the use or selection of additional criteria and indicators by which an enterprise can be included in one or another of the above risk categories.

In this regard, the purpose of this study is to develop recommendations for improving the methodology of classifying the activities of legal entities and individual entrepreneurs to a certain category of risk.

2. Methods

The methodological basis of the research is the analytical method, the method of systematization, generalization, comparison, deduction, the method of document analysis.

To develop recommendations for improving the risk-based approach, the materials-methodology for assigning employers' activities to a certain risk category were studied. [6]

To determine the criteria and indicators that influence to a greater extent in the categorization, enterprises of the following sectors of the economy were analyzed: education, electricity supply, trade and

Content from this work may be used under the terms of the Creative Commons Attribution 3.0 licence. Any further distribution of this work must maintain attribution to the author(s) and the title of the work, journal citation and DOI.
Published under licence by IOP Publishing Ltd

manufacturing. The data provided by the labor protection service of each enterprise were studied, namely, information on injuries over the past 3 years, the average number of employees, information on the presence of wage arrears and administrative penalties.

3. Results

In the course of the work, a comparative analysis of the risk categories of enterprises of various economic activities was carried out, the difference in the considered indicators was carried out.

The comparative analysis revealed the following:

1. The industries of education and electricity supply are assigned into the high risk category, while trade and manufacturing are considered as moderate risk. Although from the point of view of threats to life and health, harmful and dangerous factors of the labor process, these industries can be related in a different way. If in the first case the coincidence of the risk category is explained by the presence of accidents, in the second, when the enterprises fully comply with labor legislation and there is no injury, there is a discrepancy between the assigned risk categories.

2. If there has been at least one minor accident, excluding severe and fatal, the potential risk indicator all (P), based on the value of which is assigned to the risk category, still increases regardless of the economic sector.

3. The definition of the risk category depends on the number of employees in a particular area. However, this should not affect the categorization, since the number of employees does not reduce the impact of occupational risks in really potentially dangerous areas for the life and health of employees.

Based on the comparative analysis, a detailed analysis of the calculations and factors affecting the final result was carried out, the shortcomings of the method were identified and a number of recommendations that could contribute to its improvement were proposed.

Conclusions on the identified shortcomings of the method:

1. The employer cannot influence the values of such indicators as the indicator of potential harm to legally protected values in the world of work (PH) and the indicator of the scale of the spread of potentially negative consequences (S), since they are already presented in numerical form in the regulatory document.

2. The product of the PH and S indicators may ultimately give such value (1 or more) that it makes no sense to consider the remaining coefficients. The indicator of potential risk already automatically takes on a value that assigns the enterprise to the high risk category.

3. The coefficient of the fact of the presence of administrative penalties for violations (Cadm) takes the value of either 0.1 or 0, but it does not depend on the number and significance of violations of mandatory requirements in legislation and the sphere of labor protection, but simply on the presence of an administrative penalty. In fact, there can be an uncountable number of violations, as well as violations may differ in degree of significance, but this does not affect the risk categories, and, consequently, scheduled inspections.

4. The risk category depends on the number of employees, excluding working conditions.

During the analysis of the methodology used for categorizing the enterprise, data that are not taken into account in the calculation were identified. Accordingly, the developed recommendations for improving the methodology include three areas:

1. Classification of violations.
2. Results of special assessment of working conditions – SAWC.
3. Risk assessment.

To determine the risk category of the enterprise in the considered methodology, such an individual coefficient, which shows the presence of the fact of administrative punishments imposed for a number of violations of mandatory requirements in the world of labor (Cadm) is used. In the presence of punishments, the coefficient is 0.1, in the absence it is equal to 0. However, this coefficient does not take into account the significance, the severity of the consequences that may arise based on a particular violation.

The first recommendation is to rank the coefficient of administrative punishments (Cadm) according to high, low and medium risk of violations of mandatory requirements. The introduction of this coefficient will help to identify the significance of the violations committed and clearly demonstrate the employer the obligation to comply with the requirements of labor legislation and legislation in the field of labor protection.

The algorithm for determining Cadm can be represented in the form of a block diagram shown in figure 1.

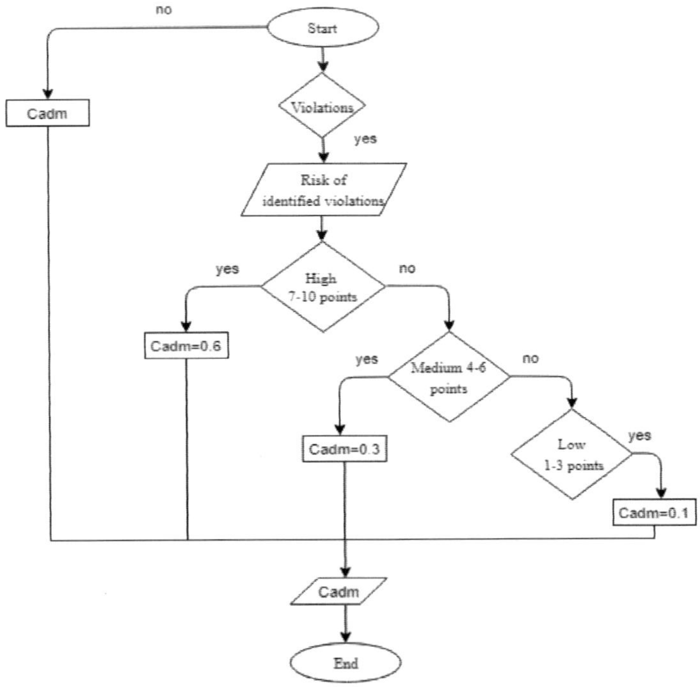

Figure 1. Algorithm for calculating the coefficient of administrative penalties.

The calculation results, taking into account ideal conditions and taking into account the introduction of the gradation of the coefficient of administrative punishments, are presented in table 1.

Table 1. Calculation results.

Sphere	No penalties		Cadm by method			Cadm by degree of risk		
	P	Category	Cadm	P	Category	Cadm	P	Category
Trade	0.3	Moderate risk	0.1	0.4	Moderate risk	0.6	0.9	Significant risk
						0.3	0.6	Average risk
Electricity supply	0.693	Average risk	0.1	0.793	Significant risk	0.6	1.293	High risk
						0.3	0.993	Significant risk

Let's move on to the second direction of the developed recommendations.

At enterprises, situations are possible when there are no accidents at work, but according to the results of the SAWC, 3 and (or) 4 classes of working conditions have been established, respectively, at the workplaces of these enterprises employees are involved whose work is carried out in harmful and (or) dangerous factors, affecting employees during the production process in the form of acquiring later or already having occupational diseases (OD).

Since the method under consideration includes an indicator of the scale of the spread of negative consequences (S), which is the average number of employees in the enterprise, an additional coefficient can be introduced to this indicator. It will consist of the presence of workers at work involved in harmful and (or) dangerous working conditions.

The algorithm for determining this coefficient is presented in the form of a block diagram in figure 2.

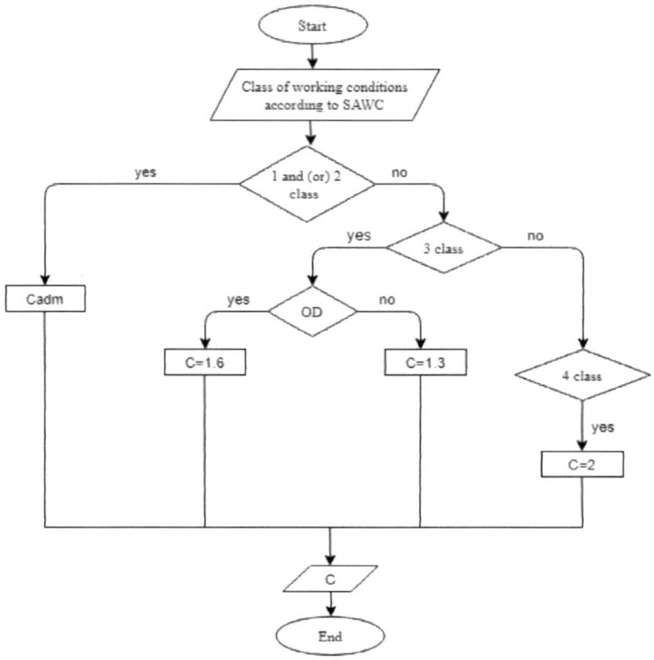

Figure 2. Algorithm for calculating the coefficient, taking into account the presence of workers involved in harmful and (or) dangerous working conditions.

The calculation results, taking into account the introduction of this coefficient, are presented in table 2.

Table 2. Calculation results.

before the introduction of the coefficient					after the introduction of the coefficient				
Sphere	PH	S	P	Category	PH	S	C	P	Category
Trade	0.2	1.5	0.3	Moderate risk	0.2	1.5	1	0.3	Moderate risk
Electricity supply	0.99	0.7	0.693	Average risk	0.99	0.7	1.6	1.1088	High risk

It is worth noting that the proposals mentioned above for improving the coefficients that affect the categorization of the enterprise do not take into account the number of identified workplaces with different classes of working conditions. In calculating the coefficient, only one workplace with the highest hazard class is used.

Thus, the introduction of this coefficient will focus the attention of the supervisory authorities on those workplaces (and, accordingly, enterprises) where it is possible to implement the impact of harmful and (or) hazardous factors that affect workers in the future in the form of occupational diseases.

One of the following recommendations for improving the risk-based approach is to use the information obtained through professional risk assessment. However, in the absence of an approved methodology, this approach cannot be implemented. This is another argument in favor of the need to develop and implement a common risk assessment methodology for all enterprises.

The use of the obtained results of a unified risk assessment, perhaps, should be presented in the form of a certain indicator and add it to the considered methodology as an independent coefficient. However, at the moment there is no way to say what this coefficient should be. Definitely it should be presented in numerical form, and the figure used should be convenient and understandable for the control and supervisory authorities. The value of this indicator should fully reflect the results of the risk assessment carried out.

4. Discussion

The introduction of a risk-based approach significantly changes the mechanism of state supervision and control in the field of labor protection. This innovation significantly reduces the load on the Rostrud system and the State Labor Inspectorate when monitoring compliance with the requirements of labor legislation and other regulatory legal acts containing labor law norms, and also allows the supervisory authorities to focus increased attention at enterprises where the risks are the highest. The use of a risk-oriented model consists in using a methodology, based on which one or another category of risk is assigned to enterprises, and the "density" of control is determined.

The developed recommendations for improving this methodology will contribute to the most complete, deep and detailed assessment of a particular enterprise, taking into account various points in terms of hazards to the life and health of employees when categorizing with the subsequent implementation of scheduled inspections, increase the indicator of the validity of the number of inspections from the activities of the enterprise.

References

[1] Klimova E V, Semeykin A Yu and Nosatova E A et al 2016 IOP *Conf. Series: Materials Sci. and Eng.* **451 (1)** 012198
[2] McLellan R K 2017 *Health Affairs* **36 (2)** pp 206-213
[3] Glebova E V, Artanova M V, Ivanova M V and Korobov A V et al 2020 *Conf. Series: Materials*
[4] Sci. and Eng. 862 062067
[5] Blagovestnikova A A 2018 Domestic jurisprudence 6 (31) 30-32
[6] Mekkodathil A, El-Menyar A and Al-Thani H 2016 International Journal of Critical Illness and Injury Science 6 (1) pp 25-32
[7] Shabanova D N and Alexandrova A V 2018 Vestnik NCBZD 3 (37) 127-133
[8] Khairullina L I, Galeeva E I and Gimranov F M 2017 Features of the Implementation of a Risk-
[9] Oriented Model of Labor Protection Management (Moscow: Mendeleev University) p 149
[10] Salakhutdinova A F and Vishnevskaya N G 2017 *Vector of Economics* **11 (17)** 76
[11] Lokteva O S and Loktev D A 2018 *Science and Technology of Transport* **1** 84-91

Analysis of height dimensions of relatively high box-parts in their drawing from variable thickness blanks by full factorial experiment implementation

A N Malyshev, S A Bysov and E N Malyshev

Bauman Moscow State Technical University, Kaluga Branch, Gagarina Street 3, Kaluga, 248000, Russia

E-mail: malen@bmstu.ru

Abstract. This article is devoted to an experimental study of the error in the contour height of relatively high box-parts with large radiuses of angular curves manufactured by drawing operation from a tailor welded blanks with variable thickness made of high-strength steel. The advantages of usage of tailor welded blanks with variable thickness are a reduction in the cost of the blanks by reducing material consumption compared to the" classic " uniform blanks, the ability to create a design of the stamped part in accordance with the requirements for operational properties, and a reduction in the mass of the stamped part as a whole.

1. Introduction

Currently the implementation of some new progressive types of metal sheet blanks (for example tailored welded sheet blanks (TWB) of variable thickness) in the development of the ways to intensify the sheet stamping process during the production of the detail parts and components for a number of sectors of industry and other branches of the economy is constantly increasing. For example in order to reduce the weight and the cost of the parts and also to manage the strength characteristics of the car [1] to improve its passive safety the modern automotive industry actively develops a new direction of the TWB with variable thickness [2, 3, 4]. At the moment due to the constantly changes of the body design, casings, aprons, tanks and other items produced during the "drawing" process the need to use TWB of variable thickness increases as well as in the other sectors of the national economy including the aerospace industry [5, 6].

2. Materials and Methods

The accuracy is a complex indicator which gives an idea about the perfection of the technological process and shows the specific features of one or other stamping process in general and sheet metal stamping in particular. For example the accuracy assessment of the sheet-blank stamping method of "drawing" is based on the study of errors of the initial blanks, the inaccuracy of the technological process, the features of force conditions and the mode of deformation.

The factors influencing the inaccuracy with regard to the height of the box-parts during the drawing process can be classified according to the nature of their manifestation:

1. The factors determine some occasional errors which mainly affect the value of the instantaneous scattering field of a random variable of the box's height dimension as follows:

a) errors of the geometric characteristics of the initial blanks;

b) fluctuation of some physical and mechanical properties of the initial blanks' material because of the variability of the chemical composition and because of the heterogeneity of the raw material structure of the blanks;

c) errors arising from the changes of the friction conditions in the die cavity and assessed through the changes in the value of the contact friction coefficient;

Content from this work may be used under the terms of the Creative Commons Attribution 3.0 licence. Any further distribution of this work must maintain attribution to the author(s) and the title of the work, journal citation and DOI.
Published under licence by IOP Publishing Ltd

d) errors arising from the incorrect placing of the initial blank in the die;
2. The factors defining some systematic constant errors/inaccuracies in the production of the kit of parts comprise the geometric error of the real dimensions of the die tooling, including the inhomogeneity of the roughness of the die surfaces involved in the formation process and the anisotropy of mechanical properties as well [7]. In some cases, for example, during the automatic stamping the geometric dimensions of the die forming surfaces and the friction conditions can change due to the action of the temperature factor. If so, the mentioned factors determine the errors which can act within a certain period as systematic and naturally changing ones.

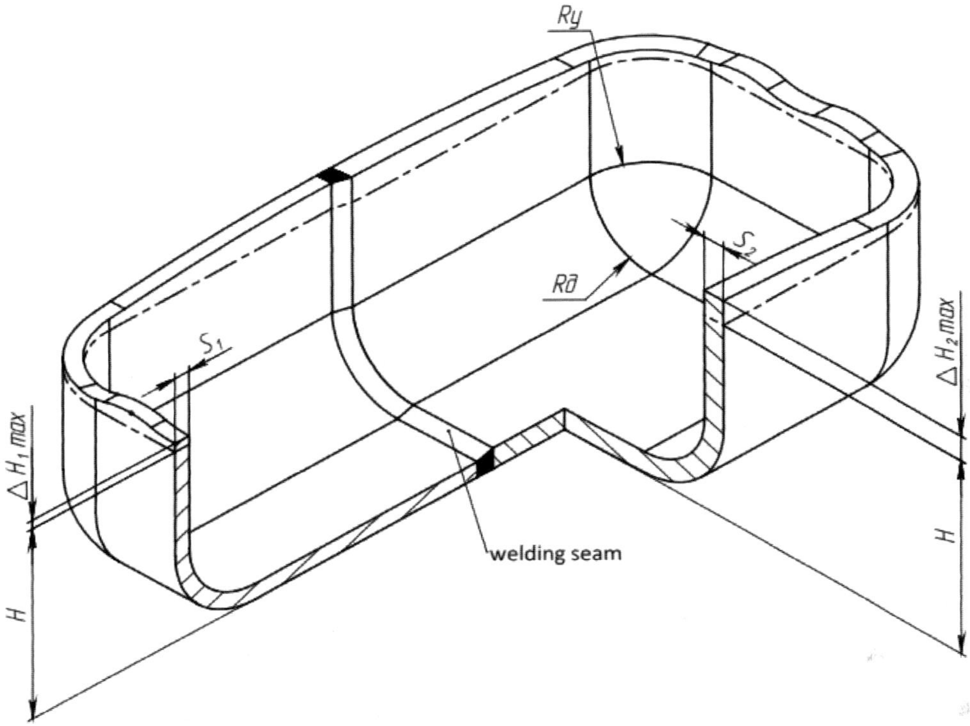

Figure 1. The sketch of the final box-part after "drawing" operation

Figure 1 shows a sketch of the final box-part and the geometric parameters of which are shown in Table 1. The initial sheet material for the experiments is high-strength steel HX260PD from a rage of RS-steels which presents a high-strength and rolled product made of rephosphorized steel. The shape formation of the initial sheet blanks was performed according to the well-known method [8] for relatively high boxes with relatively large radiuses of angular rounding.

Composite blanks were used for making experiments which parts were butt-welded by laser; the quality control of welding was carried out according to the standard [9]. The experiments [10] were performed in the laboratory of the Department MK1 "Machine-Building technologies" of Kaluga branch of Bauman Moscow State Technical University (MSTU named after N.E. Bauman). The drawing of box-parts was implemented in an experimental die without using of any additional force to the surface of the initial sheet blank by "upper" and "lower" blank holder and without any lubrication. Two types of initial TWB (1-st type of blanks: S_1=1.9mm; S_2=2.4 mm; thickness difference ΔS=0.5 mm; 2-nd type of blanks: S_1=1.4 mm; S_2=2.4 mm; thickness difference ΔS=1.0 mm) and 8 sizes were used three times for all experiments while changing the calculated (rated) height H of the box-part from 23.15 mm up to 30.15 mm. The welded seam was at the center of the sheet blank, see Figure 1. The results of all experiments are presented at pictures Figure 2 and Figure 3.

Table 1. Geometric parameters of the final box-part after the "drawing" operation

№	Geometric parameters of the box-part
1	thickness of material: 1-st type of blanks - $S_1 = 1.9mm$; $S_2 = 2.4\ mm$; thickness difference $\Delta S = 0.5\ mm$; 2-nd type of blanks: $S_1 = 1.4\ mm$; $S_2 = 2.4\ mm$; thickness difference $\Delta S = 1.0\ mm$
2	calculated (rated) height of the box-part: H, mm
3	radius of corner curvature: $R_y = 10$ mm
4	radius of bottom curvature: $R_\partial = 10$ mm
5	minimum deviation (error) of the calculated (rated) height of the box with regard to the part of the contour of thinner material : $\Delta H_{1\ min}$, mm
6	maximum deviation (error) of the calculated (rated) height of the box with regard to the part of the contour of thinner material: ΔH_{1max}, mm
7	minimum deviation (error) of the calculated (rated) box height with regard to the part of the contour of thicker material: $\Delta H_{2\ min}$, mm
8	maximum deviation (error) of the calculated (rated) height of the box with regard to the part of the contour of thicker material: ΔH_{2max}, mm
9	the width of the box-part: A = 40mm
10	the length of the box-part: B= 80mm

The arithmetic mean ΔH_i is calculated by:

$$\Delta H_i = (H_{i\ min} + H_{i\ max})/2 \quad (1)$$

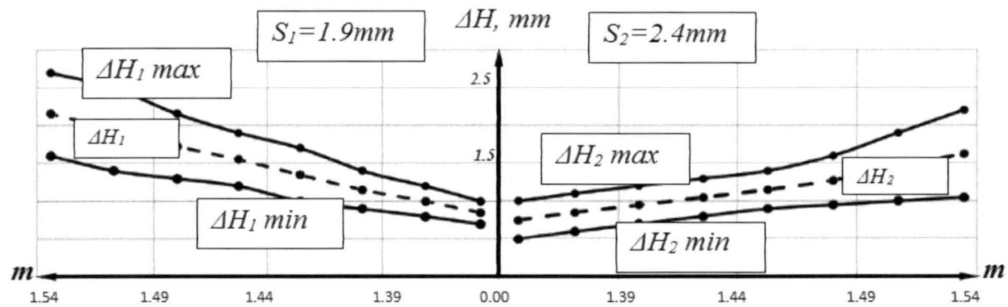

Figure 2. Experimental curves reg. the dependence of errors ΔH in the height of the box-part contour (S_1=1.9 mm, S_2=2.4 mm, ΔS=0.5 mm) on the deformation degree m

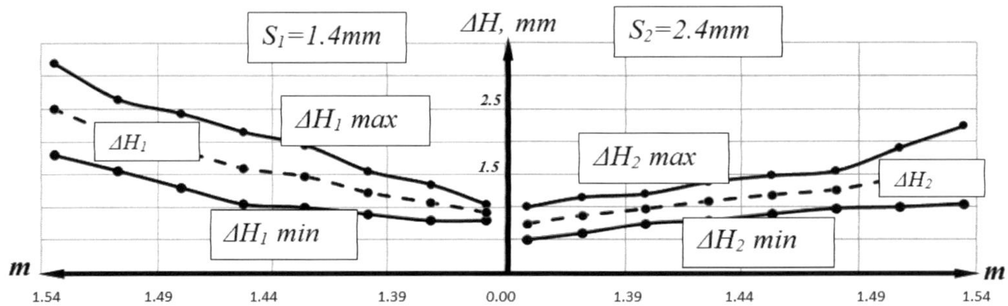

Figure 3. Experimental curves reg. the dependence of errors ΔH in the height of the box-part contour (S_1=1.4 mm, S_2=2.4 mm, ΔS=1.0 mm) on the deformation degree m

3. Results and discussions

The full factorial experiment 2^2 matrix, its results and also the levels of factors at the natural and normalized scales are presented in Table 2 and respectively in Table 3 where ΔS (thickness difference) and m (deformation degree) are the investigated factors at the coded scale; $\overline{\Delta H}$ is the average value of the errors/inaccuracies reg. the height in every i-th experiment; S_i^2 is the dispersion in i-th experiment.

Table 2. Factors level at natural and normalized scales.

Factors	Factors level			Interval
	+1	0	-1	
m	1.53	1.435	1.34	0.095
ΔS, mm	1.0	0.75	0.5	0.25

Table 3. The full factorial experiment 2^2 matrix and results.

№	m	ΔS, mm	ΔH, mm			$\overline{\Delta H}$, mm	S_i^2
			1-st exp.	2-nd exp.	3-d exp.		
1	+	+	2.50	2.46	2.54	2.50	0.0016
2	-	+	0.89	0.95	1.01	0.95	0.0036
3	+	-	2.11	2.19	2.15	2.15	0.0016
4	-	-	0.85	0.88	0.82	0.85	0.0009
							$\sum S_i^2 = 0.0077$

To create a model (2) of regression analysis we used a generally accepted method [11], when an adequate regression model was obtained at the normalized scale of factor levels in the statistical processing of the results of full factorial experiment 2^2:

$$y(\Delta H) = 1.61 + 0.71 \cdot m + 0.11 \cdot \Delta S + 0.06 \cdot m \cdot \Delta S \quad (2)$$

The value and the sign of the coefficients (plus–minus) in the regression equation indicate the strength and the character of influence of the investigated factors on the relative error/inaccuracy of the box-part's height.

From the analysis of the regression model it follows that the deformation degree m has the greatest influence on ΔH (the average value of the error reg. the height of box-part). The effect of the first one is almost 7 times more than the effect of thickness difference ΔS.

The analysis of variance is used to select in a more objective way some factors which have the strongest impact on the response (ΔH) and makes it possible to determine the contribution of factors and their interactions to the variance (variation) of the response (ΔH). So the deformation degree m has 96.84% and makes the greatest contribution of to the overall response variance, the thickness difference ΔS has 2.41% and the paired effect of the factors m and ΔS is 0.75%.

4. Conclusions

The experimental results indicate the need to make certain corrections in the calculation of the optimum contour tailor welded blanks of variable thickness for drawing box-shaped parts as confirmed by well-known researches [12, 13] aiming at some recommendations for calculation of the contour of the initial tailor welded blanks to avoid cutting of the box-part's contour after drawing process and to get a relatively smooth contour of the item where some inherent roughness or errors will be kept within tolerance on the height of the box-part.

References

[1] Gaysin S V 2016 *Trudy NAMI* **3(266)** 6–14
[2] Klimychev S B et al 2004 *J. Forg. and Stamp. Prod. Material Working by Pressure* **7** 22-7
[3] Hayrullin T V and Stolbov V I 2012 *J. Welding and Diagnostic* **2** 42-4
[4] Merklein M, Johannes M, Lechner M and Kuppert A 2014 *J. Materials Processing Technology* **214** 151-64

[5] Hayrullin T V and Stolbov V I 2011 *J. Bulletin of SSAU* **3(27)** 20-7
[6] Sinke J, Iacono C and Zadpoor A A 2010 *Int. J. of Material Forming* **3** 849-52
[7] Barlat F et al 2005 *Int. J. Plasticity* **21** 1009-39
[8] Romanovsky V P 1979 *Reference cold forming* (Leningrad: Machine building) p 520
[9] GOST (Russian State Standard) 13919-1. 2017 *Welding. Electron beam and laser welded joints. Guidelines for assessing the quality level for defects* (Moscow: Standard inform)
[10] Malyshev A N and Bysov S A 2020 *Int. Conf. Modern problems of mechanical engineering* (Tomsk: TPU) pp 128-129
[11] Box G E, Hunter W G and Hunter J S 2005 *Statistics for Experimenters: Design, Innovation, and Discovery* (Wiley) p 672
[12] Colgan M and Monaghan J 2003 *Deep drawing process: analysis and experiment, J. Materials Processing Technology* **132** 35-41
[13] Fazli A 2013 *Int. J. of Automotive Engineering* **3(2)** 435-45

Possibility of creating composite materials with self-diagnosis function based on the effect of mechanoluminescence

K V Tatmyshevskiy and D D Pavlov

Institute of Information Technology and Radio Electronics, Vladimir State University, 87 Gorky Street, Vladimir, 600000, Russia

E-mail: pavlovdd84@mail.ru

Abstract. The increasing productivity of modern information processing systems opens up new possibilities for various technical devices. The ability to very quickly analyze a large amount of information allows you to constantly improve technical systems, including autonomous ones. Sensors or primary converters are used to obtain information about the surrounding world or external influences in technical systems. However, modern technologies make it possible to create materials with desired properties - composite materials. The programmable properties of materials can be strength, rigidity, resistance to mechanical, chemical and temperature influences, but it is especially worth noting the possibility of obtaining materials with specified informative properties. These materials include composites containing mechanoluminescent substances. Such materials can simultaneously be housings of any devices or enclosing structures, they can be parts of load-bearing structures of buildings and structures and at the same time play the role of shock and destruction sensors. Using such materials, you can create panels with a self-diagnostic function.

1. Introduction

Mechanoluminescence is the radiation of solids caused by mechanical action [1,2]. Based on this physical phenomenon, it is possible to design sensors for recording various external influences (impacts, slippage, deformation, cracking, destruction). Due to the optical nature of their functioning, mechanoluminescent sensors have certain advantages over other types of sensors. These advantages are: generator mode of operation; lack of sensitivity to electromagnetic interference; solidity; the ability to apply on surfaces of different shapes and areas; the possibility of embedding in a structure or embedding in a construction material. In some works [3-6], it is proposed to create mechanoluminescent (or triboluminescent) damage sensors. The range of applications for materials with mechanoluminescent properties is very wide. They can be used in construction to monitor the state of various supports or structures [7], to monitor the integrity of the hulls of mobile objects and robots [8] and to monitor the condition of the skin of aircraft [9].

2. The mechanism of luminescence

When luminescence is excited, an atom (molecule), absorbing energy, passes from the ground energy level 1 (figure. 1) to the excited level 3. As a rule, the emission level 2 lies below level 3, part of the energy is lost during excitation into heat, and the wavelength of the emitted more light than absorbed (Stokes luminescence). Processes are also possible when an emitting atom receives additional energy from other atoms; then the emitted quantum can have a shorter wavelength (anti-Stokes luminescence) [10, 11].

Content from this work may be used under the terms of the Creative Commons Attribution 3.0 licence. Any further distribution of this work must maintain attribution to the author(s) and the title of the work, journal citation and DOI.
Published under licence by IOP Publishing Ltd

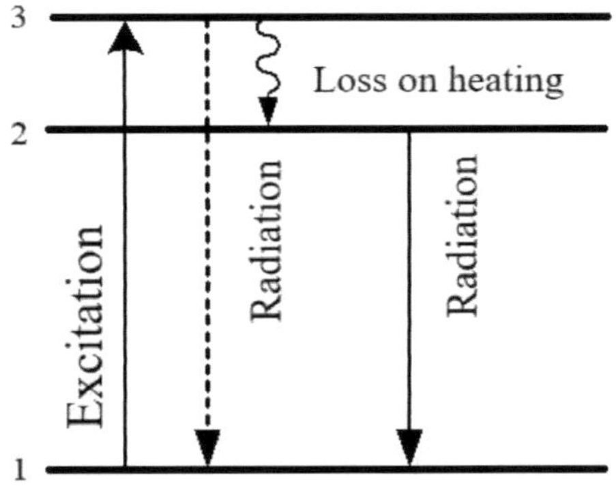

Figure 1. Scheme of quantum transitions in the elementary process of luminescence: 1-the main energy level; 2-the radiation level; 3 – the excitation level. The dotted line indicates the transition corresponding to resonant luminescence, and the wavy line indicates a non – radiative transition.

The emission level can belong to the same atom (molecule) that absorbed the excitation energy (such transitions are called intra-center), or to another particle. The transfer of energy to other atoms and molecules is carried out by electrons during electron-ion shocks, during ionization and recombination processes, by inductive resonance or exchange, when an excited atom directly collides with an unexcited one. Finally, energy transfer via conduction electrons, holes, and electron-hole pairs (excitons) becomes crucial in crystals. If the final act of energy transfer is recombination, then the luminescence accompanying this process is called recombination [10].

In the vast class of substances that have the ability to luminescence in general and to mechanoluminescence in particular, a special place is occupied by substances united under the general name "crystal phosphors" (CF). In CF, luminescence occurs due to the presence of a small amount of impurities of other substances called activators. Crystal phosphors are: sulfides (ZnS, CdS), selenides (ZnSe), silicates (Zn_2SiO_4), tungstates ($CaWO_4$), molybdates, and some other compounds [11]. Zinc sulfide CF have the highest luminosity [10] and make up more than 80 % of all industrial electroluminophores [11]. It is for this reason that the vast majority of mechanoluminescence studies are conducted on ZnS phosphors.

3. Necessary conditions for mechanoluminescence

Explorations [12] showed that mechanoluminescence in A_2B_6 compounds is a consequence of the processes of motion and multiplication of dislocations accompanying plastic deformation of crystals. It was found experimentally that dislocations in A_2B_6 semiconductors, and in particular in ZnS, have a strong electric charge [8]. In the process of plastic deformation, the luminescence centers interact with the electric field of moving charged dislocations, which leads to the excitation (ionization) of luminescence centers with their subsequent radiative transitions.

To create a material with a self-diagnostic function based on the phenomenon of mechanoluminescence, it is necessary to ensure the hardness of the material. This condition is necessary for the energy of mechanical action to reach the crystals of the mechanoluminophor. As a result of this, plastic deformation of the crystals of the luminescent substance will occur, which will allow the luminescence mechanism to be triggered at the energy level.

Another important parameter of a material will be its optical properties. The optical parameters of a mechanoluminescent sensitive element, which determine the transfer of radiation in it, primarily depend on its structure. In the most common simplest case, the SE structure is a plane-parallel light-scattering layer of particles of a powdery luminescent material in a hardened transparent binder [8]. Such a layer is deposited on a substrate of a transparent material, which ensures, firstly, the creation of a stressed state of the particles during propagation of a pressure pulse (mechanical stress waves) through the layer, and secondly, the removal of radiation to the photodetector. Epoxy, melamine formaldehyde and other resins, varnishes, polystyrene and other transparent materials are used as a binder for powder phosphors [13]. Faceplates made of various glasses, optical ceramics, fiber-optic bundles and cables, tapers can be used as a substrate [8].

4. Methods for receiving composite materials with self-diagnostic function

As described above, in order to function, the mechanoluminescent sensor element embedded in the composite panel must be in a stressed state and it must be possible to output optical radiation from the sensor. A composite panel can be a rigid structure made up of several layers (figure. 2).

IOP Conf. Series: Materials Science and Engineering 1100 (2021) 012022 doi:10.1088/1757-899X/1100/1/012022

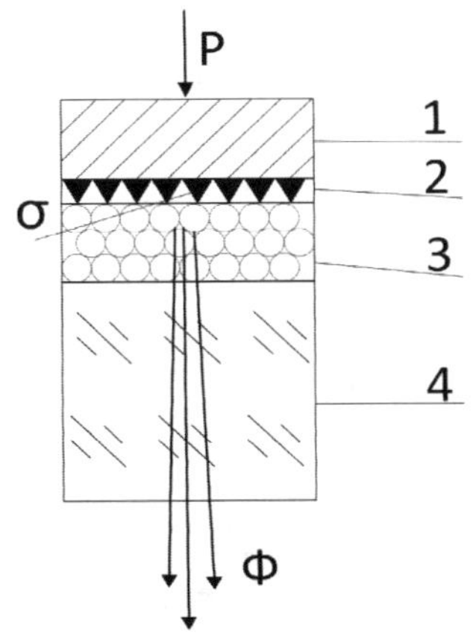

Figure 2. Structure of mechanoluminescent impact sensor for SHM, consisting of: a steel plate (item 1) simulating the skin of an object (there were used steel plates with a thickness of 1 to 6 mm); flat mechanical stress concentrator (item 2) to increase the sensitivity of the sensor; mechanoluminescent distributed sensing element (item 3); a transparent base (item 4) and a closed optical channel for image transmission.

As a result of experiments [9] with panels of this configuration, the following frames of the glow of the phosphor were obtained (figure. 3).

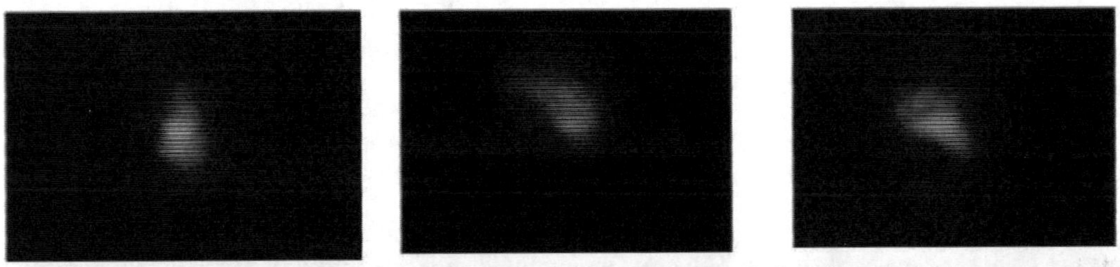

a) impact of a metal ball falling from a height of 0.5 m on a steel plate 1 mm thick;

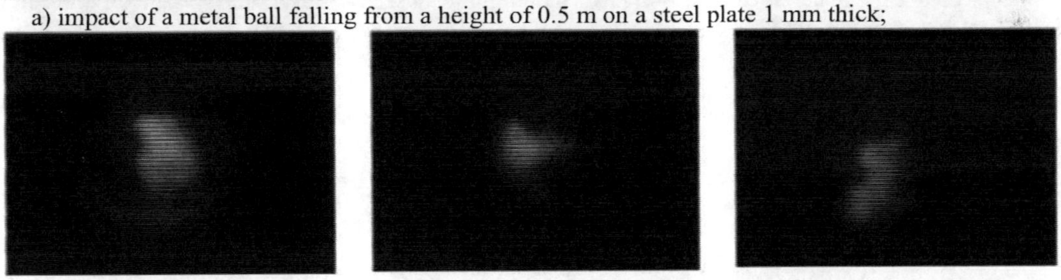

b) the impact of a metal ball falling from a height of 1 m on a steel plate 1 mm thick;

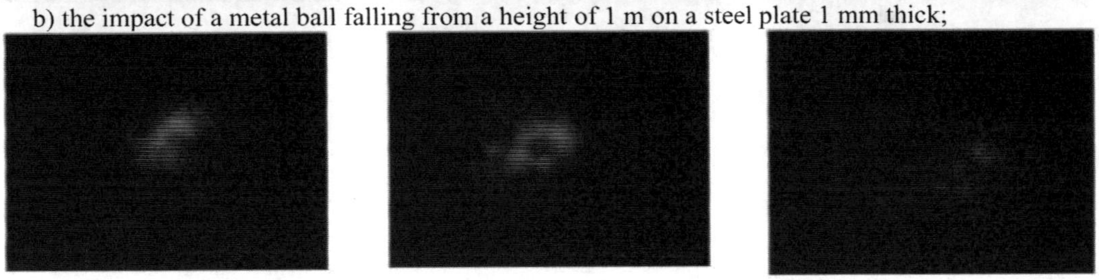

c) the impact of a metal ball falling from a height of 0.5 m on a steel plate with a thickness of 2 mm;

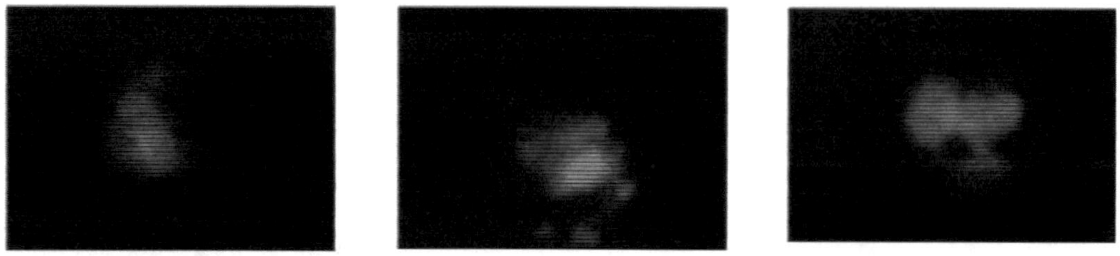

d) the impact of a metal ball falling from a height of 1 m on a steel plate with a thickness of 2 mm;

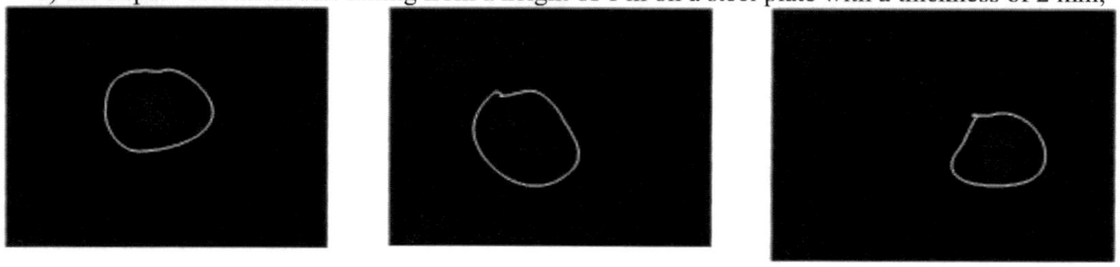

e) the impact of a metal ball falling from a height of 1 m on a steel plate with a thickness of 6 mm;

Figure 3. Luminescence patterns of the distributed sensor at different impacts and shell thicknesses.

Another way to obtain a composite panel is to apply a mechanoluminophor to a transparent base in a vacuum [14]. This method consists in blend composed of bis (isopropylxanthonenate) ((2,2 ′-bipyridyl) zinc and bis (isopropylxanthogenate) ((2,2′-bipyridyl) manganese in 1000:(1-20) molar ratio is heated at evaporator temperature of 110-130 C, vapors are transferred and thermal decomposition is carried out in vacuum of 1×10^{-5} to 1×10^{-6} torr at substrate temperature of 150-300 C.

It is also possible to create composite materials with a self-diagnostic function based on the phenomenon of mechanoluminescence by introducing optical fibers with lateral light transmission and manganese-activated zinc sulfide grains into the material. The addition of mechanoluminescent materials to the epoxy compound and the creation of a structure that provides radiation transmission through optical fibers to the photodetector system allows solving the problem of monitoring and self-diagnostics of composite structures [13, 4]. An example of a composite material with built-in mechanoluminescent sensory areas and lateral light transmission optical fibers is shown in figure 4.

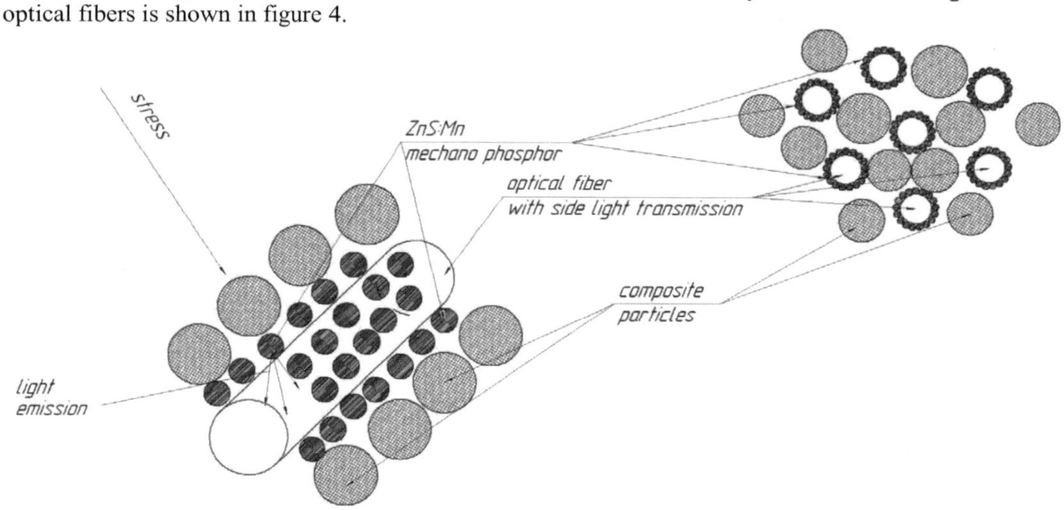

Figure 4. The implantation of a mechanoluminophor with an optical fiber with lateral radiation input into optical fibers in the composition of a composite material.

Distributed optical fiber systems and mechanoluminescent materials with different radiation spectra are used here for reliable recognition of a specific damage site [8].

5. Conclusions

A promising area of application of mechanoluminescent sensor elements is the development of composite materials and structures with the property of self-diagnostics [8,15]. On the whole, composites are materials consisting of carbon-graphite or quartz fibers embedded in epoxy compounds [12]. Such materials have high strength and stiffness with low weight and radio transparency. Currently, composites in the form of panels are widely used in mechanical engineering, aircraft and rocketry, in the construction of bridges, sports facilities, and others.

In addition to those given at the beginning of the article, as a positive property, it is worth noting the presence of a threshold nature of the dependence of the glow intensity on the applied mechanical effect. This ensures that the sensors are insensitive to minor loads and vibrations during operation. Integration of functions within a single structure allows the creation of built-in sensitive elements (sensors) that have the ability to selectively determine both the degree of damage to the composite panel and the specific location of the damage.

References

[1] Boswell L F and Alekhin V N 2013 *Academic Bulletin of the Ural SRIproject* **2** 77-81
[2] Ossipyan Y A and Schmurak S Z 1981 Defects in insulating crystals *Proc. of the Intern. conf.* (Berlin: Springer-Verl) pp 18-23
[3] Chandra B P 1985 *Nucl. Tracks and Radiat. Meas.* **1** 225-229
[4] Sage I and Bourhill G 2001 Triboluminescent materials for structural damage monitoring *J. of Mat. Chem.* **11** 231–245
[5] Mastro S, Mathur V and Jarrett A 2004 United States Patent No 6710328 appl. 27.02.2002, publ. 23.03.2004
[6] Xu C N, Zheng X G, Watanabe T and Akiyama M 1999 *Applied Physics Letters* **74** 2414-2416
[7] Tatmyshevskiy K, Pavlov D and Krylov V 2020 Functional materials based on the effect of mechanoluminescence for monitoring the state of load-bearing structures *Materials Science and Engineering* **896**:012092
[8] Tatmyshevskiy K V 2004 *Mechanoluminescent sensing elements. Fundamentals of theory, calculation, and design issues* (Vladimir: Vladimir State University) p 135
[9] Tatmyshevskiy K V and Pavlov D D 2020 To a question on the possibility of using mechanoluminescent sensor elements with area-distributed sensitivity in aerospace engineering *Materials Science and Engineering* **734**:012007
[10] Vereshchagin I K 1974 *Electroluminescence of crystals* (Moscow: Nauka) p 280
[11] Aven M and Prener J S 1967 *Physics and Chemistry of II-VI Compounds* (Amsterdam: Interscience) p 862
[12] Banishev A F, Targonsky A V, Shubny A G and Banishev A A 2020 *Izvestiya RAS. Phys.* Series **84(7)** 1010-1013
[13] Sage I, Humberstone L, Oswald I, Lloyd P and Bourhill G 2001 *Smart Mater. Struct.* **10** 332-337
[14] Belyj V I, Rastorguev A A, Ivanova E N, Larionov S V, Kirichenko V N, Bessergenev V G and Kovalevskaja Ju A 1996 Pat. of the Russian Federation No 2061015 appl. 15.10.1992, publ. 27.05.1996
[15] Tatmyshevskiy K V 2019 *IOP Conference Series: Materials Science and Engineering* **537:**022077

IOP Conf. Series: Materials Science and Engineering 1100 (2021) 012023 doi:10.1088/1757-899X/1100/1/012023

Modelling of normal pressures distribution in a polymer-made orthopaedic insole for plantar fasciitis treatment

N M Gorokhova, S Y Perepelkina and P P Kovalenko

ITMO University, 49, Kronversky str., Saint Petersburg, 197101, Russia

e-mail: gnatali1997@gmail.com

Abstract. This research is aimed at the developing an optimal design of orthopaedic insole for plantar fasciitis treatment. This article presents a brief description of the experiment performed to analyse the pressure distribution in human foot, and describes the necessary calculations made for the insole construction A mathematical model was built after the experiment and its further correction was performed, using machine learning algorithms. After the correction of the model, the polymer properties were analysed. The results of the analysis show that to perform the optimal unload of the loaded foot segments there is a need for a polymer thickness change. The respective analysis was performed and a necessary polymer thickness alteration range (1-2 mm) was introduced.

Introduction

Nowadays Plantar Fasciitis (PF) appears to be one of the most wide-spread orthopaedic diseases. Since this disease usually causes a range of problems to different age groups of people [1, 2], there is always a need for an orthopaedic device which can eliminate the source of pain and provide efficient treatment. Despite the fact that the range of treatment methodologies greatly vary [3], some methods may be insufficient for particular individuals due to their personal peculiarities related to musculoskeletal system. Unfortunately, PF can gradually deteriorate if no appropriate treatment is provided.

The main idea of the current research is to develop a specific orthopedic insole which would be able to provide the effective treatment of PF and significantly improve the life quality of people who suffer from this disease. To accomplish this task, there is a strong need for a good model which can represent the pressure distribution in the orthopedic insole and demonstrate the dynamics of this distribution. This article concentrates on the modelling of these pressures and analyses the anomalous pressures on the overloaded human foot segments. Moreover, a machine learning algorithm is used to predict "good" pressures and unload the most loaded insole parts and thus, lessen the load upon the human foot arc in general. The received data is used to change the shape of the insole relief along z axis by changing the thickness of a polymer, and to use this information in further orthopedic insole development. As a result, the provided algorithm changes the insole thickness automatically by calculating the most overloaded parts of the insole and increases the polymer thickness of particular foot segments to decrease the overall load upon the human foot.

The research methodology and experiment description

The current research is based on the theoretical calculations, where the insole is considered as a beam, divided into three equal segments with the respective percentage of the pressure distribution. The total pressures were calculated via weight coefficients, demonstrating a percentage of the pressure distribution and the initial "theoretical" pressure values calculated by [4].

Content from this work may be used under the terms of the Creative Commons Attribution 3.0 licence. Any further distribution of this work must maintain attribution to the author(s) and the title of the work, journal citation and DOI.
Published under licence by IOP Publishing Ltd

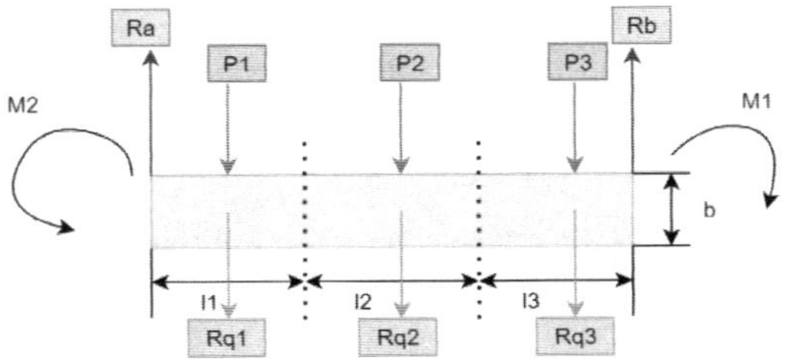

Figure 1. Simplified scheme of the insole for further calculation, where M1, M2 – acting rotational moments, P1… P3 – distributed weight, RA, RB – bearing reaction, Rq1, Rq3 – distributed forces acting inside the insole.

The three segments l_1, l_2, l_3 are equal. According to the theoretical data, the acting pressures in each segment are $\sigma_1, \sigma_2, \sigma_3$ and equal to 52.6 ×10³ (Pa), 23.5 ×10³ (Pa), 135.7 ×10³ (Pa) respectively [5]. In addition, weight coefficients were used in the calculation: c_1, c_2, c_3 ,equal to 1, 0.5 and 1. Considering these parameters, it is possible to perform the total pressure calculation:

$$\sigma_1 * c_1 = 52.6\times10^3 * 1 = 52.6\times10^3 \text{ (Pa)}, \quad (1)$$

$$\sigma_2 * c_2 = 23.5\times10^3 * 0.5 = 11.75\times10^3 \text{ (Pa)}, \quad (2)$$

$$\sigma_3 * c_3 = 135.7\times10^3 * 1 = 135.7\times 10^3 \text{ (Pa)}. \quad (3)$$

The weight coefficient c_1 illustrates the pressure distribution part in forefoot, c_2 – the pressure distribution part in middle foot, and c_3 – the pressure distribution part in rearfoot.

Based on the calculated pressure values, a model of the respective pressures distribution was built in MATLAB. Figure 2 represents a graph, where each red mark displays the pressure value in human foot, acting in each 1cm2-sized area of the specific platform [6], which will be described below.

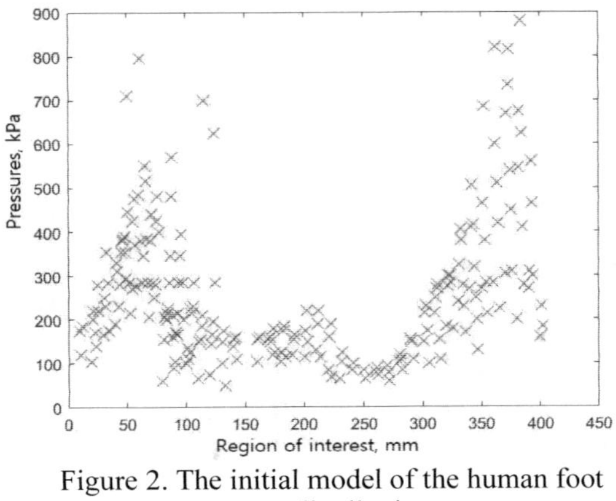

Figure 2. The initial model of the human foot pressure distribution.

Since the graph appears to be nonlinear, it is necessary to perform an approximation of the received data using non-linear algorithms. In this research an algorithm of the fifth-order polynomial regression was used [7]. The polynomial regression equation can be formed using the linear regression cost function equation which will be further minimized:

$$J(\theta) = \frac{1}{2m}\sum_{i=1}^{m}(h_\theta(x^{(i)}) - y^{(i)})^2 \quad (4)$$

where $h_\theta(x) = \theta^T x = \theta_0 + \theta_1 x_1 + \cdots + \theta_n x_n$ is the hypothesis given by the linear model.

However, the input argument is an only one parameter – the pressure values in the region of interest, which is not enough to create a sufficient machine learning algorithm and perform a good approximation. Therefore, there is a need to add more features into the model to form a fifth-order polynomial regression algorithm to provide a correct algorithm. 4 other arguments were added: the arguments were calculated using power function from 1 to 5. These features were added the following way:

$$x_1 = X, \dots x_5 = X^5 \tag{5}$$

where X is the input matrix with the argument values. Thus, the polynomial regression hypothesis for the model is:

$$h_\theta(x) = \theta^T x = \theta_0 + \theta_1 x_1 + \theta_2 x_1^2 + \theta_3 x_1^3 + \theta_4 x_1^4 + \theta_5 x_1^5 \tag{6}$$

Due to the difference in the order of the features, the normalization procedure was also performed to provide correct scaling.

$$x_{n\_norm} = \frac{x_n - \frac{1}{m}\sum_{i=1}^{m} x_n^{(i)}}{x_{n\_max} - x_{n\_min}} \tag{7}$$

where $\frac{1}{m}\sum_{i=1}^{m} x_n^{(i)}$ is the average values of x_n feature values and $x_{n\_max} - x_{n\_min}$ is the range of x_n feature values.

The cost function for the model is similar to the one applied for the linear regression.

Considering the remarked calculation, a new graph was plotted. Figure 3 shows the graph with the "learning data" – the initial data, and the resulting approximation.

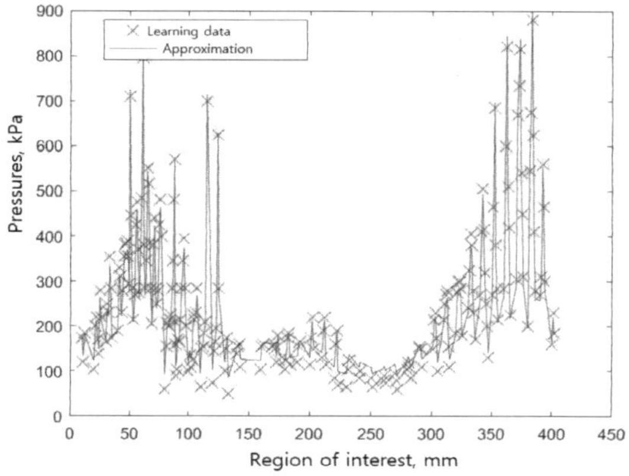

Figure 3. The model with the approximation.

Most of the research is based on the experiment, where human foot pressure distributions were investigated. The experiment can be briefly described as following. An individual was standing on the special pedographic platform by Emed for a few seconds and the sensors in the platform were registering foot pressures in each area of the platform sized 1 cm^2. A few subjects participated in the experiment, and each of them had an individual foot position and different lifestyle. The experiment is illustrated with one of the examples in Figure 4. The red lines show the foot areas and each of them contains maximum value which was used in calculations. Red and pink colors illustrate overloaded parts of the human foot, colder color show the less loaded areas respectively. The three images show the pressure distribution during statics – standing on the platform for 2-3 seconds and during dynamics – human walking and running.

Figure 4. The pressure measurements in the foot during statics (the left picture) and dynamics (the middle and the right pictures).

The calculations of the total pressures were performed via the similar method introduced above. To calculate the pressures, special weight coefficients, which define the percentage of the load distribution inside the insole, were added. The loads in dynamics are significantly higher than in statics, and therefore, there is a need to consider this in the further insole construction. The model correction and further values prediction should be performed, considering the maximum pressure values in the insole to prevent any material deformation.

Prediction model description and analysis
As a result of the polynomial regression algorithm implementation, "predicted" optimal pressure values in the insole were received. The most crucial point in the model construction was the critical pressure values in the insole. Based on each critical values, the predicted value was calculated. Figure 5 demonstrates the initial values, marked via red and predicted values, marked via blue.

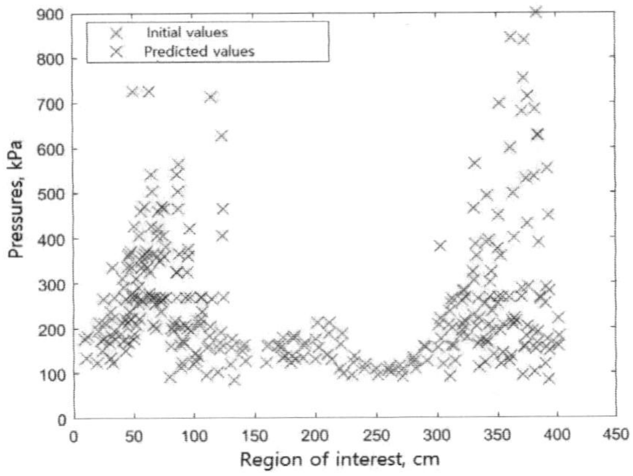

Figure 5. The model of the initial and predicted pressure values distribution

Moreover, based on the received data, the difference model was built, where the differences between the initial and predicted values are depicted. The green values on the Figure 6 illustrate these differences. These calculations may be used for further calculation of forces, acting on the insole area, sized 1 cm^2. These force values demonstrate which force a person with particular weight characteristics can produce on the insole, and how it affects different insole areas.

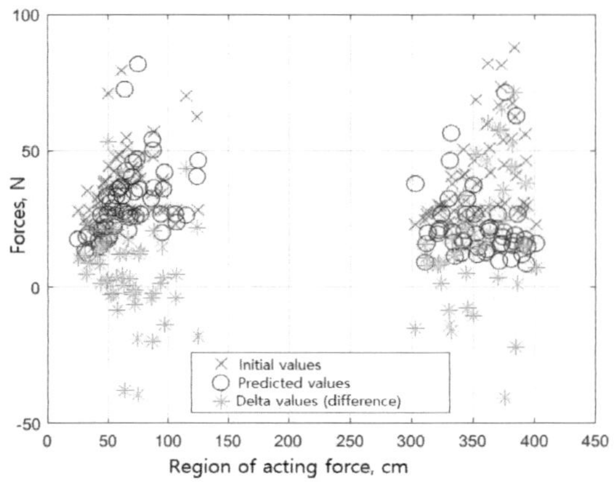

Figure 6. The model of the initial, predicted, and difference pressure values distribution

However, during the development of the insole, it is necessary to consider friction forces within polymer, which is used for the insole construction, on the intermolecular level [8]. To overcome the forces of inner friction inside the polymer, a polymer thickness alteration along z axis within 1-2 mm should be performed.

Polymer thickness alteration

The values of inner intermolecular forces of the polymer are 57.8 MPa [8]. In order to overcome these forces, it is essential to increase the force acting on the material since during the device construction a work with real material is performed, and the real material may contain not only defects but also impurities. To calculate the polymer thickness change, a few following equations should be used:

$$\sigma = \varepsilon * E \quad (8)$$

$$\varepsilon = \frac{\Delta l}{l_0} \quad (9)$$

$$\Delta l = \frac{l_0 * \sigma}{E} \quad (10)$$

The equations above demonstrate the relation between pressure and Young modulus, and the relation between the polymer height change along z axis to the initial height of the polymer. The resulting values Δl were calculated for each area, sized 1 cm^2, considering the pressure values, mentioned above. Then the received data were multiplied by the safety factor coefficient which is necessary for correct insole development. The polymer alteration is illustrated via the Figure 7.

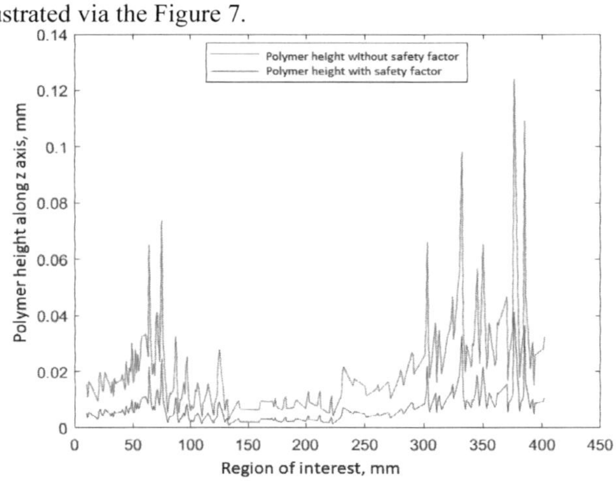

Figure 7. The thickness values with and without safety factor coefficient

The graph demonstrates the difference between the polymer thickness with and without safety factor coefficient, which equals to 3 in the calculations. The red line shows the maximum thickness of the polymer to provide an optimal unload of the insole by lifting its relief. Both lines depict so-called borders for polymer alteration to provide optimal unload of the foot arc unload. Last but not least, it is important to note that this alteration should be provided automatically within the device.

Conclusion

Thus, a mathematical model of pressure distribution was introduced and its further approximation and correction was performed. Since the main material used for the insole construction appears to be a polymer (PLA), there is a need to make a range of considerations during the device development. First and foremost, physical qualities of the polymer should be analyzed to prevent any polymer deformation. Considering this, a special model with approximation was developed, which included the initial values, predicted values and the differences between them. Secondly, necessary polymer thickness alterations were made to provide the optimal unload of especially loaded human foot segments. The range of the thickness alteration is 1-2 mm, which significantly lessens the total load influence on the human foot arc.

References

[1] Thompson, J. V., Saini, S. S., Reb, C. W., & Daniel, J. N., 2014. Diagnosis and management of plantar fasciitis. *The Journal of the American Osteopathic Association*, 114(12), 900-901
[2] Schwartz, E. N., & Su, J. (2014). Plantar fasciitis: a concise review. *The Permanente Journal*, 18(1), e105
[3] Plantar Fasciitis Treatment. Available from: https://www.transparencymarketresearch.com/plantar-fasciitis-market.html, [Accessed 15 November 2020]
[4] Cavanagh, P. R., Morag, E., Boulton, A. J. M., Young, M. J., Deffner, K. T., & Pammer, S. E. 1997. The relationship of static foot structure to dynamic foot function. *Journal of biomechanics*, 30(3), 243-250
[5] Hall S. J., Basic Biomechanics / S. J. Hall, PhD, 2015 (College of Health Sciences, University of Delaware). – Seventh Edition. McGraw-Hill Education, 556p
[6] Emed: Pressure distribution under the foot. Available from: https://www.novel.de/products/emed/, [Accessed 15 November 2020]
[7] Müller, A. C., & Guido, S. 2016. Introduction to machine learning with Python: a guide for data scientists. (O'Reilly Media, Inc.)
[8] Overview of materials for Polylactic Acid (PLA) Biopolymer. Available from: http://www.matweb.com/search/DataSheet.aspx?MatGUID=ab96a4c0655c4018a8785ac4031b9278&ckck=1 [Accessed 15 November 2020]

IOP Conf. Series: Materials Science and Engineering 1100 (2021) 012024 doi:10.1088/1757-899X/1100/1/012024

General principles of studying regularities of biological populations' dynamics for environmental protection and reclaim

V V Kiselev, I N Aleksandrova and I N Tikhonova

Ministry of Education of Stavropol Territory, branch of the budgetary educational institution of higher education "Stavropol State Pedagogical University" in Essentuki, 7 Dolina Roz street, Essentuki, 357635, Russia

E-mail: stasjairustix@mail.ru

Abstract. Currently the ecological problems, including those arising in the process of environmental protection and sustainable use of resources, are solved within the paradigm of dynamic stability of biological communities in the natural environment. This paradigm is canonic in the ecology, it was formed during almost a century. Materials on the studies of various species populations and their generalizations are presented in books and monographs on the ecology of plants, animals, general ecology. It has almost become a rule in the scientific literature to include the discussion of mathematic modeling methods into the parts on dynamics of various organisms populations. On this way we can find a lot of monographs of domestic and foreign scientists. Populations of forest insects have become especially popular with the scientists completing their comprehensive research which includes building mathematical models and providing suggestions on various corporate actions aimed at environmental protection and reclaim. The forest insects are a rewarding object not only due to their numerosity which makes it possible to apply the apparatus of differential calculus to describe the population (density), discrete mathematics to study the structure of the populations (generations, age groups, sex dimorphism, etc). Different groups of xylophages and phytophagans demonstrate breeding grounds development, pandemic irruptions with unfailing regularity, allow the researcher making controlled experiments and assessing quantitative characteristics of the population and parameters of the animal units cooperation which are so necessary and indispensible for building the mathematical models.

1. Introduction
The living systems possess a number of remarkable, functionally and structurally effectual, properties. In their systematic totality these properties are referred to as self-regulation mechanisms by the academic community. Depending on the reversibility or irreversibility of events these self-regulation mechanisms are also called adaptation or homeostasis. Despite the surprising variety and unique character of the functions of the self-regulation mechanisms they are all based on the highly similar principles of the object state regulation at all organization levels of a living being (organism, population, biogeocenotical, biospheric, including the sublevels) [1,2,3].

2. Results and discussion
When solving the problem of biocommunities stability and finding the criteria for "norm" and "pathology" as far as the structure and functioning of variously ranked natural associations are concerned the scientific research in various environmental areas brings us to the conclusion that the primary elements of the regulation mechanism ensuring the dynamic stable state of populations are a complex of reactions and acts of collective and individual behavior of organisms and their populations caused by endogenous population and exogenous environmental changes. They are very diverse by their nature: experience gained; instinctive; congenial – sexual, eating, defensive; physiological; behavioral; adaptive; reflex; with or without time lag and other possessing certain stabilizing properties and systems principles of both deviation and advance regulation. This mechanism ensures correlating the biological characteristics, properties and parameters of a population (magnitude or density, age and sexual structure, reproduction capability, survivorship rate, mortality,

Content from this work may be used under the terms of the Creative Commons Attribution 3.0 licence. Any further distribution of this work must maintain attribution to the author(s) and the title of the work, journal citation and DOI.
Published under licence by IOP Publishing Ltd

IOP Conf. Series: Materials Science and Engineering 1100 (2021) 012024 doi:10.1088/1757-899X/1100/1/012024

capability for emigration, immigration and expansion, the pronouncedness of collective and individual behavior, nurture and protection of the offsprings from the unfavorable biotic and abiotic environmental factors and so on) and the biogeocenotic capacity of habitats in the natural environment.

The suitable regulation mechanisms allow achieving such qualitative composition of organisms or species diversity and magnitude that is optimal in the asymptotic for various characteristic times or spaces and corresponds to the environmental and resource gradient [4,5,6].

The regularities of the biological community functioning and its habitus in the habitats (even land, highland, deciduous or coniferous forests, moor, steppe, river-basin, other landscape units) are a certainty, an unbiased fact [1,2].

Various interactions between the components of the community (mutually useful, useful-neutral, useful-harmful, harmful-neutral, mutually harmful) and their quantitative values which the researcher always wants to take into consideration when he or she analyses the populations to estimate the state of the community and to forecast its spatiotemporal dynamics [7], inevitably make him or her face two sets of problems.

The first one is the "curse of dimensionality" (a large amount of parameters and variables). The researcher analyses qualitative and quantitative traits and properties which are essentially incomparable. Their values are fixed on the various axes of the multidimensional space. The space is not homogenous and isotropic. Fluctuations of the value levels of all factors and resources (biotic and abiotic) are the objective reality.

The second problem is associated with the system properties of the biological community. From the viewpoint of the system analysis it is the problem of the non-linearity of the biocommunity components interaction, emergence, synergy, equifinality, robustness, opportunity of bifurcation points formation, manifestation of collective behavior and other qualities and properties that "make" the living system remain such a unit with its immanent structure and functions within some natural-territorial complex at the intervals of its characteristic time and space.

Any naturalist understands that these two problems – the "curse of dimensionality" and the system properties of the biological community – make the analysis of the biological community too difficult or even unrealizable. Multidimensional and combinatorial problems are impossible to solve in the general way. Only in some special cases the methods of the probabilistic-statistical pattern recognition with application of digital techniques and computational procedures give reasonable estimates. The complexity and hardness of research grows exponentially depending on the number of parameters and variables being analyzed. Then some idealized images of communities which are lesser dimension but significantly impoverished serve as an object of theoretical cognition of things and phenomena immanent to the communities. These images being some constituent parts of the real natural environment represent the communities.

Such examples are the research works concerning the population dynamics of forest insects [1,2,8]. The established modern concepts about the regularities of spatiotemporal development of "forest-insects" system and about the breeding grounds and pandemic irruptions of needle- and leaf-eating phytophagans and xylophages within this complex biological community were preceded by various theories of population dynamics. Let us mention some of them, those, which names allow understanding the essence of each theory: the weather theory, the trophic resilience theory; statical theory; synthetic theory; integral theory; biogeocenotic theory; theories based on the atmosphere circulation or solar activity and other [8]. The conclusion about the effects of the meteorological conditions and climate in the weathers upon the propagation of insects seems indisputable. All processes in the biosphere including the resistance of trees and that of the tree stand in general depend on these conditions. The physiological state of the woody plants determines the concentration of protective agents with insect-repellent and toxic properties which affects the population density and prevents insect epidemics [2]. The consumer eating the fodder changes its quality as the green mass accumulates the antibiotic active ingredients. The change of fodder quality and quantity makes some insects fall into dormancy, especially those faring on fruits and seeds. Population explosion is an endogenetic succession of biogeocenosis with age structure and composition of species typical for it [1].

We believe that all theories in total complement each other and do not contradict to the modern views upon the classification of the major factors of population dynamics.

We have empirically established the dependences between the density of differently resistant trees colonization with some species of insects in diverse forest types with their taxation characteristics in various geographic areas of Siberia and the most important parameters of population. The constant values and the functional dependences are taken into consideration when building various mathematical models (parametric, non-parametric, statistical, imitational, phenomenological and others) and phase portraits of the dynamics of the interacting populations [1,2,3].

The first most important fact with the far-reaching implications for estimating of the populations dynamics is that insects-xylophages have collective behavior when choosing their habitat and colonizing an individual tree depending on their resistance. So, the density of the larch bark beetle colonies on the tree trunks is limited at the top and at the bottom, its value is within 0.3 to 6.5 families per 1 dm^2 of the trunk surface.

Each following nest of the bark beetle family on the tree trunk during the xylophages emergence period is built not closer than 2-3 cm and not farther than 40-50 cm from the previously built nests being placed with as little density as possible up to the minimal one. Thus, emergence of nests on the tree trunk resembles a "progressing" wave of density. Finally the correlation between the fodder supply and the number of insects determines 1) – spatial holding of the fodder base and 2) – density of the population.

Depending on the density of the colony the empirical correlations (linear and non-linear) were established with such characteristics of species' activity (larch bark beetle, Monochamus sutor, Altaian larch borer, larch bud midge, line looper, gypsy moth, Siberian silk moth and other) as the amount of eggs laid by the female insects, survival of various stages of offspring, net reproduction, distance of insects departure from their hatch places, sex ratio and other.

When studying the location of breeding grounds of different insect species in the Siberian taiga forests which area equals hundreds of thousands hectares we defined landscape-ecological morphostructures with various risks of breeding grounds emergence. Such ecosystems, differing from each other, are seen on the aerospace photographs. Besides, hundreds of parameters and variables, causal relationships and functional dependences in the "forest-insects" system including those for particular landscapes, areas, stows, tree stands, trees colonized by the insects to a different extent and also having different physiological state reflecting their resistance to the insects were linked to the deciphering characteristics which made the basis of aerial photography materials produced at different scales. These indirect indicators were used to build the forecast system for the population dynamics of the most dangerous species for different lead time: very-long-term (for 10 years and longer), long-term (3 years minimum) and medium-term (annual). Thus, application of distance methods of forest areas sounding allowed passing from the real object to its image in the form of aerial or space image filled with various information. This information correlates to the population characteristics of forest insects, resistance of tree stands and individual trees, allows estimating the capacity of populations habitats and using for building mathematical models of breeding grounds in time and space.

The highest degree of the achieved understanding and knowledge of an object is its mathematical model including all causal relations. Although in practice even when building the imitational models the researchers leave behind many properties of the system or simplify them greatly [2]. All parametric models are simplified independently of whether they are built for one or several interacting populations.

Analysis of qualitative models is usually considered on a plane. But even under such significant limitation there are questions left concerning the amount of quasi-equilibrium states, type of critical points (node, focus, saddle, center or other critical point), mechanisms of bifurcation points emergence. In all cases an important question is at what times and spaces the structure of the community and its functions remain invariant within the activity of a higher living system [1, 8, 9].

3. Conclusion

When integrating or synthesizing the knowledge about the certain components of the community when the segregates and the studied components are united into a cohesive whole we face a difficult task to make a correct model of the whole community's behavior with consideration not only to the mentioned above specific system particularities but also to the philosophical category of quality. A good model implies integrating new knowledge into it, allows calculating the consequences of such integration and estimating the quality of the original more precisely and forecast its behavior in various situations.

The biological communities are self-oscillating systems. They are able to oscillate with the frequency of the external periodic cyclic action. Synchronization of some interacting populations densities with or without a time shift is also possible.

Solving the problems of the population dynamics may become foundational when formulating rules, regularities, laws, principles and concepts of formation and development of many biological communities interacting with each other in a living system of any level. They can be used in the practices of planning and organizing environmental protection and reclaim.

References

[1] Isaev A S, Khlebopros R G, Kiselev V V, Kondakov Y P, Nedorezov L V and Soukhovolsky V G 2009 *Euroasian Entomogical journal.* **8(2)** 3-115

[2] Isaev A S, Rozhkov A S and Kiselev V V 1988 *Fir Sawyer Beetle* (Novosibirsk:«Nauka» Publishing House Siberian Branch 271)

[3] Kiselev V V, Alexandrova I N and Tikhonova I N 2020 *General ecology part 1* (Pyatigorsk: RIA «KMV» 184)
[4] Kashkarov D N 1944 *Basics of animal ecology* (Leningrad: Publishing House Commissariat RSFSR 384)
[5] Macfadyen A 1965 *Animal ecology: Aims and Methods* (London: sir Ysaac Pirman and Sons LTD 376)
[6] Larcher V 1978 *Plant ecology* (Moscow: «Mir» 384)
[7] Odum E P 1971 *Fundamentals of Ecology* (Philadelfia-London-Toronto: Third Edition W B Saunders Company 743)
[8] *Population Dynamics of Forest Insects* 2001 red. Isaev A S (Moscow: «Nauka» 374)
[9] Volterra V 1976 *Mathematical Theory of the Struggle for Existence* (Moscow: «Nauka» 287)

IOP Conf. Series: Materials Science and Engineering 1100 (2021) 012025 doi:10.1088/1757-899X/1100/1/012025

Superconducting lines cable inputs. Thermal processes modelling

V P Firsov, A A Yakovlev and D V Golovanov

[1]New Technologies Research Center, Moscow Aviation institute (National Research University), 4 Volokolamskoe shosse, 125993 Moscow Russian Federation

E-mail: tempero.m@gmail.com

Abstract. The article considers the thermal state in the basic elements of a cryogenic energy transfer line with high-temperature superconductors. The temperature fields distribution in the heat-stressed elements of the cryogenic energy transmission line with high-temperature superconductors depends on the methods of supplying the cryogenic coolant to cable cryostats and current leads. A large contribution to the total heat flux, especially in cable power lines of short length, falls at the current leads. Computations of complex real current lead structures consist in the joint solution of the hydrodynamics and heat transfer equations for the flux, and the energy equation for the wall. The article considers a three-dimensional non-stationary motion of weakly compressible media (gas and liquid) with a free surface in a semi-closed volume at Mach numbers up to 0.3. The mathematical model is based on the Navier-Stokes equations and total enthalpy. These equations are being solved in conjunction with the equation of state for gas or liquid, the equations for the low Reynolds k-ε turbulence model, and the three-dimensional energy equation for the wall. Recently, FlowVision application package, or specially designed software are being used for complex three-dimensional thermal calculations.

1. Introduction

The thermal state in the heat-stressed elements of a cryogenic power transmission line with high-temperature superconductors (HTSC) depends on the methods of supplying cryogenic coolant to cable cryostats and current leads. A large contribution to the total heat input, especially in cable power lines of small length, is accounted for by current leads.

Current leads are designed for electrical connection of current-carrying elements of the HTSC power electric cable located in the low-temperature area with current-carrying elements of the cable industrial network. Current leads must provide the necessary current load in the power cable with minimal heat input from the environment to the low-temperature area. According to the total value of heat flows, the cooling capacity of the cryo-refrigerator in the cryostat loop is been selected.

The value of AC or DC current passing through the current-carrying elements of the current lead and cable usually does not exceed 3500 A at a high voltage (more than 20 kV) and the operating temperature of the current-carrying elements of the industrial cable network from 223 to 323 K. The modern cryostats uses liquid nitrogen or liquid hydrogen as a heat carrier. The temperature range in cryostats with liquid nitrogen varies from 65 to 78 K, and in hybrid power lines with liquid hydrogen-from 19 to 35 K.

The main coolant heating sources in the path of a closed cryopreservation system for extended power cables when transmitting AC and DC currents are:

- heat flows to the low-temperature zone through current leads determined by the design features of the current leads (the value of heat inflows on a single current lead can vary from 100 to 350 W), and not depended on the length of the cable);
- heat flows through passive thermal insulation of cryostats (in modern flexible nitrogen cryostats - from 1 to 3 W/m));
- •heat generation due to hysteresis losses during AC passage (in first generation HTSC conductors they are from 1 to 3 W / m per phase in the presence of a superconducting screen);

Content from this work may be used under the terms of the Creative Commons Attribution 3.0 licence. Any further distribution of this work must maintain attribution to the author(s) and the title of the work, journal citation and DOI.
Published under licence by IOP Publishing Ltd

- heat generation in the cable path due to irreversible hydraulic losses (channel walls friction, local resistance losses), which are determined by the cryostat mass flow rate, liquid temperature, flow path design features (roughness of the channel wall, turbulence of the flow due to twisting, local hydraulic resistances presence, etc.);
- heat release in the pumping units path (determined by the pump efficiency, mass flow rate, amount of pressure generated by the pump).

The most urgent problem is to determine the values of heat flows through current leads calculated numerically or experimentally [1].

2. Results and Discussion

The complex real structures of current leads calculations consist in the joint solution of hydrodynamics and heat exchange equations for the flow and the energy wall equation. In power HTSC systems, liquid nitrogen or hydrogen is used as a heat carrier. Recently, FlowVision and ANSYS application packages or specialized programs have been used to calculate heat flows.

A three-dimensional mathematical model was created to calculate the values of heat flows from the environment to liquid nitrogen and hydrogen passing through the flow paths of the current lead and cryostat system. The model provides solutions to the thermal conductivity equations for the wall and the Navier-Stokes equations for a flow with variable thermal properties. Three-dimensional unsteady motion of weakly compressible media (gas and liquid) with a free surface in a semi-closed volume is considered for Mach numbers up to 0.3.

The mathematical model is based on the Navier–Stokes and total enthalpy equations:

$$\frac{\partial(\rho V)}{\partial t} + \nabla(\rho V \otimes V) = -\nabla P + \nabla\big((\mu + \mu_t)(\nabla V + (\nabla V)^T)\big); \tag{1}$$

$$\frac{\partial \rho}{\partial t} + \nabla \rho V = 0; \tag{2}$$

$$\frac{\partial(\rho H)}{\partial t} + \nabla(\rho V H) = \frac{\partial P}{\partial t} + \nabla\left(\left(\frac{\lambda}{C_p} + \frac{\mu_t}{Pr_t}\right)\nabla H\right); \tag{3}$$

These equations are solved together with the equation of state for a gas or liquid, the equations for the low-Reynolds k-ε turbulence model, and the three-dimensional energy equation for the wall. A solid-state three-dimensional "volumes" geometry corresponding to the current lead actual geometry was developed to perform calculations .

The FlowVision software package allows to study the processes in the current lead and get a movement detailed view in the current leads and cables flow elements. Figure 1 shows the current lead characteristic design for the HTSC industrial frequency AC cable line (50 Hz). The structure and operating principle are discussed in detail in [1].

Figure 2 shows the upper part characteristic design area of the current lead. The vertical axis of the current lead is combined with the y axis. The FlowVision numerical algorithm is implemented on a rectangular grid with local adaptation to the third level and complex geometry subgrid resolution. This approach allows to perform calculations efficiently with a sufficient degree of accuracy, using minimal computing resources.

A multi-level adaptation of the finite-difference grid is applied to ensure the necessary accuracy of calculations . The calculation is carried out by a non-stationary method with a gradual thickening of the finite-difference grid near the input and output holes and in the zone bounded by the gradient of the Mach number change.

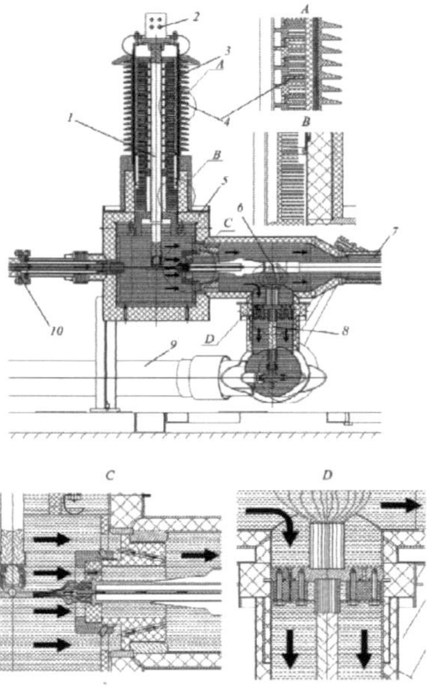

Figure 1. The HTSC current lead design.

The following initial data were used: working fluid-nitrogen (or hydrogen); working fluid initial pressure-1 MPa; working fluid initial temperature -75 K; working fluid mass flow rate -1.2 kg/s; ambient temperature -300 K. The calculations took into account the change in thermal conductivity from temperature and the convective cooling processes. The calculation results are presented in a plane passing through the axis of symmetry of the current lead.

Figure 3 shows a temperature distribution diagram over the height of the Central copper current lead and in the annular cavity of the current lead. Heat is supplied from the environment to the external side current lead cylindrical surface due to the convection of nitrogen gas located in a closed annular cavity between the current lead external surface and the electrical insulator internal surface, as well as through the upper end surface. Heat removal is provided by convection from the current lead lower part in contact with liquid nitrogen. The temperature varies from the upper boundary ambient temperature (300 K) to the liquid nitrogen temperature at lower boundary (75 K). It can be seen the gas temperature varies from 270 K in the upper section to 160 K in the lower section of the current lead. The asymmetry of the temperature field relative to the current lead axis is noted.

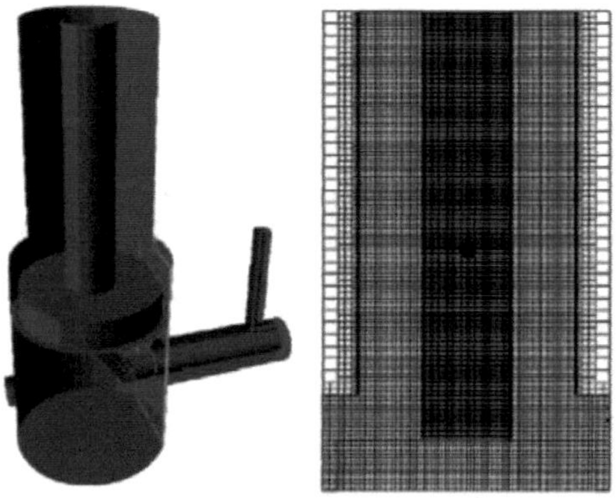

Figure 2. The current-carrying geometry in the section for hydrodynamics and heat transfer processes mathematical modeling and the design area (upper part).

Figure 4 shows a visualization of the "flash" flow. It can be noted that a system of three-dimensional vortices is formed in the annular cavity of the current lead. A helical vortex is formed in the lower zone of the current lead annular cavity. The speed of steam movement does not exceed 0.1 m / s. Non-stationary vortices transmit significant convective heat fluxes.

The value of the cross-section of the current-carrying element is determined by the results of calculations using the developed mathematical verified current lead model [3]. To solve the problem, a non-stationary one-dimensional heat equation is used, taking into account:

- current passing heat generation;
- the third kind boundary condition in the upper section (the heat transfer coefficient is set) and the first kind boundary condition in the current lead lower section (liquid and current lead temperatures equality);
- heat drains from the current lead side surface (the heat transfer coefficient is set);
- electrical resistivity and thermal properties of the material dependences on temperature.

Figure 3. The current line temperature change along the height in the Park area depending on the current load for hydrogen.

Figure 4. The "flash" stream flow visualization.

The nitrogen or hydrogen boiling curve (heat transfer and temperature pressure dependence) is used when calculating the current line section immersed in a liquid . The program was verified in experiments on nitrogen. Typical current leads calculation results for hydrogen are shown in figure 5.

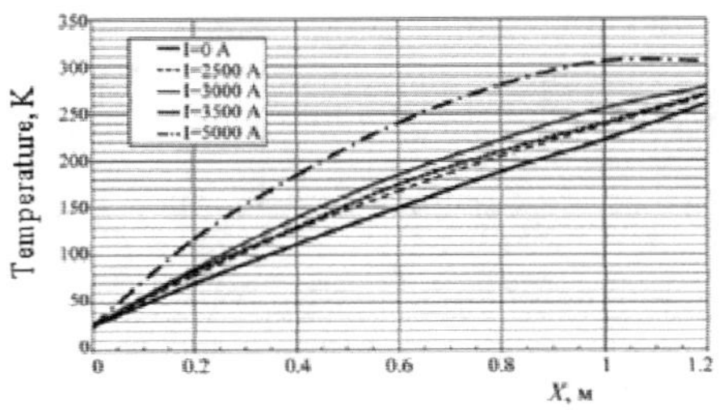

Figure 5. The current line temperature change along the height in the Park area depending on the current load for hydrogen.

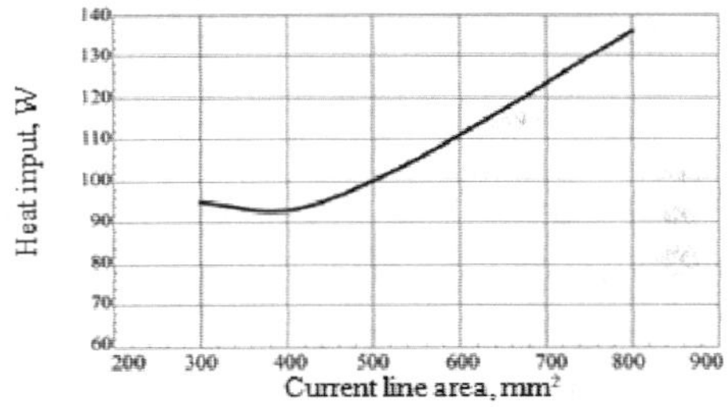

Figure 6. The number of partitions on the heat flow Influence to the liquid through the current channel.

It can be seen that if an electric current is applied, the temperature change law along the current line length changes from linear to parabolic. There is a maximum temperature and its displacement with increasing load current to the lower current lead cross-section. If a load current of 2.5 kA, the maximum current lead temperature is 265 Km.

Figure 6 shows the heat input to liquid nitrogen dependence on the current line cross-section. It can be seen there is an optimal current line cross-section. At a load current of 2.5 kA, the heat input is 145 W (through the current line directly -110 W, screen-vacuum insulation-10 W, due to convection-25 W). The heat input value is 70 W in the absence of a current load. The heat input to liquid hydrogen at a load current of 2.5 kA is 270 watts. Directly on the current-carrying element-170 W, due to convection-100 W. In the absence of a current load, the heat input value is 80 W.

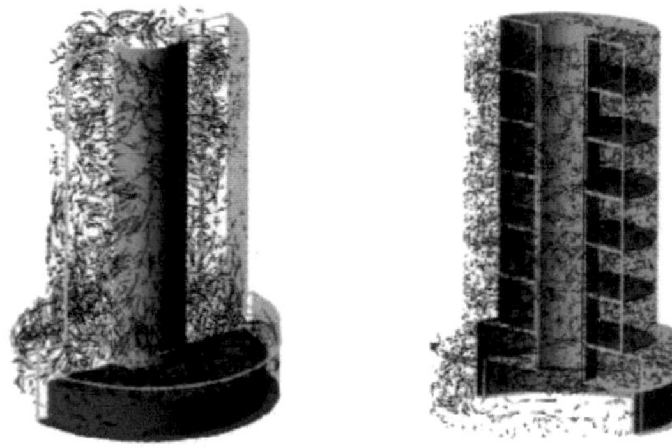

Figure 7. The convective steam movement visualization in the annular cavity of the current isolator by "flashes".

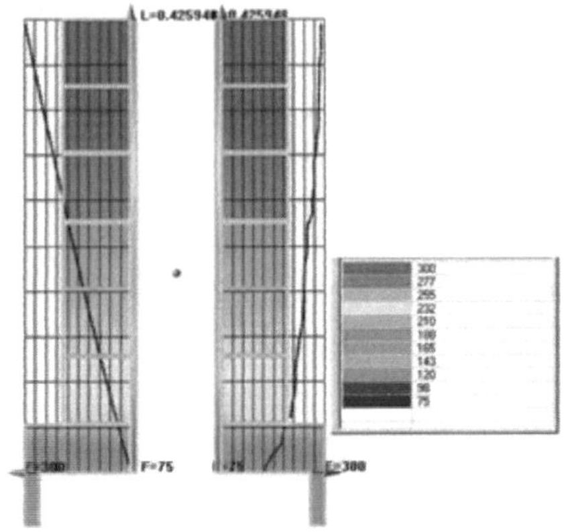

Figure 8. Temperature distribution over the Central current flow (left) and the annular gas gap (right).

The caprolon diaphragm transverse partitions installation in the steam annular current insulator cavity along the current line height significantly reduces the convective heat transfer by steam between the current lead upper section and the phase interface. Transverse partitions are installed along the current-carrying element height with no more than 1.5 mm gap between the current line and the partition. They reduce the three-dimensional non-stationary vortices scale by splitting the insulator volume annular cavity into lower height volumes. In this case, the resulting vortex structures scale in hydrogen gas is proportional to the distance between the partitions. When dividing the annular chamber space by partitions, a significant reduction in the unevenness of the temperature fields and the velocities of vortex gas flows along the current line is provided. As a result of using horizontally installed dielectric partitions, the heat flow to the device low-temperature zone is reduced by 15%. The changes in the convective steam movement nature are shown in figure 7.

Figure 8 shows the temperature distribution over the Central current lead (left) and the annular gas gap (right).

The total heat flow calculating results to the liquid through a current-flow channel with six partitions are shown in figure 9. It can be seen that heat flow is reduced by 50% with six partitions, compared to the original version. Thus, the partitions installation in the current leads allows to effectively reduce heat flows to the low-temperature zone of the HTSC power cable.

The thermal processes modeling in hydrogen current leads was also performed using the FlowVision package with gradual thickening of the finite-difference grid near the characteristic elements of the highway path. The calculation results for hydrogen current leads shows the same heat exchange features as for nitrogen ones.

The main heat input to the current lead low-temperature zone between the cable and liquid hydrogen comes from thermal conductivity through the copper current-carrying element 1 (see figure 1) and due to heat generation occurs when an electric current passes (100-110 W). Additional heat flows are supplied through the polyimide current lead current insulator housing and the current insulator 3 (see figure 1) (27-30 W). Hydrogen gas natural convection occurs in a closed current lead annular cavity without partitions between the electrical insulator inner surface, the current lead upper flange, the current-carrying element surface and the liquid, provides a heat flow of more than 250 W, since hydrogen gas has a significant thermal conductivity (≈ 0.1 W/(MK) at 20 K). The total heat flow through the current according to the calculation results will be 370-390 watts.

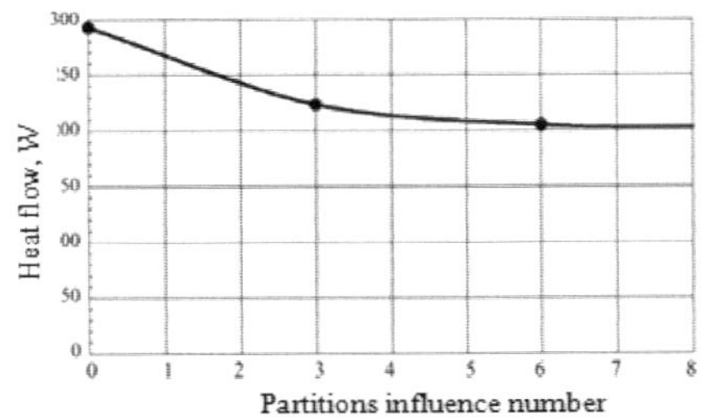

Figure 9. The number of partitions influence on the heat flow to the liquid through the current channel.

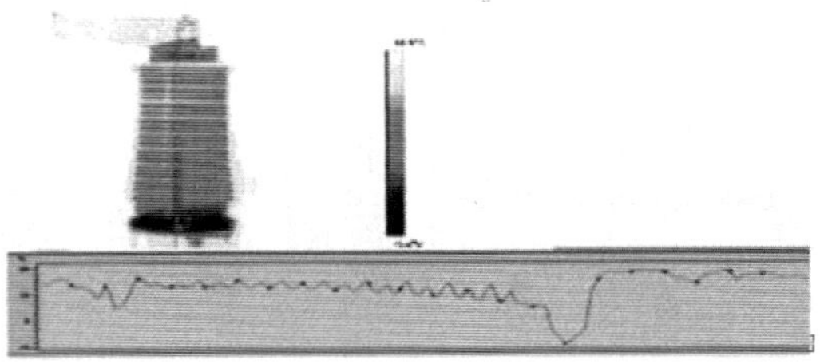

Figure 10. The temperature field measured on the silicone high-voltage insulator outer surface and obtained using the ThermaCAM P60 "FLIR systems" thermal imager.

As for nitrogen current leads, when the electric current increases, the temperature change law along the current line length changes from linear to parabolic. There is a movement of the maximum heating zone to the cold section. The current line maximum temperature reaches 50 C with a rated load current of 2.5 kA. According to the calculation results, the total heat input at a current load of 2.5 kA through the current lead with partitions will be 370-390 watts. The total heat input without electric current is 320-350 watts. Figure 10 shows the temperature field measured on the silicone high-voltage insulator outer surface and obtained using the ThermaCAM P60 "FLIR systems" thermal imager. It can be seen that the temperature of the silicone shell exceeds -10°C.

The selected design solutions to ensure electrical strength in an extended cryostat and current leads (the use of lining, etc.) excluded the possibility of breakdown both in areas with liquid hydrogen and gaseous. The test results confirmed the created structural elements high electrical strength at a voltage of up to 60 kV.

3. Conclusion

Mathematical models were developed for calculating extended high-temperature superconducting cables and their current leads. The proposed current leads mathematical models were used for calculations using the FlowVision software package and for creating an experimental device.

The caprolon diaphragm transverse partitions installation in the current insulator steam annular cavity along the current line height significantly reduces the convective heat transfer by steam between the current lead upper section and the phase interface. The heat flow to the device low-temperature zone is reduced as a result of using horizontally installed dielectric partitions,.

The selected design solutions for ensuring electrical strength in an extended cryostat and current leads eliminated the possibility of breakdown in both liquid and gaseous zones. The test results confirmed the high electrical strength of the created structural elements.

Acknowledgments

The work was supported by the Russian science Foundation, grant 19-79-10206.

References

[1] Antyukhov I V, Marinin K S, Nosov A A, Potanina L V, Firsov V P and Yakovlev A A 2018, *Kabeli i provoda – Cables and Wires*, **6(374)**, 20–30. In Russ.

[2] Kostyuk V V, Antyukhov I V, Blagov E V, Vysotsky V S, Katorgin B I, Nosov A A, Fetisov S S and Firsov V P 2012, *Technical Physics Letters*, **38(3)**, 279–282. https://doi.org/10.1134/S106378501203025X

[3] Vysotsky V, Nosov A, Fetisov S, Svalov G, Kostyuk V, Blagov E, Antyukhov I, Firsov V and Katorgin B 2013, *IEEE Transactions on Applied Superconductivity*, **vol. 23, no. 3.** https://doi.org/10.1109/TASC.2013.2238574

[4] Kostyuk V V, Blagov E V, Antyukhov I V, Firsov V P, Vysotsky V S, Nosov A A, Fetisov S S, Zanegin S Yu, Svalov G G, Rachuk V S and Katorgin B I 2015, *Cryogenics*, **66**, pp. 34–42. https://doi.org/ 10.1016/j.cryogenics.2014.11.010

[5] Molchanov A, Yanyshev D and Bykov L 2017, *IOP Conf. Series: Journal of Physics: Conf. Series* **891** 012051 doi :10.1088/1742-6596/891/1/012051

IOP Conf. Series: Materials Science and Engineering 1100 (2021) 012026 doi:10.1088/1757-899X/1100/1/012026

Development of an epoxy inhibiting primer using barium manganate (IV) as a pigment

M R Ziganshina[1], I A Ziganshin[1], V E Katnov[1,2] and A M Hannanov[1]

[1]Kazan National Research Technological University, 68 Karl Marx street, Kazan, 420015, Republic of Tatarstan, Russia
[2]Kazan Federal University, 18 Kremlyovskaya street, Kazan, 420008 Republic of Tatarstan, Russia

E-mail: vkatnov@yandex.ru

Abstract
The properties of anticorrosive coatings based on epoxy resin pigmented with barium manganate (IV) produced by ceramic method have been investigated. Electrochemical tests of coatings based on the developed primer showed inhibiting properties with respect to corrosion processes and also demonstrated that the coatings retain a high adhesion even after exposure to an electrolyte solution for 1000 hours. It has been established that the inclusion of the synthesized pigment in the coatings increases their ability to inhibit sub-film corrosion in steel.

1. Introduction
One the most common method of protecting metals against corrosion is the use of polymer anticorrosion coatings [1-3]. The main role in ensuring the protective properties of such coatings is played by anticorrosive pigments [4]. The disadvantages of pigments that give modern metal primers high anti-corrosion properties include toxicity due to the highly oxidised content of chromium [5]. The search for low-toxicity pigments comparable in protective effect to those of chromate pigments, such as strontium and zinc chromates, remains a current challenge [6-11]. Many substances may act as substitutes for chromate pigments, but zinc phosphate and related compounds [12] are currently the most common, but they are not without their disadvantages and do not provide the necessary level of anticorrosive protection at similar concentrations. For this reason, less toxic manganese (IV) compounds with oxidative and complex-forming properties, as well as chromium (VI), are of great interest from the point of view of protective action similar to chromate pigments.

In a number of previous works [13,14] we have shown that the regulation of the chemical composition of manganese pigments, including nanoscale pigments [15,16], and the correct selection of formulations for paints and coatings allow us to obtain coatings with high protective properties.
The aim was to develop and optimise the composition of an epoxy primer to protect metals using synthetic barium manganate (IV) as an anticorrosive pigment.

2. Experimental part
The application compositions were dispersed in a laboratory bead mill to a grinding degree of 25 μm, which was determined using a grindometer. The development of the primer was based on the pigment part of EP-0191 primer (TU 6-10-16-92-86) containing zinc tetraoxychromate, zinc oxide, talcum and calcite. E-41 epoxy-diane resin (TU 6-10-1316-84) cured with hardener No. 11 (TU 2332-098-21743165-97), which is a polyamide resin solution in a mixture of organic solvents, was used as a film forming base. The calculation of the composition of the primer (MiniTab programme) is presented in Table 1, and a diagram of the calculation results is shown in Figure 1.

Content from this work may be used under the terms of the Creative Commons Attribution 3.0 licence. Any further distribution of this work must maintain attribution to the author(s) and the title of the work, journal citation and DOI.
Published under licence by IOP Publishing Ltd

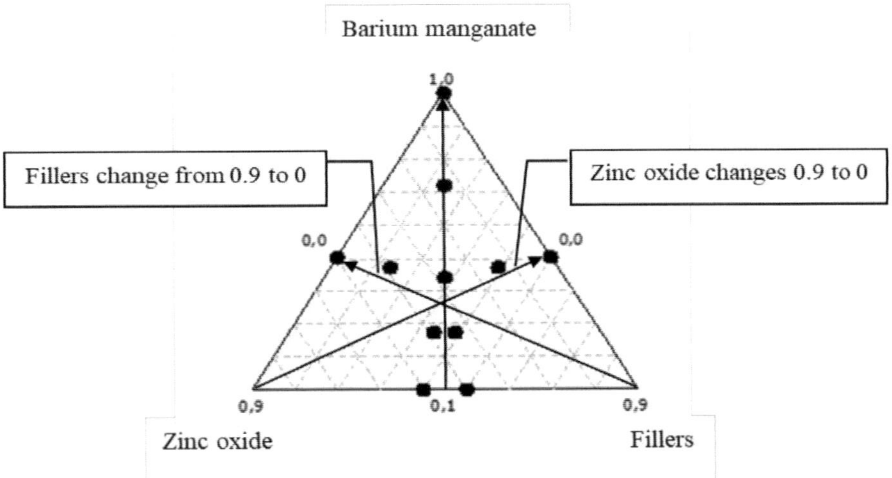

Figure 1 - Schematic presentation of the results of the primer composition calculation

Table 1. Compositions of E-41 resin-based primers received

Number	Quantity of paint composition component, %					
	Barium manganate	Zinc oxide	Talc	Calcit	E-41	Hardener
1	4	20	3.2	12.8	134.7	47.15
2	40	0	0	0	134.7	47.15
3	20	0	4	16	134.7	47.15
4	10.8	13,6	3,12	12.48	134.7	47.15
5	18.8	15,6	1.12	4.48	134.7	47.15
6	18.8	5.6	3.12	12.48	134.7	47.15
7	4	16	4	16	134.7	47.15
8	10.8	15.6	2.72	10.88	134.7	47.15
9	20	20	0	0	134.7	47.15
10	17.6	11.2	2.24	8.96	134.7	47.15
11	28.8	5.6	1.12	4.48	134.7	47.15

Coatings from the obtained primers were applied in three layer on 08kp steel samples in accordance with GOST 1050-2013 and cured for 72 hours without heat supply. The thickness of the formed coating was 30±5 microns. Primer samples in the form of free films were applied to a fluoroplastic sheet with subsequent delamination from it after the films were cured under similar conditions as described above.

Extracts of filled epoxy films were studied by linear polarization. To carry out electrochemical tests on coatings according to the method described in [17,18], an electrochemical cell was formed on their surface by gluing a glass cylinder into which a 3% aqueous sodium chloride solution was poured. Electrical capacity and corrosion potential were monitored during electrochemical tests. Coating adhesion was determined by the stripping method in accordance with GOST 32299-2013 (ISO 4624: 2002).

3. Results and discussion

In parallel, corrosion studies of water extracts from calculated compositions were carried out using the method of low linear polarization. The data obtained is shown in Figure 2, according to which the curves obtained for some compositions were close.

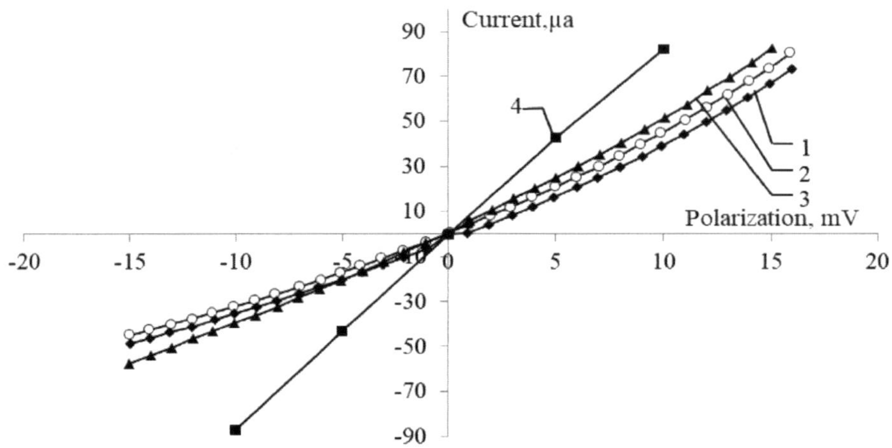

Figure 2. Small linear polarization curves of extracts of filled epoxy films: 1 - for samples 2,3,9,11; 2 - for samples 4,5,6,8,10; 3 - for samples 1.7; 4 - background solution

The data presented in Figure 2 also confirms the inhibitory effect of the synthesized pigment.

The formed coatings were electrochemically tested for 1000 hours in 3% aqueous sodium chloride solution. Contour diagrams were drawn based on the results obtained (Figure 3).

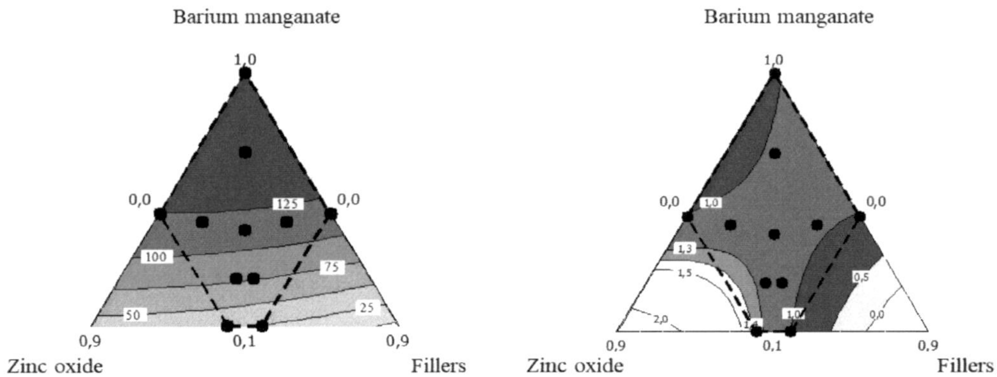

(a) (b)

Figure 3. Contour diagram of the values of electrical capacity (a) and corrosion potential from the composition of the pigment part in E-41 epoxy resin after 1000 hours of testing

In selecting the optimal pigment part composition for further research, we proceeded from the need to minimize the content of the most expensive component - the anti-corrosion pigment. The contour diagram of electrical capacitance (Figure 3a) and corrosion potential (Figure 3b) shows the optimum areas in the form of an increase in the grey color intensity corresponding to the improvement of these characteristics.

The analysis of the results shows that the optimum areas of the investigated characteristics do not coincide with each other. For the convenience of determining the optimum composition of the pigment part, values of electrical capacitance and corrosion potential have been superimposed within given limits. The results of approximation for the compositions under study are shown in Figure 4.

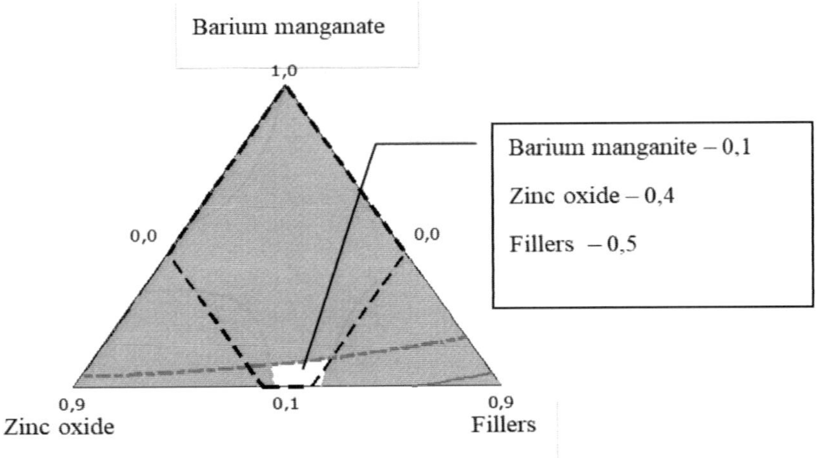

Figure 4. Areas of optimal ratio of pigment part components in E-41 epoxy resin after 1000 hours tests

The area of proportions of the pigment part, in which the pigment content is minimal but the protective properties of the coatings are preserved, is highlighted in white. Following these tests, a comprehensive assessment of the condition of the samples was carried out, the results of which are shown in Table 2.

Table 2. Comprehensive evaluation of samples after testing within 1000 hours

Number	Capacity C, nF	Corrosion potential E, mV	Corrosion area, %	Adhesion by separation, MPa before the tests	Adhesion by separation, MPa after the tests
1	1.282	17	0	4.810	4.249
2	0.938	150		3.873	3.124
3	1.153	123		5.061	4.157
4	1.215	99		6.370	4.499
5	1.261	110		6.685	4.813
6	0.955	93		7.250	4.249
7	1.369	25		4.186	3.937
8	1.196	100		6.216	4.651
9	1.411	126		5.435	4.300
10	1.454	120		5.373	5.373
11	1.411	141		4.155	3.524

The corrosion potential values for all compositions given in Table 2 are in the positive area, which also confirms the inhibitory properties of the obtained pigment.

4. Conclusion

Studies have shown that synthetic barium manganate (IV) can be used as an anti-corrosion pigment in primers to protect metals. Electrochemical tests revealed the high barrier and inhibitory properties of coatings containing the synthesized pigment. In addition, no corrosion damage was detected under the coatings after the tests, and the adhesion for some formulations remained almost unchanged. The optimum ratio between pigments and fillers was determined during the formulation of the pigment part of the primer.

References

[1] Montemor M F 2014 *Surface and Coatings Technology* **258** 17–37
[2] Grundmeier G, Schmidt W and Stratmann M 2000 *Electrochimica Acta* **45** 2515–33
[3] Vakhitov T R, Katnov V E, Grishin P V, Stepin S N and Grigoriev D O 2017 *Proceedings of the Royal Society A: Mathematical, Physical and Engineering Sciences* **473** 20160827

[4] Stepin S N, Garifullina E I, Katnov V E, Usmanov I V and Tolstosheyeva S I 2018 Properties of anti-corrosive ferrite pigment synthesized with the use of production waste *International Multidisciplinary Scientific GeoConference: SGEM* **18(6.1)** 417-23
[5] Twite R L and Bierwagen G P 1998 Progress in Organic Coatings 33 91–100
[6] Sinko J 2001 *Progress in Organic Coatings* **42** 267–82
[7] Bethencourt M, Botana F J, Marcos M, Osuna R M and Sánchez-Amaya J M 2003 *Progress in Organic Coatings* **46** 280–7
[8] Chico B, Simancas J, Vega J M, Granizo N, Díaz I, de la Fuente D and Morcillo M 2008 *Progress in Organic Coatings* **61** 283–90
[9] Forsyth M, Seter M, Hinton B, Deacon G and Junk P 2011 *Australian J. of Chemistry* **64** 812
[10] Granizo N, Vega J M, Díaz I, Chico B, de la Fuente D and Morcillo M 2011 *Progress in Organic Coatings* **70** 394–400
[11] Deflorian F, Rossi S, Fedel M, Ecco L G, Paganica R and Bastarolo M 2014 *Progress in Organic Coatings* **77** 2133–9
[12] Deyá C, Romagnoli R and del Amo B 2007 *J. of Coatings Technology and Research* **4** 167–75
[13] Ziganshina M R and Nurislamova E A 2018 *IOP Conf. Series: Materials Sci. and Eng.* **311** 012017
[14] Ziganshina M R, Peshkova M S, Stepin S N and Katnov V E 2005 *Lakokrasochnye Materialy i Ikh Primenenie* **10** 34–6
[15] Katnov V E, Stepn S N, Zganshina M R, Kayumov A A and Mendelson V A 2020 *IOP Conf. Series: Materials Sci. and Eng.* **939** 012032
[16] Ziganshina M R, Aliyew F A, Karandashov S A, Svetlakov A P and Shakurov I I 2018 Nanodimensional metal manganates (IV) in the protection coatings *International Multidisciplinary Scientific GeoConference: SGEM* **18(6.1)** 323-30
[17] Rammelt U and Reinhard G 1992 *Progress in Organic Coatings* **21** 205–26
[18] Stepin S N, Vakhin A V, Sorokov A V and Ziganshina M R 2000 *Lakokrasochnye Materialy i Ikh Primenenie* **1** 25–8

Continuous and semi-continuous industrial production of lubricants modified with graphene nanostructures

Z Alhilo[1], W Mansour[2], V Pershin[3], A Pasko[3]

[1]Post-graduate student from the Republic of Iraq, Tambov State Technical University, 106 Sovetskaya Street, 392000, Tambov, Russian Federation
[2]Post-graduate student from the Syrian Arab Republic, Tambov State Technical University, 106 Sovetskaya Street, 392000, Tambov, Russian Federation
[3]Department of Technology and Methods of Nanoproducts Manufacturing, Tambov State Technical University, 106 Sovetskaya Street, 392000, Tambov, Russian FederationRussian Federation

E-mail: pershin.home@mail.ru

Abstract. This article proposes one of the options for the production of graphene-containing suspensions and their modification of lubricants. It is based on the technology of obtaining graphene nanostructures by the method of liquid-phase shear exfoliation of graphite. The proposed technological scheme can operate in both semi-continuous and continuous modes. The main advantages of this technology are simplicity and reliability of equipment, as well as environmental friendliness of production.

1. Introduction

Currently, lubricants have high requirements regarding antifriction and anti-wear characteristics in friction units of many mechanisms and machines operating at variable and elevated speeds, pressures and under various temperature conditions. Various modifiers are used to improve the performance of lubricants. Many researchers argue that the addition of nanoparticles to lubricants, such as metal [1], metal oxide [2], metal sulfides [3], carbonate [4], borate [5], carbon materials [6, 7], organic materials [8] and rare earth compounds [9] are effective in reducing both friction and wear [10]. Friction reduction and antiwear properties are improved due to the individual characteristics of nanoparticles, for example, their

Content from this work may be used under the terms of the Creative Commons Attribution 3.0 licence. Any further distribution of this work must maintain attribution to the author(s) and the title of the work, journal citation and DOI.
Published under licence by IOP Publishing Ltd

size, shape, and physicochemical nature [11]. One of the most promising modifiers is graphene plates [12].

The use of graphene plates as modifiers of lubricants on an industrial scale is constrained by the high cost and complexity of the technology for its production. At present, graphene plates, more precisely, partially recovered graphene oxide, are produced using one or another modification of the Hamers-Ofemann method. This technology has many disadvantages. In our opinion, the main disadvantages are environmentally hazardous production, since strong acids and high energy consumption are used, since ultrasound is used to exfoliate graphite oxide. In addition, using this technology, an aqueous suspension is obtained, which contains a mixture of graphene plates with a different number of layers. To obtain the modifier, an oil is added to this suspension and then the water is removed (evaporated). As a result, an oil-based modifier is obtained, which contains 10-15% graphene plates.

The method of liquid-phase exfoliation of graphite was developed relatively recently [13]. This method is environmentally friendly, since it does not use acids and has a low cost price, since it does not use ultrasonic processing of graphite. However, the problem of switching from an aqueous suspension to an oily suspension remains. We proposed to implement liquid-phase shear exfoliation of graphite in oil [14, 15]. Thus, it is advisable to continue improving the technology for the production of a lubricant modifier based on graphene plates obtained by liquid-phase shear exfoliation of graphite in oil.

The industry uses the following operating modes of equipment: periodic; continuous; semi-continuous. The choice of the mode mainly depends on the required performance and the specifics of the technological process. As many industries continue to accelerate the current paradigm shift from batch to continuous production of materials, options for continuous production of graphene wafers must now be considered to produce new materials with improved performance.

2. Technology of continuous and semi-continuous production of lubricants modified with graphene nanostructures

2.1 Functioning principle

By graphene nanostructures we mean a mixture of single-layer, two-layer, few-layer and multilayer graphene. There are two main options for modifying lubricants with graphene nanostructures. The essence of the first option is that first, an oil suspension with graphene structures is obtained and then, on this basis, you implement the entire production cycle of a particular lubricant. In the second variant, it is necessary to make a graphene concentrate in oil and then modify the finished lubricant.

From an economic point of view, at the first stage, it is advisable to implement the second option, since there is no need to make changes to the existing lines for the production of lubricants. Moreover, the equipment that will be used in the implementation of the second option can be fully used in the first option.

Previously, we developed a technological scheme for the preparation of graphene concentrate and modification of lubricants [15]. The results of further studies have shown a number of disadvantages, which are eliminated in the scheme, which is shown in figure 1.

This technology is implemented as follows. Powder of crystalline graphite by the metering device 1 is continuously fed into the rod drum mill 2. At the same time, the metering device 3 feeds into the mill either pure oil, which is the basis for the lubricant to be modified, or a clarified suspension from the filter 9. Mass concentration of graphite is from 10 to 20%. In the mill, graphite undergoes mechanical activation and partial exfoliation, that is, exfoliation of graphite particles into graphene nanostructures. After the mill, all particles have at least one size less than 100 nanometres. From the mill, the mixture enters tank 4, where either pure oil or clarified suspension from filter 9 is fed simultaneously by metering unit 5. Tank 4 has a mixer for distributing graphene nanoparticles and graphite particles throughout the volume of this tank. From tank 4 by pump 6, the suspension is fed into the rotary apparatus (exfoliator) 7. From the exfoliator 7, the suspension enters the exfoliator 8, etc. To organize continuous production in an industrial technological scheme, more than 10 exfoliators can be used (in figure 1, these exfoliators are not conventionally shown). After the cascade of exfoliators, the suspension enters the filter 9, where untreated graphite particles and large graphene particles are separated. The liquid part of the suspension enters the filter 10, where the graphene nanoparticles and oil are separated. The clarified part of the suspension, containing a very small amount of graphene, is fed to the dispenser 3. The precipitate from the filter 10 is fed to the mixer 11, where the modified lubricant is also supplied. From the mixer 11, the mixture of lubricant and graphene nanostructures is fed into the rotary homogenizer 12. After processing the mixture in the homogenizer, the lubricant modified with graphene nanostructures is fed to the packing section.

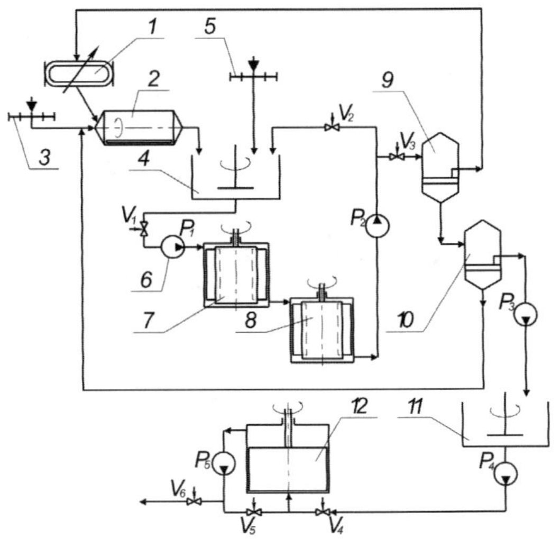

Figure 1. Functional diagram of the production of graphene nanostructures and the modification of lubricants

Despite the external complexity of the technological line for the modification of lubricants with graphene nanomaterials, all processes for processing the starting materials are purely mechanical and there are no pronounced chemical influences. During the mechanical activation of graphite in a drum rod mill 2, as well as during the movement of intermediate intermediates in high-gradient flows in the exfoliator 7 and rotary homogenizer 12, the planes of graphene plates and especially at their edges are defects. Among other things, dangling bonds can arise, due to which new chemical bonds can be formed, but there are much fewer of them than in the case of targeted functionalization of graphene, and even more so partially reduced graphene oxide.

2.2 Weight continuous feeding of graphite powder

In the presented technological scheme, in contrast to the one proposed earlier, a continuous weighing batcher is used to supply graphite powder to a rod drum mill. As in our work, we used a two-stage dosing technology. The essence of this method is that at the first stage, individual portions of bulk material with a certain weight ΔP are formed and at regular intervals ΔT are fed into a cylindrical tray, which performs circular oscillations about a longitudinal axis inclined to the horizon at an angle of 1 to 5 degrees. In the second stage, the portions are converted into a continuous flow due to the vibration of the tray. Figure 2 shows the conversion of individual batches to a continuous stream. Immediately after being fed into the tray, the portion of material has the shape shown in figure 2a. Under the action of vibration, the portion moves along the lok and simultaneously changes its shape (figure 2b). Before pouring from the vibrating tray, the portion turns into a uniform flow (figure 2c). The next portion is loaded into the tray when the previous portion is not completely emptied from the tray. The portions are connected to each other and thus a uniform continuous flow of bulk material with a given productivity is formed. The capacity of such a feeder Q is calculated as follows:

a) b) c)

Figure 2. Form transformation of a single portion in a vibrating tray

The results of experimental studies have shown that the error of continuous weight dosing of graphite powder at a productivity of 120 g / h does not exceed 0.25%.

3. Results and discussion

As noted earlier, in a real technological scheme, it is supposed to use several exfoliators sequentially. To check the effect of the number of exfoliators on the kinetics of the process and productivity of the line, we used two containers, similar to container 4 shown in figure 1. From the first container, the initial suspension was pumped with a fixed capacity into the first exfoliator (position 7 in figure 1) and then entered in the second exfoliator (item 8 in figure 1). From the second exfoliator, the treated suspension entered the second container. After the end of the suspension in the first container and the treated suspension stopped flowing into the second container, the exfoliator drives were turned off, i.e. one cycle of suspension processing ended. The treated suspension was pumped into the first tank by an additional pump and then the processing cycle was repeated. Every ten cycles, the concentration of graphene nanostructures in the suspension was determined and the productivity was calculated in terms of dry graphene. Similar experiments were performed with one exfoliator.

Figure 3 shows the dependence of the mass concentration of graphene nanostructures on the number of processing cycles. In this case, the initial concentration of graphite was 10%, the rotor speed was 10,000 rpm, the flow rate of the suspension by the pump when fed into the first exfoliator was 5 l / min. It can be seen from the graphs that when using two exfoliators (curve 2 in Fig. 3), a concentration of more than 6 mg / ml is reached approximately two times faster than when using one exfoliator (curve 1 in Fig. 3). With 30 exfoliators, this concentration can be achieved in one cycle.

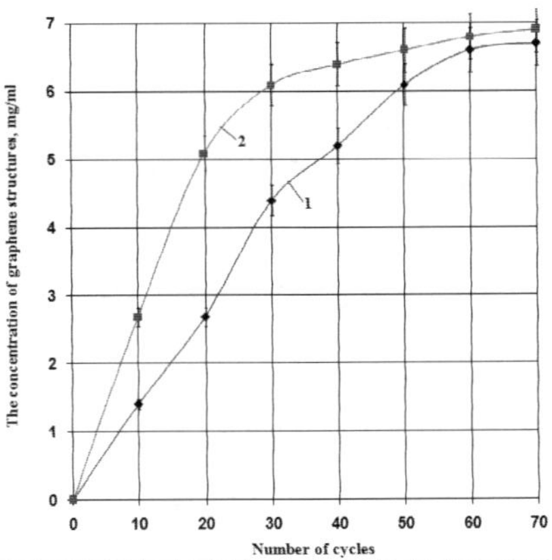

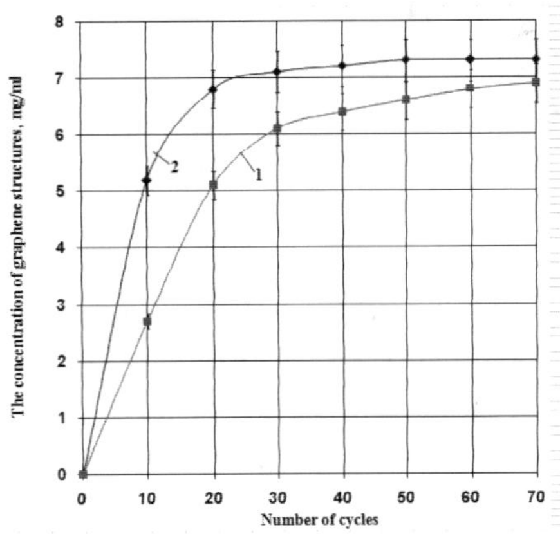

Figure 3. Dependence of the mass concentration of graphene nanostructures on the number of processing cycles, with a different number of exfoliators

Figure 4. Dependence of the mass concentration of graphene nanostructures on the number of processing cycles at different pump flow rates

If the pump power is reduced by 2 times, the residence time of the suspension in each exfoliator will approximately double. In fig. 4 shows the dependences of the mass concentration of graphene nanostructures on the number of processing cycles at a pump capacity of 5 l/min (curve 1) and 2.5 l/min. It can be seen from the graphs that to achieve a certain concentration, for example 6mg/ml, with a pump capacity of 2.5 l/min, suspension processing cycles are required approximately 2 times less than with a capacity of 5 l/min. It should be especially noted that with an increase in the residence time of the suspension in each exfoliator, productivity will not increase. This technique allows to reduce the number of exfoliators in the production line, while ensuring the required performance for graphene.

Fewer exfoliators can be used with semi-continuous operation than with continuous operation. In this case, the outlet from the last exfoliator is connected to the inlet of the first exfoliator and until a certain concentration is reached, the suspension moves in a closed loop through all exfoliators. Thus, in a continuous mode, all devices of the circuit operate continuously, and in a semi-continuous mode, the suspension is processed continuously, and the lubricant is modified periodically.

4. Conclusion

The advanced technology for modifying lubricants with graphene nanostructures can operate in both continuous and semi-continuous modes. In addition, the graphene-containing suspension after removal of graphite particles and large graphene particles (after filter 9 in Fig. 1) can be used as a basis for the production of grease. In this case, it is not necessary to modify the existing production lines, but the end product is a grease modified with graphene nanostructures.

Acknowledgments
The reported study was funded by RFBR, project number 19-38-90159.

References

[1] Liu G, Li X, Qin B, Xing D, Guo Y and Fan R. 2004 Tribol Lett 17(4) p 961
[2] Battez A H et al 2008 Wear 265(3-4) p 422
[3] Chen S, Liu W and Yu L 1998 Wear 218(2) p153
[4] Rapoport L et al. 1999 Wear 225-229(Part 2) p 975
[5] Hu Z S and Dong J X 1998 Wear 216(1) p 87
[6] Huang H D, Tu J P, Gan L P, Li C Z 2006 261(2) p 140
[7] Al-Saadi D, Pershin V, Salimov B and Montaev S 2017 J. Frict. Wear 38(5) p 355
[8] Rico E F, Minondo I and Cuervo D G 2007 Wear 262(11-12) p 1399
[9] Zhang Z, Yu L, Liu W and Xue Q 2001 Tribol Int 34(2) p 83
[10] Zhang B et al 2011 Tribol Int 44(7-8) p 878
[11] Wu Y Y, Tsui W C and Liu T C 2007 Wear 262(7-8) p 819

[12] Berman D, Erdemir A and Sumant V 2014 Materials Today, Volume 17, Number 1 p 31
[13] Paton K et al 2014 Nat. mater. 13(6), 3944
[14] Pershin V et al 2019 AIP Conf. Proc. 2101, 020011
[15] Alhilo Z, Zhumagalieva G and Pasko T 2020 MATEC Web of Conferences 315, 06005

Simulation the motion of a viscous fluid containing graphene plates in the gap between static and rotating stepped discs

Z Alhilo[1], A Baranov[2] and V Pershin[2]

[1]Post-graduate student from the Republic of Iraq, Tambov State Technical University, 106 Sovetskaya Street, 392000, Tambov, Russian

[2]Department of Technique and Technology of Production of Nanoproducts, Tambov State Technical University, 1 Leningradskaya Street, Tambov 392000, Russian Federation

E-mail: pershin.home@mail.ru

Abstract. The motion of a viscous fluid containing graphene nanostructures in a thin gap between a stationary stepped stator and a rotating stepped rotor is considered. Dependences are obtained for calculating the parameters of motion of an elementary volume of a viscous liquid. Using the Maple mathematical package, characteristic trajectories of movement of elementary volumes of fluid in the gap are obtained. A similar idea of the movement trajectories of an elementary volume was obtained using computational fluid dynamics using the FlowVision package (TESIS, Russia). The transition from a mathematical description to a computer model makes it possible to take into account changes in the viscosity of a liquid caused by a change in the concentration of graphite in the initial suspension. The use of a computer model opens the way to modeling the mixing of graphene nanostructures with a viscous liquid.

1. Introduction

The main functional purpose of lubricants is to reduce the coefficient of friction between surfaces moving relative to each other and to reduce their wear. The analysis of promising developments and studies reviewed by scientists from different countries and presented in the journal Friction and Wear over the past 35 years is given in the review [1]. The addition of nanoparticles to lubricating oils and greases significantly reduces interfacial friction and increases the load-carrying capacity of parts, which is considered a great potential as lubricating additives. Few-layer and multilayer graphene are very promising nanoadditives for lubricants. The study [2] presents the macrotribological behavior of graphene-containing oil during lubrication of steel / steel contacts. It has been experimentally proven that graphene plates as an additive to the base oil can, especially in the boundary lubrication mode, reduce friction in steel / steel contacts by up to 44%. It was shown that the predominant mechanism for lubricating graphene plates is the formation of a protective tribofilm on the contacting surfaces.

Not only low-layer and multilayer graphene, but also graphene derivatives are used as additives to lubricants. Three-dimensional hierarchical porous graphene as an additive to lithium grease was investigated by measuring the morphology and volume loss of the worn surface in three modes of motion [3]. The results show that the addition of graphene can significantly improve the friction reduction and anti-wear properties of the lubricant and thus results in a smoother wear surface. Specifically, a grease with 0.3 wt% graphene can be reduced by 52.0% and 20.3% in wear volume loss and coefficient of friction, respectively, compared to the original grease.

Interest in few-layer and multi-layer increased after they learned how to obtain these products by liquid-phase shear exfoliation of graphite, i.e. without the use of strong acids. Liquid phase exfoliated graphene sheets were modified by oleic acid and dispersed in lubricant oils as additives [4]. The tribological behaviours of graphene-contained oils were investigated using a four-ball tribometer. The lubricant with optimized graphene concentrations of 0.02–0.06 wt% showed enhanced friction and anti-wear performance, with friction coefficient and wear scar diameter reduced by 17% and 14%, respectively. The titanium

Content from this work may be used under the terms of the Creative Commons Attribution 3.0 licence. Any further distribution of this work must maintain attribution to the author(s) and the title of the work, journal citation and DOI.
Published under licence by IOP Publishing Ltd

complex grease was prepared by dispersing graphene (GN) with a proper concentration into the titanium complex grease [5]. In order to uniformly disperse it into the base grease, the GN-oil dispersion was prepared; then the GN-oil dispersion was added into the reaction mixture with the weight concentration of GN 0.02, 0.04, 0.06, 0.08, 0.1%; and then, based on the variations of the average friction coefficient (AFC) and wear scar diameter (WSD) with the additive concentrations, the grease containing the optimum concentration of GN was determined carried out with the four-ball tester. It was the titanium complex GN grease that consisted of base grease and GN with the optimum additive concen. When the addition was 0.02 wt% dosage, the AFC of Grease A and B decreased by about 13.8% and 5.0%. When the concentration was 0.06 wt%, the AFC was minimum, being reduced by 21.99% and 21.25% compared with the base greases A and B, respectively. As the amount of GN further increased, the AFC slightly increased. It indicated that the antifriction effect was weakened when the dosage of GN was excessive.

We get graphene directly in oil, which is the basis of the lubricant. The results of our experimental studies showed that the diameter of the wear spot of a lithium complex lubricant for threading from a pipe plant is 50% larger than when using the same lubricant modified with 0.1% multilayer graphene (graphene plates) [6]. Disk devices that use the energy of shear deformations in the gap between the rotating and fixed discs hold great promise for dispersing and homogenizing nanoparticles in a liquid base. For the transition from laboratory to industrial installations, a mathematical model was developed for the movement of a viscous fluid in a small gap between a moving and rotating discs [7]. This model makes it possible to determine the basic geometric and operating parameters of a device for homogenizing viscous liquids, in particular lubricants. Subsequent studies have shown that increasing the homogeneity of the mixture of the initial lubricant and the graphene modifier, it is necessary to increase the path length of this viscous liquid in the gap between the discs. The easiest way to solve this problem is to increase the geometric dimensions of the device, but this is an extensive way. We solved this problem by changing the disk configuration.

2. Simulation object and problem statement

In the previous study [7], we considered the scheme presented in figure 1. In this article, we consider an installation with a stepped rotor, the diagram of which is shown in figure 2.

Figure 1. Geometric interpretation of movement in the gap between the disks: front view with a cut (a), top view (b).

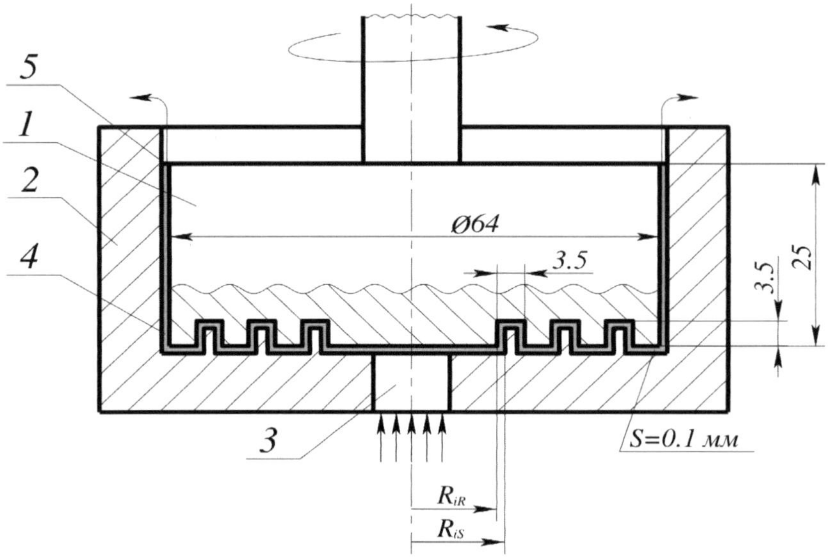

Figure 2. Geometric interpretation of movement in the gap between the disks: front view with a cut (a), top view (b).

The installation contains a cylindrical rotor 1, which is located inside the cylindrical stator 2. The lower plane of the rotor and the bottom of the stator are stepped. Lubricant containing graphite or multilayer graphene is pumped into hole 3. When the rotor rotates, the lubricant passes between the stationary stator and the moving rotor along the gaps 4 and then goes up through the gap 5. In the vertical sections, the gap between the stator and the rotor is 0.1 mm, and in horizontal sections, the gap can be changed by moving the rotor in the vertical direction. Since the speed of movement varies along the height of the gap, it practically changes from zero, near the housing 1, to the peripheral speed of the rotor 2, the lubricant layers are shifted relative to each other, homogenization occurs.

The modeling problem can be formulated as follows: find the velocity of the distribution of a viscous fluid in a stationary mode in the gap between the body and the rotor, as a function of the coordinates of the elementary volume of this fluid. Knowledge of the distribution of velocities allows not only to start modeling the homogenization process, but also to calculate the geometric and operating parameters of the industrial installation to ensure the required mode of motion.

The use of a stepped stator and rotor made it possible, without changing the overall dimensions of the device, to almost double the path of movement of the lubricant in a thin gap between the stationary stator and the rotating rotor. In addition, periodic changes in the direction of movement, during transitions from horizontal to vertical sections and vice versa, not only intensify the homogenization process due to the occurrence of turbulent flows, but also contribute to the exfoliation of graphite or multilayer graphene.

3. Mathematical description of motion

For the installation, the diagram of which is shown in figure 1, we used a cylindrical coordinate system and obtained the following dependences to describe the trajectory of motion of an elementary volume of a lubricant and graphene nanostructures that are in this volume [7]:

$$r(\tau) = \frac{\sqrt{\pi \cdot h \cdot (V \cdot \tau + r_0^2 \cdot \pi \cdot h)}}{\pi \cdot h} \quad (1)$$

$$\varphi(\tau) = \varphi_0 + \frac{\omega \cdot x \cdot \tau}{h} \quad (2)$$

where h - gap size (m), V - volumetric capacity (m^3/s), τ - current time (s), r_0 - radius for lubricant hole (m), x - distance from the considered elementary volume to the stator, m

It should be especially noted that the change in the angular coordinate (dependence 2) depends on the position of the elementary volume in the gap between the stator and the rotor. Parameter X can vary from 0 to h. For the installation, the diagram of which is shown in figure 2, the stator and rotor are stepped, therefore there are two types of sections with different patterns of viscous fluid movement. The trajectories and nature of movement on horizontal sections can be determined by dependencies (1) and (2), i.e. the movement of elementary volumes is carried out in a spiral. In vertical sections, elementary volumes move along a helical path. It should be especially noted that the pitch of the spiral decreases with distance from the axis of rotation of the rotor, but the helical lines within one section have a constant pitch. The velocity of the elementary volume along the x-axis in vertical sections can be calculated using the following relationship:

$$w_x = \frac{V}{\pi(R_{iS}^2 - R_{iR}^2)} \quad (3)$$

With this in mind:

$$x(\tau) = w_x \cdot \tau = \frac{V \cdot \tau}{\pi(R_{iS}^2 - R_{iR}^2)} \quad (4)$$

R_{iS} – the radius of the stator in the section i, R_{iR} – the radius of the rotor in the section .
The angular coordinate can be calculated as follows:

$$\varphi_v(\tau) = \varphi_{0i} + \frac{\omega \cdot (R_{iS} - r) \cdot \tau}{(R_{iS}^2 - R_{iR}^2)} \quad (5)$$

φ_{0i} – the angular coordinate at which the elementary volume began to move in the vertical section (the numbering of the vertical sections starts from the rotor axis), r - current radius at which the elementary volume is located in the vertical section

4. Results and discussion

Using solutions (1, 2) and (4, 5), it is easy to obtain the characteristic trajectories of the elementary volumes of a viscous fluid and graphene particles that are in these volumes, in the gap between a stationary step stator and a rotating step rotor with different flow characteristics and device geometry. In addition, the residence time of these elementary volumes in the gap and the path length can be calculated. Figure 3 shows a typical trajectory of movement of an elementary volume. This trajectory is calculated at a productivity $V = 0.1\times10^{-6}$ m^3/s, an angular rotor speed $\omega = 52.36$ rad/s (500 rpm) and geometric parameters that are shown in Figure 2, i.e. we used the same parameters as in the previous study [7]. It should be noted that the path length of the elementary volume of the lubricant in the gap increased by about 2 times. The movement of a viscous fluid in a small gap between a stationary stepped stator and a rotating stepped rotor was modeled by computational fluid dynamics using the FlowVision package (TESIS, Russia). For this, a three-dimensional model was imported into the program, which describes the "liquid region". In addition, the parameters corresponding to Figure 2 were corrected. Using a laminar fluid model and setting the boundary conditions at the inlet and outlet ($V = 0.1\times10^{-6}$ m^3/s) and the boundary condition of the wall on a stationary and rotating wall (stator and rotor), the velocity and pressure fields were calculated and the trajectories of the elementary volume were visualized. Figure 3 shows that in the horizontal sections the pitch between the turns decreases with increasing distance from the axis of rotation of the rotor, and in the vertical sections the pitch of the spiral within the boundaries of each section remains constant. Experimental studies on a laboratory setup with a rotor diameter of 62 mm and speeds from 300 to 1500 rpm showed satisfactory convergence of the calculated and experimental values. The choice of optimal geometric and operating parameters can be done after experimental studies of the influence of these parameters on the performance characteristics of the finished product. If we consider the process of modifying lubricants with graphene nanostructures, then the main performance characteristics are the coefficient of friction and wear.

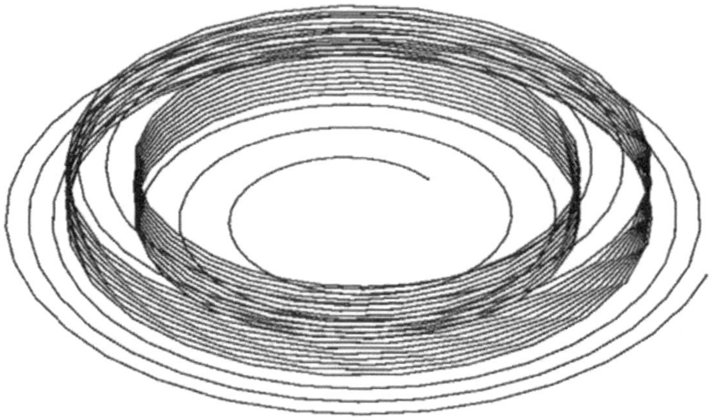

Figure 3. The trajectory of movement of an elementary volume of fluid between a stationary stepped stator and a rotating stepped rotor

5. Conclusion

A mathematical description of the motion of a viscous fluid with graphene nanostructures in the gap between a stationary stepped stator and a rotating stepped rotor has been developed. Using the Maple mathematical package, characteristic trajectories of movement of elementary volumes of fluid in the gap are obtained. A similar idea of the movement trajectories of an elementary volume was obtained using computational fluid dynamics using the FlowVision package (TESIS, Russia). The transition from a mathematical description to a computer model makes it possible to take into account changes in the viscosity of a liquid caused by a change in the concentration of graphite in the initial suspension. The use of a computer model opens the way to modeling the mixing of graphene nanostructures with a viscous liquid.

Acknowledgments
The reported study was funded by RFBR, project number 19-38-90159.

References

[1] Abdullah S F 2009 *J. Energy Environ.* 1 p 34
[2] Kogovšeka J and Kalina M 2019 Tribology International 138 p 59
[3] Ouyang T et al. 2020 Tribology International 144 106118
[4] Zhang W et al. 2011 J. Phys. D: Appl. Phys. 44 205303
[5] Niu M Qu J and Gu L 2019 Tribol Int 140:105815.
[6] Pershin V F Ovchinnikov K Alhilo Z Stolyarov R and Memetov N 2018 Nanotechnologies in Russia 13 (5–6) p 340
[7] Pershin V et al 2019 J. Phys.: Conf. Ser. 1278 012025

The study of the possibility of X-ray inspection of fiber-optic sensors embedded into the structure of a polymer composite material

G S Shipunov[1*], A S Nikiforov[1], M A Baranov[1], A A Tihonova[1] and A A Tretyakov[1]

[1]Perm National Research Polytechnic University, 29, Komsomolski av., 614990, Perm, Russian Federation

[*]E-mail: gsshipunov@gmail.com

Annotation. Currently, during the operation of structures, it is important to use the so-called systems for monitoring the state of structures, which in real time allow you to monitor changes in temperature and deformation fields. To diagnose the condition in such systems, various sensors are used, for example, strain gauges, fiber-optic, piezoluminescent. Most modern monitoring systems use fiber-optic sensors because they have more advantages and fewer disadvantages than other sensors. When monitoring the state of a structure, fiber sensors are mounted on the surface of the structure, however, when monitoring structures made of composite materials, it is possible to introduce fiber lines directly into the structure of the composite material at the stage of manufacturing the structure. When fiber lines are introduced into the structure of a composite material, a number of difficulties arise, namely, due to the fragility of fiber lines, breaks are possible both at the exit from the structure and inside. When considering polymer composite materials, the problem of the location of fiber sensors arises due to the resin flow during the process of forming structures. The solution to the problems described above can be the creation of the so-called Smart-layer, in which the fiber lines will be rigidly fixed, which will protect the fiber from breaking, and limit the movement of the fiber during the formation of structures. Thus, within the framework of this work, we investigate the issue associated with the assessment of the location of fiber sensors in the Smart layer after the formation mode. The location of the sensors after the structure is formed is investigated by radiation methods of non-destructive testing.

1. Introduction

At present, during the operation of structures, special attention in the literature is paid to the creation and implementation of the so-called Smart-structures [1-3], which in real time record the change and evolution of deformation and temperature fields, as well as the origin and evolution of defects of various types. Strain gauges, fiber-optic and piezoluminescent are used as sensors in such monitoring systems. Most modern monitoring systems use fiber-optic sensors, since they have a number of advantages: small size, flexibility, heat resistance, ease of implementation in multilayer composites, the ability to integrate into a single development network.

The Sandia National Laboratories in collaboration with Structural Monitoring System, Sykorsky, Boeing, AEM, Acellent and others, which use fiber-optic sensors and strain gauges as sensors, are implementing systems for monitoring the state of structures in carbon fiber reinforced plastic (CFRP) [4]. Sandia National Laboratories are embedding sensors in commercial aircraft to detect incipient defects since 2001, in mining equipment since 2005, and since 2016 are embedding fiber lines to US bridges to detect cracks, corrosion, and structural health monitoring. However, Sandia National Laboratories are currently using Acellent technology for embedded monitoring. Acellent has developed a monitoring system based on strain gauges imprinted in plastic. This technology makes it possible to protect sensors at the stage of implementation,

Content from this work may be used under the terms of the Creative Commons Attribution 3.0 licence. Any further distribution of this work must maintain attribution to the author(s) and the title of the work, journal citation and DOI.
Published under licence by IOP Publishing Ltd

however, the issue related to the location of sensors in structures made of carbon fiber reinforced plastic (CFRP) remains open (due to the increased temperatures of molding products).

As part of the implementation of the state assignment «Development of a system for monitoring the state of structures made of polymer composite materials for modern aircraft engines based on built-in fiber-optic sensors», the team of PNRPU developed a prototype of a system for monitoring the state of a structure, which includes: the object of monitoring; fiber-optic sensors (FOS); interrogator required to obtain data on the deformation of the monitoring object; software that processes data from FOS. When the developed monitoring system was introduced into the outlet guide vane by radiation methods of non-destructive testing, it was found that the fiber lines change their location relative to the initial one during the molding process. This is due to the fact that the vane pen is curvilinear and different in thickness, and during the molding process, due to the movement of the resin, the fiber-optic lines are displaced (Figure 1). It should be noted that the introduction of FOS into the structure was difficult due to the fragility of the fiber during the insertion process, as a result of which one of the ends of the fiber lines broke off.

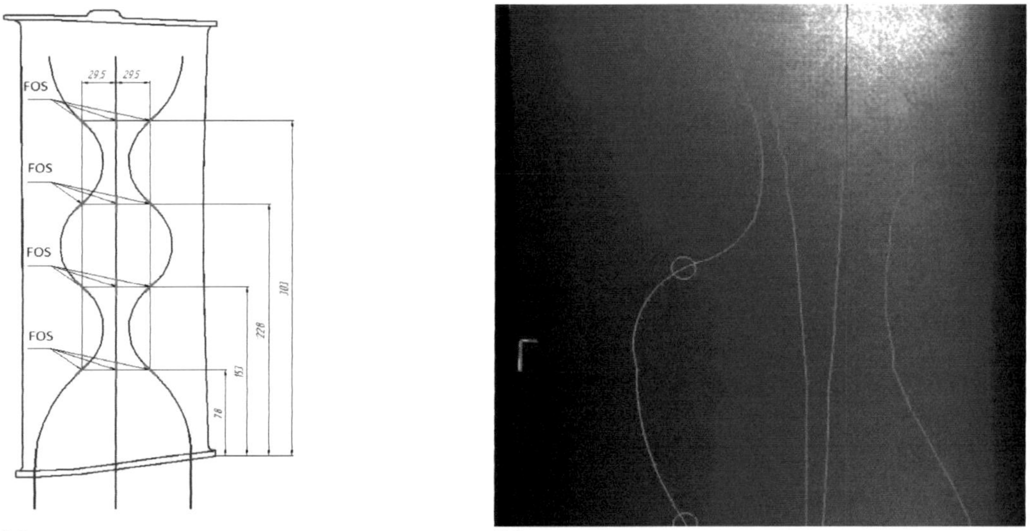

(a) (b)

Figure 1. Initial FOS placement (a), FOS placement after forming mode (b).

Thus, it was established that it is necessary to control the location and protection of fiber sensors when introduced into a CFRP structure. When solving the described problems, prototypes of Smart-layers of various configurations were created, which make it possible to protect the optical fiber and outputs during the laying of the fiber into the structure. Within the framework of this work, the question of the location of the Smart-layer with fiber lines after the mode of forming the structure from CFRP was investigated. A plate with the curvature of the corresponding outlet guide vane was chosen as the object of research.

2. Making a curved plate

To conduct research to identify the location of fiber lines in the CFRP, two Smart-layers were made. The first configuration included a polyurethane-based Smart layer (melting point 200 °C) into which a fiber line was embedded. The second layer consisted of polyamide (with a melting point of 220 °C). When choosing materials, it was taken into account that the molding temperature of CFRP structures is 180 °C, therefore, the material from which the Smart layer is made should have a melting point above this mark.

It should be noted that at this stage, the Smart layer manufacturing technology has a significant drawback - the displacement of the optical fiber relative to the central position inside the layer, which is associated with insufficient tension and fixation of the fiber during the formation of the layer. Insufficient tension and fixation of the fiber in the tooling contribute to the longitudinal displacement of the fiber during the movement of the melt (in the case of using polyamide and polyurethane) during the formation of the layer, which leads to a wave-like deformation of the optical fiber. It is planned to solve this problem by modifying the technological equipment, a system of pre-tensioning of the fiber-optic line.

A curved plate with a curvature corresponding to the feather outlet guide vane was made of carbon fiber on the basis of a unique scientific installation «research complex for scientific and technological research in the field of creating products from polymer composite materials». At the manufacturing stage, Smart-layers were laid between layers 5 and 6, as shown in Figure 2a. Such stacking of Smart-layers is justified by the area of greatest curvature, in which the maximum displacement of fiber lines in the vane was recorded. The plate after the forming mode is shown in Figure 2b.

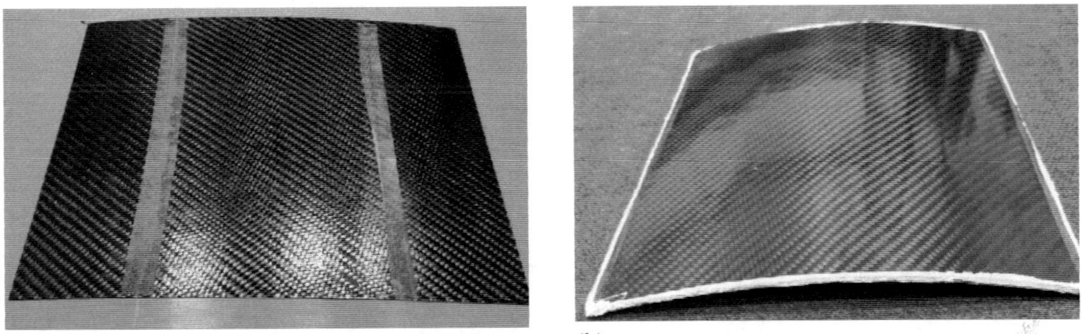

(a) (b)

Figure 2. Making a curved plate: (a) - embedding Smart-layers, (b) - plate after the molding mode.

The next stage of the study was to conduct radiation non-destructive testing in order to identify the location of fiber lines after the forming mode.

3. X-ray non-destructive testing

Radiation non-destructive testing was carried out on a portable microfocus X-ray unit. The X-ray source is made in a monoblock design with a focal spot size not exceeding 50 microns. A flat-panel indirect conversion detector with a pixel size of 140 μm was used as an X-ray image detector. The detector consists of a matrix of thin-film transistors and a scintillator (phosphor) layer based on cesium iodide with a thickness of 500 μm. Figure 3 shows a general view of the X-ray unit with the investigated curved plate.

Figure 3. General view of the X-ray unit with the studied plate.

Figure 4 shows an X-ray image of the investigated curved plate with embedded Smart layers.

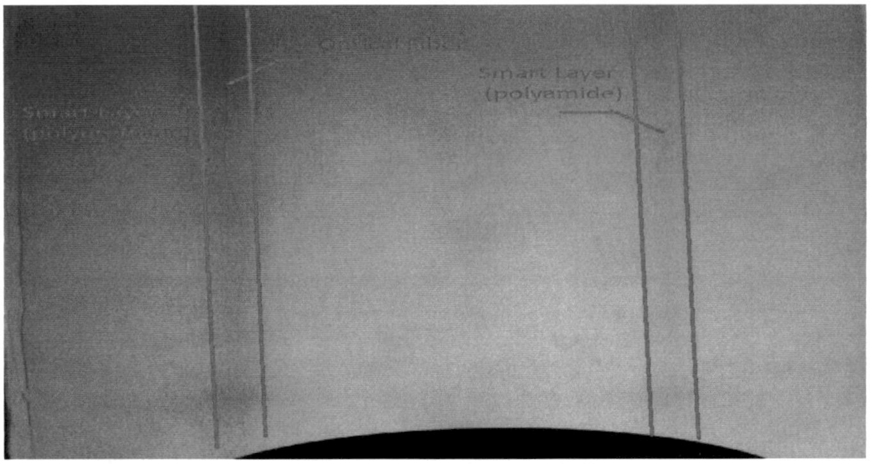

Figure 4. X-ray image of a curved plate.

Thus, in the course of X-ray non-destructive testing, it was found that the developed Smart layers make it possible to solve issues related to the location of optical fibers when they are introduced into a curved structure directly in the Smart layer. It should be noted that the displacement of the optical fiber relative to the central position inside the layer is associated with insufficient tension and fixation of the fiber during the formation of the layer, which will be eliminated in the near future.

4. Reference

As part of the study, Smart-layers were manufactured and embedded in a CFRP sample, radiation non-destructive testing of a curved CFRP plate with embedded fiber optic sensors, with a curvature that corresponds to the outlet guide vane pen, was carried out. Within the framework of radiation monitoring, it was found that the embedded Smart layers do not change their position relative to the original packing in a curved plate. The next stage of the study is the refinement and implementation of the Smart-layer directly into the outlet guide vane structure made of CFRP.

Thus, when introducing FOS into a structure, it is recommended to use Smart-layers, since they allow safe (without fiber damage) implementation and ensure the required location of sensors in the structure to control the area under study.

Acknowledgments

The results were obtained during the fulfillment of the state task of the Ministry of Science and Higher Education of the Russian Federation for the implementation of fundamental scientific research (project No. FSNM-2020-0026).

References

[1] Fu T, Liu Y, Li Q and Leng J 2009 *J.* Optics and Lasers in Engineering 47 1056-1062.
[2] Motwani P, Perogamvros N, Taylor S, Sonebi M, Laskar A and Murphy A 2020 *J.* Sensors and Actuators A: Physical 303 1-13.
[3] Guemes A, Fernandez-Lopez A and Soller B 2010 *J.* Structural Health Monitoring 233-245.
[4] Qing X, Beard S, Armita K and Chan H 2006 *J.* Composites Science and Technology 66 1694-1702.

The use of polymer composite materials in cylindrical gears

V M Medunetsky, V A Zenkov, M V Abramchuk and S Yu Perepelkina

ITMO University, Kronverksky Prospekt, 49, St. Petersburg, 197101, Russia

E-mail: vm57med@yandex.ru

Abstract. The paper describes the use of PCM-polymer composite materials in the construction of cylindrical gears. Various types of composite materials and their characteristics for gear links of cylindrical gears, in particular with an asymmetric profile, as well as shell forming matrices for their manufacture from PCM are considered. The main issues of ensuring the accuracy of PCM gears for assessing their quality level are presented.

Keywords: polymer composite materials (PCM), gears, shell forming matrices, precision gears made of PCM.

1. Introduction

Currently, there is a steady trend of using structural polymer materials (PCM) for the manufacture of gears in the instrument mechanisms of mechatronic devices for various purposes. The advantages and disadvantages of composite and plastic gears are well known [1-4]. The advantage of polymer composite products is to reduce friction, and as a result, reduce wear, as well as reduce weight, increase durability and reliability, which directly affects the cost of the product. This is reflected in detail in [5-9].

The advantage of toothed links made of composite materials over metal ones is the fact that they have less friction losses. Gears made of structural polymer materials are lower noise, retain their inherent lubricity, are smaller in weight and are corrosion-resistant.

It can be noted that in many structures of mechanisms, gears made of such materials are used with reverse, and the so-called dead stroke occurs due to the presence of gaps in the joints and elastic deformations [10]. Thus, there is an objective need to significantly reduce the dead stroke of the transmission.

In some cases, there is a need to increase the load capacity of gears from PCM.

2. Polymer composite materials. Properties, applications, databases

Composite materials (composites) – multicomponent materials consisting, as a rule, of a plastic base (matrix) reinforced with fillers that have high strength, rigidity, etc. The combination of dissimilar substances leads to the creation of a new material whose properties differ quantitatively and qualitatively from the properties of each of its components. By varying the composition of the matrix and filler, their ratio, and the orientation of the filler, a wide range of materials with the required set of properties is obtained. Composites, in which the matrix is a polymer material, are one of the most numerous and diverse types of materials. If you combine the matrix and filler, calculate different proportions, you can easily get materials much better than "traditional", while they are generally much easier than ordinary reinforcement. The properties of polymer composite materials (PCM) are becoming more accessible and are already being used not only in the field of space technologies, but also as a more accessible basis for household appliances.

Composite components can be plastics, almost all metals, polymer fibers, etc. There are even more complex composites – polymatric ones, if they use several different polymers for the matrix, as well as hybrid variants, in which reinforcing fibers are combined.

The density of PCM varies widely from 400 to 2800 kg/m^3 and averages 1400 kg/m^3. Metals are much heavier, for example, the density of steel – 7800 kg/m^3, copper – 8900 kg/m^3, aluminum – 2700 kg/m^3. Thus,

Content from this work may be used under the terms of the Creative Commons Attribution 3.0 licence. Any further distribution of this work must maintain attribution to the author(s) and the title of the work, journal citation and DOI.
Published under licence by IOP Publishing Ltd

the density of the polymer composite is 5-6 times lower than that of ferrous and non-ferrous metals and alloys, and 2 times lower than that of aluminum.

PCM based on glass fiber and basalt fiber are excellent dielectrics. The electrical resistivity of fiberglass is 1.0×10^{10} Ohm·m. The specific resistance of steel at normal temperatures from 13 to 30×10^{-8} Ohm·m, which is 18 orders of magnitude less. At the same time, carbon plastics conduct an electric current, which can also be used in engineering.

Polymer composite materials with a glass or basalt filler as dielectrics do not break down in electrolytically conducting media, i.e. they do not corrode in the ground, water, under the influence of atmospheric precipitation, etc.

With its low density, PCM has high physical and mechanical characteristics.

So, the tensile strength of carbon steels is 240 MPa, aluminum alloys – from 50 to 440 MPa, polymer composites from 70 MPa (fiberglass with a minimum content of glass fiber) to 1800 MPa (high-strength carbon fiber). The coefficient of thermal expansion of polymer composites is significantly lower than that of metals and simple polymers, which allows the use of PCM products in a wide temperature range.

PCM refers to materials with low thermal conductivity. So, the thermal conductivity of steel is 64 W/m °C, aluminum – 105-200 W/m °C, and fiberglass – 0.75 W/m °C.

The variety of PCM in composition and properties has led to the creation of many local databases in practice, and the complexity of forming materials and developing technologies for their manufacture (in the context of designing specific products) has led to the creation of numerous modeling programs based on mathematical models. There are also regulatory documents to help you work. For example: "GOST R 56806-2015 Polymer Composites. Identification of polymer composites in electronic databases".

Today there are extensive databases of polymer composites:

1) Total Materia of the Swiss company Key to Metals AG [11];

2) Material Data Center of the German developed by M-Base Engineering + Software GmbH [12];

3) Plastinfo.ru localized for processors in Russia and CIS countries material Data Center database of the Russian Plastinfo LLC.ru [13].

International database of polymer materials Material Data Center. It combines the full range of characteristics of polymer materials offered worldwide. The data in the database includes characteristics such as fairly detailed curves of properties depending on temperature and / or time. It is also possible to superimpose such curves for different materials on a single diagram, which provides a visual comparison of the manufacturability of materials or their thermal and mechanical properties [14].

3. Use of polymer materials in gears

It is advisable to consider the characteristics of such materials, on the basis of which it is possible to form composites for use in gears.

The most commonly used materials for the manufacture of gears are ABS and PLA plastics. With the help of additive technologies, you can get products with the specified parameters, but the accuracy of printing leaves much to be desired.

The main disadvantage of this technology is a small resolution, both horizontally and vertically, which leads to a more or less noticeable layering of the surface of the manufactured model. It is also worth noting that traditional software solutions do not fully meet the quality requirements, since various defects can be formed during the printing process, which negatively affect the subsequent operation of finished products.

Casting in forming matrices allows you to get products of almost any complexity and avoids the occurrence of porosity and the formation of surface defects. In this way, you can make complex shaped cast parts, including cone-shaped gears. Finished products can have various physical and mechanical properties-rigid, imitating the properties of ABS, semi-rigid, elastic, heat-resistant, transparent, as well as combined.

The condition for reliable operation of a gear train made of composite polymer materials is the maximum compliance of operational requirements with the properties of the selected material. The high load determines the choice of hard rigid plastics, in particular reinforced thermoplastics.

Research shows that polymer materials are very actively used in various fields of technology [15].

Currently, various technological, physical-mechanical, tribological and chemical properties of polymer materials are used for the manufacture of gears, such as polyamide and its variety – polyamide 6, better known as caprolon, and fluoroplast, both pure and with various fillers.

Polyamides have high strength, hardness, elasticity, wear resistance, heat resistance.

Caprolon is in addition to all the above has a high tensile strength. This material has a low coefficient of friction in a pair with any metals, it is well and quickly worked out, 6-7 times lighter than bronze and steel, instead of which it can be used. It is highly technologically advanced.

Composite materials based on them are also known: glass-filled, mineral-filled.

Polypropylene belongs to the class of polyolefins. Melting point 160-176 °C; thermal degradation begins at 300 °C; density 0.90-0.92 g/cm3; shrinkage 1.3-2.4%. Distinctive properties of polypropylene: resistant to water (up to 130°C), to acids and alkalis, increased impact resistance, resistance to bending loads; good wear resistance, the maximum operating temperature of polypropylene products is 120-140 C. [16]

The second most common polymer - fluoropolymer

Fluoroplast has an extremely low surface energy and can therefore be used as an anti-adhesive material.

It is resistant to aging under normal conditions and has high anti-friction properties, an exceptionally low coefficient of friction (in certain conditions and pairs, the coefficient of friction is up to 0.02). [17]

Compositions based on fluoroplastics are also widely used when fillers are introduced into the fluoropolymer that increase the wear resistance, strength, hardness or elasticity of products made of fluoroplastics.

As fillers for fluoroplast compositions, materials that can withstand the sintering temperature of the fluoroplast are used. The most common fillers can be divided into the following groups:

a) powdery;

b) fibrous (reinforcing fillers);

c) reinforcing fillers of the frame type.

When using glass fiber, silica, asbestos fabric, and metal wool as fillers, the composite stiffness increases, and the relative deformation decreases at low friction coefficients [18].

Gears can be made from polymer materials with improved properties. Whether they are used as part of combined wheels with metal or combined bushings, polymer materials after high-precision processing allow you to get a long service life without lubrication in the packaging, food, medical and semiconductor industries.

The advantage of gears made of composite polymers over metal ones is not only reduced noise level, but also greater efficiency (efficiency) due to lower friction losses.

The design of the polymer gear largely determines the performance of the gear. The chosen ratio of the crown dimensions, the reliability of the interfaces, and the location of the reinforcing elements depend on the manufacturing accuracy and strength of the transmission. Polymer gears are divided into all-polymer and reinforced according to their design [19].

4. PCM in non-standard gears

To increase the load capacity of the gear transmission, as well as to increase its service life, the following addition to their design can be proposed - to make the profile of the teeth of the rims asymmetric. This option should be used only if the torques are different when changing the direction of rotation [20]. This is illustrated in figure 1, which shows asymmetric involute teeth formed from two main circles dba and dbb.

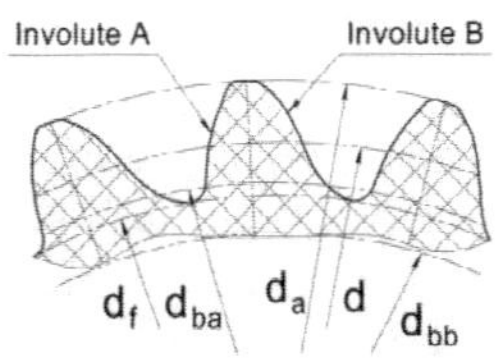

Figure 1. Cross-section of a gearwheel with an asymmetric tooth profile.

The studies have shown that to ensure a high-quality working surface of the proposed gears, it is advisable to make them from composite thermoplastics by casting them into shell-forming matrices [21, 22], and to fill the matrix with reactoplastics, for example, from phenol-formaldehyde and epoxy polymers. Thermosets have a fairly great stability of physical and mechanical properties when changing influences from the environment. Due to their strength and electrical insulation properties, as well as the ability to operate at high temperatures and in any climatic conditions, phenopolasts are successfully used for the manufacture of structural, friction and anti-friction products, housings [23]. In this case, the internal surfaces that form the cavity should be made of metal in the form of a shell.

This can be done if the toothed crown of metal is made from a thin ribbon by bending it along the contour of the master model. In this capacity, it is recommended to use cold-rolled tape made of low-carbon steel, which is often used when bending parts. It is worth noting that on the inner (adjacent to the workpiece) side of the metal tape, it is necessary to create a microrelief with the specified functional parameters, on the other surface - with sufficient roughness for a strong connection with the material of the forming matrix. An increase in roughness leads to an increase in the adhesive and cohesive strength.

Currently, it is advisable to make a master model for forming the shell on a 3D printer. This design of the forming matrix is shown in figure 2. It should be noted that in order to increase the bending strength of the teeth, in some variants, reinforcing disks with a metal toothed rim of gear should be used. This option is shown in figure 3.

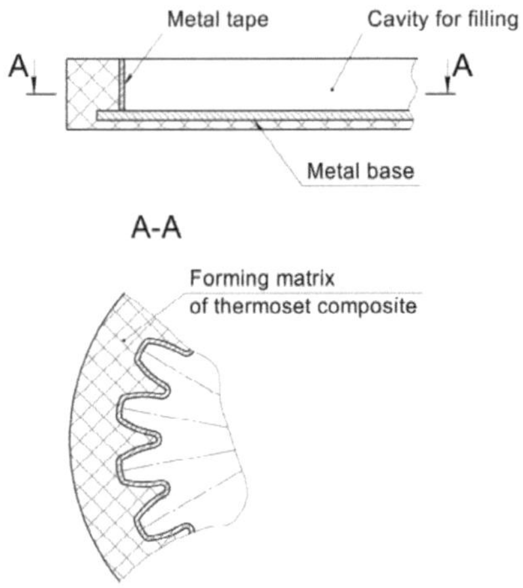

Figure 2. Fragment of the forming matrix.

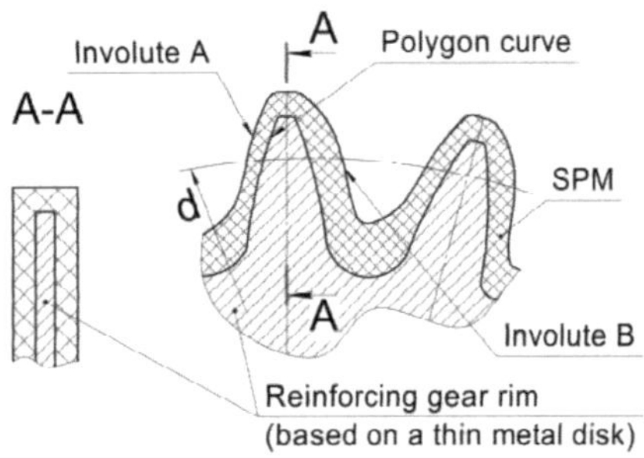

Figure 3. Cross-section of a toothed reinforced rim of gear with an asymmetric external profile (the internal reinforcing rim of gear is symmetrical).

In the above figure 3, the teeth of the rim of gear are made of composite material and are asymmetric in the profile, and the reinforcing metal rim of gear is made symmetrical.

5. Evaluation of the accuracy of non-standard gears from PCM

Given the fact that PCM products have the property of shrinkage and warping (uneven shrinkage), accuracy should be highlighted among the quality indicators for evaluating gears. For gears, accuracy is the main indicator due to the limited materials used for manufacturing the transmission and the thermochemical treatment of gears.

In the Russian Federation, according to the GOST 1643-81 standard, there are 12 degrees of accuracy, 6 types of coupling are set for gears with a tooth modulus m ≥ 1 mm: A, B, C, D, E, H and 8 types of T_{jn} tolerance for the backlash: x, y, z, a, b, c, d, h. The accuracy of manufacturing the gear and transmission is set using the degree of accuracy, and the requirements for the backlash are set by the type of coupling according to the standards of the backlash.

For cylindrical gears that are mainly involute, taking into account mounting errors, the values of the minimum and maximum kinematic error are determined by the expressions:

– for degrees of accuracy 3–6: $F'_{i0\min} = 0,62K_s\left(F'_{i1} + F'_{i2}\right)$.

– for degrees of accuracy 7–8: $F'_{i0\min} = 0,71K_s\left(F'_{i1} + F'_{i2}\right)$.

For all degrees of accuracy (the same applies to other gears)

$$F'_{i0\max} = K\left[\sqrt{\left(F'_{i1}\right)^2 + E^2_{\Sigma M_1}} + \sqrt{\left(F'_{i2}\right)^2 + E^2_{\Sigma M_2}}\right],$$

where K, K_s – phase compensation coefficients, i.e. coefficients that take into account the degree of change in the tangential composite deviation from the initial position of the links; F'_{i1}, F'_{i2} – tangential composite deviations of the driving and driven gears, respectively; $E_{\Sigma M_1}$, $E_{\Sigma M_2}$ – the total reduced installation error (in GOST 21098-82 is determined by Appendix 2).

IOP Conf. Series: Materials Science and Engineering **1100** (2021) 012030 doi:10.1088/1757-899X/1100/1/012030

Tolerance for the tangential composite deviation of the gear: $F_i' = F_P + f_f$,

where F_p is the tolerance for the accumulated error of the gear pitch, f_f is the tolerance for the error of the tooth profile.

The value of the minimum kinematic backlash of the gear transmission:

$$j_{t\min} = \frac{j_{n\min}}{\cos\alpha\cos\beta},$$

where $j_{\text{n min}}$ is the allowance backlash (normal); α is the pressure angle of the standard basic rack tooth profile; β is the helix angle.

The value of the maximum kinematic backlash for a cylindrical gear transmission:

$$j_{t\min} = 0,7\left(E_{Hs_1} + E_{Hs2}\right) + \sqrt{0,5\left(T_{H_1}^2 + T_{H_2}^2 + 2f_a^2 + G_{r_1}^2 + G_{r_2}^2\right)},$$

where $j_{\text{t max}}$ – the maximum circular backlash in the gear transmission, E_{Hs} is the smallest additional bias source circuit; T_H – tolerance on the basic racks tooth profile modification; $\pm f_a$ is the center distance limit error, G_r is the radial clearance in the bearings of rotation, is equal to the radial runout (in GOST 21098-82 is defined in Annex 2).

You can also use the probabilistic method for calculating the accuracy - the Monte Carlo method, in which the components of the tangential composite deviation and backlash will be random values in certain intervals [25-26].

6. Conclusion

Thus, the considered options and features of the manufacture of non-standard cylindrical gears from polymer structural materials, for example, such as those with asymmetric reinforced wheel teeth, make it possible to improve their quality in practice. The given characteristics of various materials for the considered structures of cylindrical gears provide an opportunity to rationally design them and form technologies for their manufacture, in particular, by using shell shaping matrices. The above questions of the accuracy of PCM gears make it possible to assess their quality level.

References

[1] Ghosh R, Ghosh S, Srivastava T, & Barman R N 2017 *International Journal of Engineering and Technology* **9(2)** 1438-51.

[2] Chatterjee S, & Gupta K 2018 *Journal homepage: http://iieta.org/Journals/AMA/AMA_C* **73(3)** 79-83.

[3] Pawar P B, Abhay A Utpat 2015 *Materials Today: Proceedings* **2(4-5)** 2968–74.

[4] Thirugnanam A, Sathish J, & Rakesh L 2014 *Middle-East Journal of Scientific Research* **20(8)** 966-68

[5] Mao K, Greenwood D, Ramakrishnan R, Goodship V, Shrouti C, Chetwynd D & Langlois+ P 2019 *Wear* **426** 1033-39.

[6] Catera P G, Mundo D, Gagliardi F, & Treviso A 2020 *International Journal on Interactive Design and Manufacturing (IJIDeM)* 1-16.

[7] Catera P G, Mundo D, Treviso A, Gagliardi F & Visrolia A 2019 *Applied Composite Materials*, **26(3)** 817-33.

[8] Gauntt S M, & Campbell R L 2019 *AIAA Scitech 2019 Forum* 0146.

[9] Singh A K, Yadav S & Singh P K 2019 *Advances in Engineering Design* 269-77

[10] Dong Y Q & Yang X C 2006 *Journal of Nanchang University (Engineering & Technology)*, 3.

[11] Total Materia [Internet]. The most extensive database of materials in the world. Available at: https://www.totalmateria.com/page.aspx?ID=Home&LN=RU. Accessed: 28 Oct 2020.

[12] Material Data Center [Internet]. Data on materials and their applications. Available at: https://www.materialdatacenter.com/mb/. Accessed: 28 Oct 2020.

[13] Base of polymer materials [Internet]. Material Data Center. Available at: https://plastinfo.ru/m-base/. Accessed: 28 Oct 2020.

[14] Kryazhev Yu A, Andreev M V, Shituk A A 2018 *Innovations in mechanical engineering* 491-97. (in Russian).
[15] González-Henríquez C M, Sarabia-Vallejos M A & Rodriguez-Hernandez J 2019 *Materials, methodologies, and biomedical applications. Progress in Polymer Science* **94** 57-116
[16] Goh G D, Yap Y L, Tan H K J, Sing S L, Goh G L & Yeong W Y 2020 *Critical Reviews in Solid State and Materials Sciences* **45(2)** 113-33
[17] Friedrich K & Walter R 2020 *Structure and Properties of Additive Manufactured Polymer Components* ed Klaus Friedrich and Rolf Walter (Woodhead Publishing) p 435
[18] Directory of polymers [Internet]. Polymer materials: products, equipment, technologies. Available at: http://www.polymerbranch.com/catalogp.html. Accessed: 28 Oct 2020.
[19] Karapetyan A N, Scharr G 2007 *Proceedings of the NAS RA and SEUA: Technical Sciences* **60(2)** 266-71. (in russian)
[20] Starzhinsky V E, Shalobaev E V, Shilko S V 2012 *Elements of Drive Devices. Calculation, Design, Technology* (Minks: Belaruskaya Navuka). 769 p. (in Russian).
[21] Starzhinsky V E, Shilko S V, Shalobaev E V 2018 *Polymer Materials and Technologies* **4(2),** 6-31 (in Russian).
[22] Starzhinsky V E, Shalobaev E V 2005 *Progressive technologies in machine and instrument engineering NSTU-API* 28-29. (in Russian)
[23] Vitkalova I A, Torlova A S, Pikalov E S 2018 *Scientific Review. Technical sciences* **2** 15-28. (in Russian).
[24] Medunetskiy V M 2002 *Ensuring Quality Indicators of Combined Cylindrical Gears* (St. Petersburg: Polytechnika) 160 p. (in Russian)
[25] Abramchuk M V *2014 Improvement of calculations of accuracy parameters of gears, gearings and multi-link gear mechanisms (Abstract of the thesis of Doctoral dissertation)* (Saint Petersburg: ITMO University) p 20 (in Russian)
[26] Abramchuk M V 2018 *Journal of Instrument Engineering* **61(2)** 118-22 (in Russian)

IOP Conf. Series: Materials Science and Engineering 1100 (2021) 012031 doi:10.1088/1757-899X/1100/1/012031

Impact of globalization and English language on communicative strategies of industrial advertising

A A Isakova

Tyumen Industrial University, 38 Volodarskogo Street, Tyumen, 625000, Russia
E-mail: IsakovaAA@yandex.ru

Abstract. Все статьи The following article is devoted to the influence of globalization and English on communicative strategies of the industrial advertising. Any person has to be guided in a powerful stream of information, defining their attitude to any realities. Tyumen regional industrial advertising is one of the forms of mass communication. The relevance of this study is determined by the daily life, man is obliged to a greater degree of self-oriented among urban stream media in order to adapt to the conditions of everyday reality. Thus, the aim of the research is a scientific assessment of influence of the industrial advertising text on the ordinary consciousness of the different layers and cohorts of the population of the region. It is useful and interesting for other regions and countries. The main idea of the article is prove that English is occupied by leading position in the system of European multilingualism and industrial advertising. The influence of language on the formation of tastes, ideas, industrial policy and communication of industrial advertising is also revealed. Thus, the word globalization, English language and advertising strategies influence and form industrial communication.

1. Introduction

A person is forced to independently navigate a powerful industrial information flow, defining his attitude to certain realities. The Tyumen region is a rich oil and gas territory. The decision of many economic problems of Russia depends on development of petroleum deposits. In this regard, the Tyumen region is interesting with a well-developed industrial infrastructure, a constantly growing population due to huge flows of migrants from countries near and far abroad. Since the end of the last century, the study of the communicative space of the region has become a priority thanks to the fundamental development and formation of the northern region as a "pilot" site of the country's labor resources. The state control and intensive development of information communications and urbanization processes promote the expansion of interethnic contacts and involve population in some uniform integrated process.

Dramatic changes of the socio-economic life, structure and pandemic of Tyumen region society are accompanied by previously unprecedented changes which cause serious changes in the existing communicative field. The change of population size to a higher or lower side, the change of population density, migration, birth and death, marriage and divorce, emigration and immigration have been influenced by flows of communications [1]. The process of technical communication is an interesting linguistic phenomenon. It is man's vision and his inventions in anthropocentric reflection. That is why, there is an abundance of units associated with human life, with functions in the development of the human person, in reflection of the speaker and listener language consciousness in the world of technical communication.

Communication of industrial advertising should be considered as a process of human interaction with the advertising environment, carried out on the basis of specific ways and means of including advertising in the production process.

The semantic and formative core of this process is culture as a concrete historical personification of the adaptive strategy of mankind.

Communication of industrial advertising should be considered as a process of human interaction with the advertising environment, carried out on the basis of specific ways and means of including advertising in the

Content from this work may be used under the terms of the Creative Commons Attribution 3.0 licence. Any further distribution of this work must maintain attribution to the author(s) and the title of the work, journal citation and DOI.
Published under licence by IOP Publishing Ltd

production process. . The semantic and formative core of this process is culture as a concrete historical personification of the adaptive strategy of mankind.

2. Theory and practice of industrial advertising communication and impact of globalization and English language on it

As you know, the concept of "communication" came into use in the early 70s of the last century thanks to the emergence of the book "Menschliche Kommunikation" by psychotherapists Paul Vaclavik, Don D. Jexon and John Bivin [2]. Communication is a phenomenon inherent i any human community at any stage of its development. Among the main trends in the development of society, it is now accepted to disseminate technologies that facilitate any communication, labor, and academic mobility of cultural communities [3].

Communicativeness means the semblance of the learning process and the communication process only according to the main features, which represent the motivated nature of speech activity, the presence of certain relationships between the interlocutors, discussion of topics, the use of speech means in the real process.

The purpose of our research is to comprehensively study the concept of "communication" and determine its place in the linguistic space of industrial advertising on the example of the Tyumen region.

Methodology. The methodological basis of the work is made up of fundamental studies of the philosophical concept and the relationship between culture and morality, culture and its communicative environment.

2.1. Industrial advertising communication of the Tyumen region

Language studying of industrial advertising is not only scientific, but also practical interest for researching. The linguistic situation breaks linguistic and speech norms, changing very approach to the culture of the language. But also, there is a certain speech chaos, where a new linguistic taste is formed.

Successful communication in any language can be carried out only if the inphone knows the lexical and grammatical knowledge of the learned language in conjunction with information about the relevant cultural standards of the people who speak this language. In particular, solving the initial problems is significant for verbal communication in general, for intercultural communication, A certain minimum of information about the basic elements of national linguistic consciousness is required in order to successfully communicate between participants belonging to different linguistic and cultural communities. The emergence of new objects requiring identification, economic reform, a change in the ethnopolitical situation, the revival of folk traditions, an appeal to the historical past, national cultural values have been embodied in toponymy, which led to the activation of the process of streamlining the names and renaming of toponymic objects [4]. Analysis of the language helps us to understand its speakers: and not only what they say, but most importantly, their vision of the world and, in part, culture [5]. Linguistic research of trademarks and brands is also relevant today.

The main attention is paid to consumer goods, industrial ones are practically not analyzed. An associative experiment serves as a marker of the explanation of the formation or unformed idea of advertising. At the same time, there is a process of spontaneous, unregulated influx of foreign words into the Russian language, since it is closely related to the activation of business, scientific, trade and cultural ties between peoples, introduction to the international layer of vocabulary, often created on the basis of Germanic ones (which is explained by the modern processes of globalization) and Turkic languages (as a result of close contacts with the local Turkic and newcomers from Turkey, Iran, Kazakhstan and the Caucasus).

Advertising is considered one of the main indicators of the borrowing and adaptation of foreign vocabulary in the Russian language. Regional advertising is a form of mass communication, since any person, stepping on the Tyumen land, joins in communication, looking at billboards in the city.

Various classes of industrial brands have denotations in the practical sphere of human activity. They are conventionally combined under the term "commercial naming". From a linguistic point of view, trademarks refer to a special vocabulary. Unlike other words, special words remain incomprehensible without exact knowledge of specific objects designated by them. The innovation driven by the ad unit borrowing process is gaining momentum. The question of their adaptation to the language system arises sharply.

The search of language optimizing problem requires to use the achievements of modern philology, linguistics, and psychology. Lack of knowledge of native and foreign languages limits the possibilities of professional communication in all spheres of life. As A.M. Solovyov rightly notes, "language is something that helps us understand each other, since it is, first of all, a means of communication [6]. In modern science, a large amount of advertising has been accumulated, which is extremely heterogeneous both in its origin and in education. Specificity can be traced in the brands themselves at all language levels. We can conclude that industrial brands are initially semantically "devastated," semantic filling occurs either viewing commercial advertisements or reading catalogs of industrial brands.

The study of all media is currently quite relevant, but it is a real fact that firstly many words and expressions are used in advertising, and then they are later fixed in the language and communication. Therefore, advertising reflects not only social processes, but the communication processes.

A feature of industrial advertising communication is a high informative richness and a pronounced pragmatic intention, which makes advertising texts as the richest material for studying of speech mechanisms and communicative strategies in difficult conditions of modern communication [7]

Advertising is aimed at the consumer, so its means should "force" a person to purchase a particular product. All levels of advertising work from font to verbal text. One of the tasks of the advertisers is to form a positive attitude among the potential consumer towards the advertised product, and one that would be deposited in the long-term memory of a person [8].

The industrial advertising slogan refers to an international layer of vocabulary, more often created on the basis of Germanic (which is understandable by the modern processes of globalization) and Turkic languages (as a result of close contacts with the local Turkic population). In addition, we see the expansion of these languages into a variety of language areas: from everyday communication to advertising space in the region.

A special sphere of communication with a peculiar Russian language was formed along with the development of the advertising business in Tyumen. Many Tyumen residents use the following expressions: fresh in the office (Airfresh is air conditioner), copy (instead of copying), I like the French (buying a Renault car), my Toyotochka (affectionately calling the Toyota car), etc.

In this regard, it is necessary to have the maximum possible accurate information about the real language and sociocultural situation in advertising in the Russian province, which is currently in the process of formation.

The purpose of the study is also a scientific assessment of the suggestive influence of an advertising text on the everyday consciousness of various strata and cohorts of the region's population. Consequently, the emergence of specific facts of linguistic culture can ultimately be stimulated by the cultural development of society.

However, it is necessary to take a fact that the new foreign-language name of the product is just a quasi-word for a Russian person creating advertising of a particular industrial product, so Tyumen advertisers are forced to look for ways to adapt these words in the communicative space of Tyumen, and, in particular, in the mental consciousness of Tyumen citizens.

2.2. Impact of globalization and English language on communicative strategies

It is a well-known fact that globalization is a process of unifying the whole world by having the same goods, styles, and technologies. Every day we come across the fruits of this process which is greatly influenced by such great states as the United Kingdom and the USA.

Globalization must be seen not only as a way of doing business or running financial markets, but it became a process. For example, the British service sector deals with its customers through a call center in India, or a manufacturer designs its products in Europe, makes them in south-east Asia and sells them in north America.

There are several examples when the UK and the USA have had a great influence on the world economics, politics, culture, promoting their initiatives, movements, way of life, technologies. For example, during the 1960-s Presidential campaign John Kennedy promoted the idea of “peace army” dedicated to helping the peace in the world. In March 1961, Congress of the USA set up the Peace Corps. The new program sought to build friendship between Americans and other nations. Americans spent several months learning the language, history, and culture of the country where they would serve. Peace Corps volunteers have served and are still serving in Africa, Asia, Latin America, in Russia. They bring enthusiasm and hope to people who are struggling to improve their standards of living.

Industrial advertising has always been part of American life. In early colonial times many merchants often peddled their goods in the streets. They advertised by calling out to passersby, in addition, they placed printed advertisements outside shop windows. Today the United States has the largest advertising industry in the world and advertisers spend more than 97 billion dollars each year. Today, mass advertising persuades people of many different nationalities to use the same products. Millions of people around the world drink the same soft drinks, drive the same cars wear the same clothes, and eat the same hamburgers. We should say that the European and American advertising explicit such an archetype as: the main person in the world is I, then the most important for me are my desires, my pleasure, my success, without doubt, as I am the best. Genetically modified food is actual in our diet. The first genetically modified plants appeared at the market in 1982. The main producers of new cultures are the USA, the Argentine and Canada. Among genetically modified products are corn, sugar beet, cotton and rape, resisting to pesticides, potatoes resisting the Colorado beetles.

One of the serious problems is that gene engineering thinks only about profit not about health of people. This is one of the bitter fruits of globalization [9].

But what is the role of the English language in the process of globalization? Cultural diversity of Europe means the existence of language diversity which provides the principle of equality of all the languages. The spread of languages means that any European must study and strive for comprehending of the languages of the neighboring countries. The most widely spread language still remains English – 58% of respondents think so, then French – 11%. Practically, the English language occupies the leading position, though it has been counted that 40 million of the European Union citizens use their regional language or the language of ethnical minorities. Hence, the European Union appreciates study of, at least, two foreign languages besides the native language. Different programs have been created. English plays a dominant role in media, inter-state communication and broadcasting.

So, English is considered the global language, but a lot of scientists ask the question: Will English stay the global language or become a minority tongue? Of course, there are changes taking place that may affect the global dominance of English. In the 20th century western countries helped the spread of English as there were the strongest economics. In the past English borrowed new words from other languages. Today the worlds of communication and the Internet are rich sources of new vocabulary. The most characteristic features of any language, its vocabulary is dynamic change, improvement [10]. New words appear all the time. Borrowings is seen on TV, radio, press, industry, advertising, and other sources.

New economic realities demand the usage of authentic texts containing English and American words. English and American words are taken from industry, computing, sport. Still great amount of English words in modern European languages proves the fact that the English language has not been weakened as far as the influence of English is concerned.

English and American don t lose their positions also because scientists have developed a new educational philosophy, which has become extremely popular around the world. The educational philosophy “Whole Language” started spreading in the USA for teaching adults and students. “Whole Language” includes the four categories: linguistic, programmed, commercial, pedagogical and political. Linguistic category includes teaching all kinds of speech activities through discussions of problematic situations, exchange of opinions, discussion of books, writing different kinds of composition, evaluation of actions and characters [11].

Trademarks can exist without advertising, but advertising can’ t exist without trademark. Different graphic, stylistic, morphological devices create such advertising, which has its aims to draw attention, raise interest, develop desire, initiate action to buy.

So, English and American trademarks are holding strong position in the global trade. Some classical goods in the UK and the USA have traditional motives, names and surnames, for example: names and surnamesж Ford, Lincoln, Mercedes; names of animals: Jaguar, Cheetah, Scorpion, Spider, Skyhawk; names of cosmetics containing flora: Lilac Lady, Night Fuchsia, Rosebuds; names of historical associations: Benjamin Franklin, Cardinal, Napoleon, Oracle, Caesar s Palace; geographical names: Fuji, Malibu, Seven Seas Lodge, Acapulco Pools; some trademarks show functions of goods: Comet, All, Tasty, Friend, Speedwriting; the principal trademarks of a man’s perfume in England contains the hint about strength, brutality: Man s Man, British Rogue, Black Night, Hugo Boss, James Bond, Demon, White Hunter .

Due to linguistic devices advertising forms standard associations in the whole world, usually with life and activities of a famous person:

- ✓ Mercedes-Benz – a prestigious car
- ✓ Rolls-Royce – a luxurious car
- ✓ Volvo-a safety car
- ✓ Ferrari – a sportive car
- ✓ BMW – a comfortable car
- ✓ Volkswagen – a family car
- ✓ Nescafe Classic – quality and accessibility
- ✓ Kodak – the chronicle of America.

English catchphrases are extremely popular with all the nations of the world. These catchphrases are sometimes translated into Russian languages and are used side by side with other set expressions.

1. Any time, any place, anywhere- from the famous advertisement of drink Martini, popular in 70-s, the 20th century.
2. Be my guest – the title of the book by a famous hotel owner Conrad Hilton;
3. Business as usual – a usual sign on a door of a shop after repair or fire, but became popular due to W. Churchill, who used it in his speech in 1914 "The maximum of the British people is "Business as usual".
4. The customer is always right – the phrase belongs to Gordon Selfridge, a well-known owner of department store "Selfridges".
5. Don't shoot the pianist – mentioned by O. Wilde in his book "Impressions of America" as the sign in some saloons in Rocky Mountains" Please, do not shoot the pianist. He is doing his best"
6. Good…to the last drop!" – the motto of the trade mark "Maxwell House", told by President Theodore Roosevelt.
7. Is he one of us? – the phrase was said by the former prime minister Margaret Thatcher, it belongs to a group of people.
8. Make love, not war! – a famous motto of 60-s, which was against war in Vietnam.
9. Not in front of the children – not to argue or quarrel in the presence of children.
10. Show must go on! – the phrase was born among circus artists in the 20-30s, the 20th century. The same name film was released, and I. Gershwin and G. Kern wrote the song with the same title. Later the leader of the rock-group "Queen" Freddy Mercury sang the same name song before his death.

All above-mentioned facts prove that English and American have not lost their functions as global languages, and English and American way of life influence greatly on the global processes. The motivating potential of our language is a major process of the speech activity and the communication.

3. Results

Analysis of the results of the experiment revealed some features of the perception of industrial brands and their advertising. It can be assumed that the reason for the discrepancies is the linguistic and conceptual picture of the world of respondents in verbal associations. In the minds of the audience we interviewed, there is an unconscious switch from industrial brands to consumer brands.

The promotional advertising text has a greater suggestive component than a detailed analysis of technical characteristics in specialized catalogs. We have to admit that consumer interests prevail over professional ones. The data obtained during the study provide an opportunity to confirm our hypothesis about the contamination of meanings, i.e. inadequate perception of respondents of a professional orientation.

The process of the formation of new technical nominations and advertising is an interesting linguistic phenomenon, which anthropocentrically reflects a person's vision of himself and his inventions. That is why there is an abundance of units related to everyday communication in the world of technical nomination around us. They should be studied according to their role for a person, according to their purpose in human life, according to their functions in the development of a human personality, and according to their reflection in the linguistic consciousness of the speaker and listener. The industrial advertising landscape of Tyumen is one-sided communication directed towards a person - a consumer of information, while a consumer does not have the ability to choose the language of information, his reaction can only be in the positive, negative or neutral assessment. The language of industrial nomination and advertising is the basis of generalization and a tool of thinking and means of speech communication. Language is more than just a mean of communication. It is a mean of critical discussion, a discussion as objectivity of arguments in all sphere of industry and science.

4. Conclusion

Thus, the word globalization, English language and advertising strategies influence and form industrial communication. English is occupied by leading position in the system of industrial advertising. The influence of language on the formation of tastes, ideas, industrial policy and communication of industrial advertising is also revealed. The interaction of English languages is clearly manifested in the process of naming service objects and private enterprises. Among the huge number of new nominations, the names of food products and the names of mechanisms occupy the first place. Every day, a person encounters many of them in various media: television, newspapers, magazines, catalogs, posters, video clips, ads and other samples of advertising products.

The process of technical communication is an interesting linguistic phenomenon. It is man's vision and his inventions in anthropocentric reflection. That is why, there is an abundance of units associated with human life, with functions in the development of the human person, in reflection of the speaker and listener language consciousness in the world of technical communication.

The following research is dictated by everyday life because a person is forced to adapt in everyday reality. Along with the development of the advertising business in Tyumen, a special sphere of industrial communication was formed on the basis of the advertising language of the Tyumen region.

Reference

[1] Isakova A A 2019 Computer analysis of brand functioning in Tyumen region. *Int. Sci. and Technol. Conf. "EastConf", EastConf*, no. 8725403.
[2] Aderhold J, Watzkawick P, Beavin J, Jackson D 1969 *Menschliche Kommunikation*, Bern, Huber.
[3] Sokolova K L 2016 Teoreticheskie osnovaniya rassmotreniya etnokul'turnoj kompetentnosti psihologa. *Psihologiya. Istoriko-kriticheskie obzory i sovremennye issledovaniya.* 3. pp. 28–37
[4] Isakova A A 2018 Transport nominations in cross-cultural communication training. *IOP Conf. Ser. Mater. Sci. and Engineer, p. 483(1)*
[5] Karasik V I 2008 *Religiya reklamy.* Volgograd. p. 54
[6] Goodman K S 1986. *What Is Whole in Whole Language?* New York . p. 3
[7] Vygotskij L S 2015 *Myshlenie i rech'* M.: Labirint. p. 351
[8] Praninskas J 1968 *Trade name creation. Processes and patterns.* The Hague-Paris
[9] Jackendoff R 1992 Languages of the mind. In: *Essays on mental representation.* London: MIT press. p. 205
[10] Miller, G A, Johnson-Laird P N 1976 *Language and perception.* Cambridge: Cambridge Univer. Press. p. 760
[11] Pil'gun M A 2018 Language consciousness in online communication: psycholinguistic analysis of protest media content. *Philol. and culture, 2(52).* p. 102–110.
[12] Poltorackij N P 1989 Life, *works, and worldviews*. New York. 320 p.

IOP Conf. Series: Materials Science and Engineering **1100** (2021) 012032 doi:10.1088/1757-899X/1100/1/012032

Study of the process of forming large-sized thin-walled parts

E G Demyanenko, I P Popov and A Yu Kosolapova

Samara National Research University Academician S.P. Koroleva, Moscow highway 34A, Samara, Russia, 443086

E-mail: nastas941104@mail.ru

Abstract. A method for producing large-sized parts of the bottom type, based on flanging processes, is considered. The method includes three operations for flanging different parts to be welded into a part. This allows flanging processes to be carried out using both the geometric and power characteristics of the Schuler press. The results of modeling different parts of a part with a minimum thickness difference are presented. The parameters of the flanging processes are determined.

1. Introduction

In modern conditions, the competitiveness of products is determined by a combination of the minimum production cost and high quality. There are many ways to obtain parts such as bottoms. This is due to the different functional purpose of parts,the increased requirements for their accuracy and the variety of sheet construction materials are used.

At present, a significant part of the bottoms in the industry are manufactured by cold stamping methods on presses: molding, drawing, flanging, followed by welding.

2. Investigation object

A characteristic feature of this group of bottoms is that with an increase in size, the formation of wall thinning during shaping is possible, therefore, it is necessary in each specific case to select the parameters of the stamping process by modeling the processes, finding optimal solutions.

During extraction, the loss of stability of the blank on the flange occurs in the zone free from contact with the tool. In some cases, with further shaping (after the appearance of folds) folds reach significant values. When the relative thickness of the part $S_0 / D_d < 0.005$, which usually leads to a scrap of the part.

The dimensions of the bottom part under consideration are shown in Figure 1.

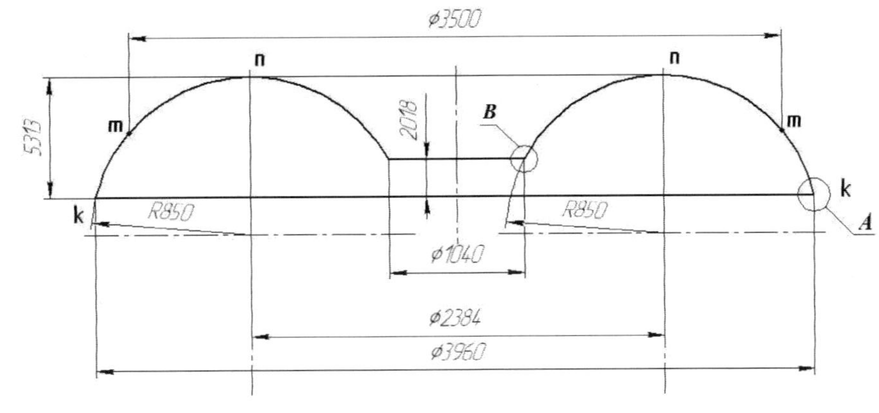

Content from this work may be used under the terms of the Creative Commons Attribution 3.0 licence. Any further distribution of this work must maintain attribution to the author(s) and the title of the work, journal citation and DOI.
Published under licence by IOP Publishing Ltd

IOP Conf. Series: Materials Science and Engineering 1100 (2021) 012032 doi:10.1088/1757-899X/1100/1/012032

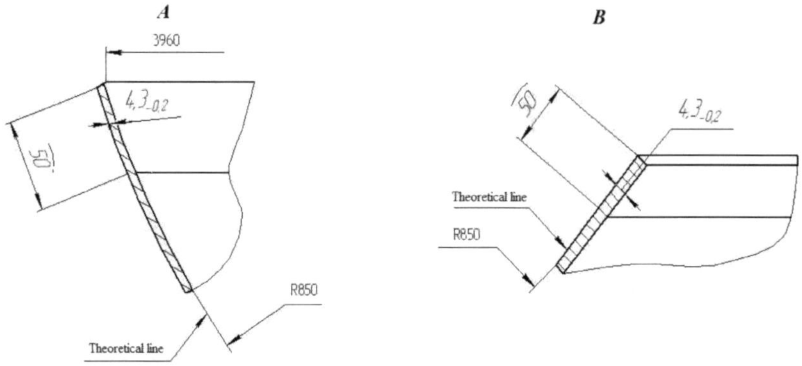

kk - section that limits the dimensions of the part; km - section that limits the outer part of the part; mn - section that bounds the middle part of the part; nn - section that bounds the inner part of the part

Figure 1. Part of the bottom (S = 5.8 × 10^{-3}m, AMg6M).

Due to its large dimensions, the part represents a relatively thin-walled shell for the processes of forming from a sheet. Therefore, its deformation must be carried out under conditions of a stressed state of tension, in order to avoid its formation.

Taking into account the existing limitations, out of several options for shaping a part from a flat blank,it is preferable to use the flanging process. The condition that ensures the stability of the flanging process is the ratio of the maximum meridional stresses.

$$\sigma_{\rho\, dr} \geq \sigma_{\rho\, fol} \tag{1}$$

where $\sigma_{\rho\, dr}$ и $\sigma_{\rho\, fol}$ - respectively, the maximum meridional stresses of the flat parts of the flange when leaving under the clamp and part of the blank for flanging.

For an ideal case, it is:

$$\sigma_{\rho fol} = \sigma_s\left(1 - \frac{r_м}{R_\sigma}\right) \tag{2}$$

The potential extraction stress of a flat flange part is ideally calculated:

$$\sigma_{\rho dr} = \sigma_s ln\left|\frac{R_{blank}}{R_\sigma}\right| \tag{3}$$

r_m - is the radius of the flange hole;
R_σ - is the radius of the section of the blank when exiting under the clamp;
R_{blank} - the radius of the edge of the blank.

Formulas (2) and (3) give estimated solutions that can be applied in the process of modeling using software products.

We divide the part into three parts, limited by sections with diameters: $D_{kk} = 3960 \cdot 10^{-3}$ m; $D_{mm} = 3500 \cdot 10^{-3}$ m; $D_{nn} = 2384 \cdot 10^{-3}$ m. Their values are selected from the condition that the diameters of the blank under the clamp, with an increase in the size of the part, should decrease due to the limitation of the press dimensions. To ensure the immovability of the flange during flanging, the following conditions must be met:

$$\frac{r_{nn}}{R_{nn}} \geq \frac{r_{mm}}{R_{mm}} \geq \frac{r_{\kappa\kappa}}{R_{\kappa\kappa}}$$

$r_{nn}, r_{mm}, r_{\kappa\kappa}$ - respectively, the radius of the flange holes for parts limited by the dimensions of the joint along the sections with diameters: $D_{nn} = 2R_{nn}$; $D_{\kappa\kappa} = 2R_{\kappa\kappa}$; $D_{mm} = 2R_{mm}$.

Figure 2 shows the scheme of forming the central part of the bottom.

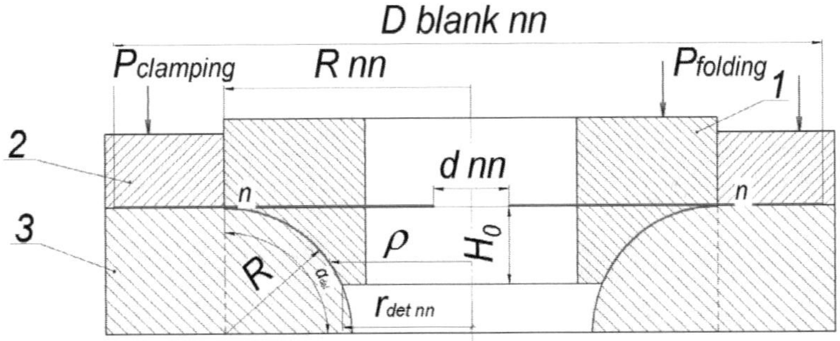

$d_{nn} = 720 \cdot 10^{-3} м; R_{nn} = 1192 \cdot 10^{-3} м; R_{mm} = 1750 \cdot 10^{-3} м; R = 850 \cdot 10^{-3} м; r_{det\,nn} = 487 \cdot 10^{-3} м$

1 - punch; 2 - clamp; 3 - matrix; R_{nn} - radius of part limited by section nn; d_{nn} - diameter of the flare hole, limited by the section nn; H_0 - height of the molded part; α_{det} - angle of the length of the forming central molded part; ρ - current coordinates of the angle and radius of the central part; $P_{clamping}, P_{flanging}$ - clamping force and flanging; R - radius of rounding of the forming central part; $r_{det\,nn}$ - radius of the finished section nn

Figure 2. Scheme of forming the Central part of the bottom.

The scheme of flanging of the Central part of the bottom (see figure 3) is considered, taking into account the allowance for cutting the flanged end face: let's take the height. The diameter of the flanging hole is in the first approximation from the equality of the lengths forming the flanged part of the part and the length of the forming blank with the hole free from clamping:

$$R_{nn} - r_{nn} = R \cdot \alpha_{det}$$
$$r_{nn} = R_{nn} - R \cdot \alpha - 3^{-3}{}_{det} \quad (4)$$

It is assumed that $r_{nn} = 360 \cdot 10^{-3}$ m and equating equations (2) and (3) we get:

$$R_{blankn n} \geq R_{nn} \cdot exp\left(1 - \frac{r_{nn}}{R_{nn}}\right) \geq \frac{2384 \cdot 10^{-3}}{2} \cdot exp\left(1 - \frac{360 \cdot 10^{-3}}{1192 \cdot 10^{-3}}\right) = 2384 \cdot 10^{-3} m \quad (5)$$

Let's take the maximum possible diameter of the blank $D_{blank\,nn} = 4990 \cdot 10^{-3}$ m

By formula (6) we find:

$$R_{blankmm} = R_{mm} \cdot exp\left(1 - \frac{R_{nn}}{R_{mm}}\right) = \frac{3500 \cdot 10^{-3}}{2} \cdot exp\left(1 - \frac{1192 \cdot 10^{-3}}{1750 \cdot 10^{-3}}\right) = 2410 \cdot 10^{-3} m \quad (6)$$

We accept the dimensions of the middle part of the workpiece with a diameter $D_{blank\,mm} = 4990 \cdot 10^{-3}$ m, and the hole diameter $d_{mm} = 2200 \cdot 10^{-3}$ m

The edge part of the bottom is limited by sections with a diameter $D_{kk} = 3960 \cdot 10^{-3}$ m and $D_{mm} = 3500 \cdot 10^{-3}$ m.

Using formula (6), we define:

$$R_{blankkk} = R_{kk} \cdot exp\left(1 - \frac{R_{mm}}{R_{kk}}\right) = \frac{3960 \cdot 10^{-3}}{2} \cdot exp\left(1 - \frac{1750 \cdot 10^{-3}}{1980 \cdot 10^{-3}}\right) = 2257.3 \cdot 10^{-3} m \quad (7)$$

The accepted dimensions of the edge part of the blank are $D_{kk} = 4990 \cdot 10^{-3}$ m, $d_{kk} = 3100 \cdot 10^{-3}$ m.

The number of intermediate anneals is calculated based on the possible breakage of the hole edge, assuming the process of deformation of the edge is similar to the process of linear stretching.

$$e_\theta = \delta_p \cdot n \text{ or} \quad (8)$$
$$r_{det\,nn} = r_{nn} \cdot e^{\delta_p N} \quad (9)$$

where *N* is a number of anneals.

For the central part of the bottom: $N_{nn} = \frac{ln\frac{r_{det\,nn}}{r_{nn}}}{\delta_\rho} = \frac{ln\left[\frac{478}{360}\right]}{0.13} = 2.3 \rightarrow 3.$

For the middle part: $N_{mm} = \frac{ln\frac{r_{det\,mm}}{r_{mm}}}{\delta_\rho} = \frac{ln\left[\frac{1192}{1100}\right]}{0.13} = 0{,}61 \rightarrow 1.$

For the edge part: $N_{kk} = \frac{ln\frac{r_{det\,kk}}{r_{kk}}}{\delta_\rho} = \frac{ln\left[\frac{1750}{1550}\right]}{0.13} = 0.93 \to 1.$

3. Modelling results

In rather time-consuming tasks, the use of modern automation tools makes it possible to exclude a significant part of routine work and to present the simulation results in a visual form [3].

To determine the stress-strain state and energy-power parameters, in the processes of metal forming by pressure, software systems based on the finite element method are widely used. The modeling uses a material with an anisotropy coefficient of a transversely isotropic body, with indicators of mechanical properties $\sigma_{02} = 15.5\ MPa; \sigma_в = 31.5\ MPa; \delta_p = 0.13\ MPa$ elastic modulus $E = 7.1 \cdot 10^4\ MPa$; the coefficient of friction is assumed to be $f = 0.2$. The process includes an adaptation program that reduces the elementary mesh when their neighboring elements deviate by one degree. The stain rate is 100 mm/sec. The blank is obtained by argon-arc welding and the blank is obtained by argon-arc welding from sheets with a width of $2500 \cdot 10^{-3}$ms with a nominal thickness of $6 \cdot 10^{-3} m$. Weld width is $20 \cdot 10^{-3} m$ Characteristics accepted for it:$\sigma_{02} = 15.5\ MPa$; $\sigma_в = 27\ MPa$; $\delta_p = 0.13$.

The blank has axial symmetry and 1/4 of the volume enclosed between the XOY and YOZ coordinate planes was used for the calculation (the OY axis was aligned with the symmetry axis) [3, 4, 5].

The hardening curve of the AMg6M aluminum alloy [3] is shown in Figure 3.

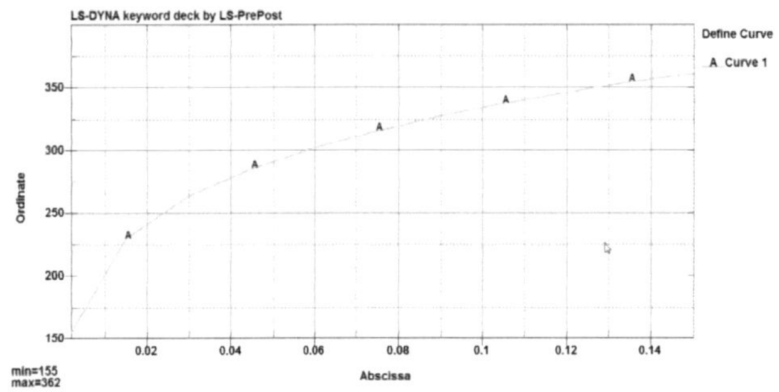

Figure 3. The hardening curve of the AMg6M alloy.

Thus, only the central part requires three intermediate anneals.

Modeling data - geometry, of course - element mesh, graphs of changes in thickness of characteristic elements, which are presented in (Figures 4 - 9) [6,7,8]:

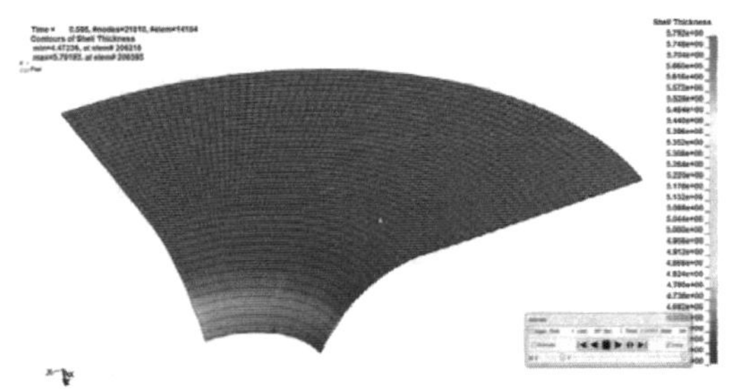

Figure 4. Change in the thickness of the semi-finished product 1st transition, mm.

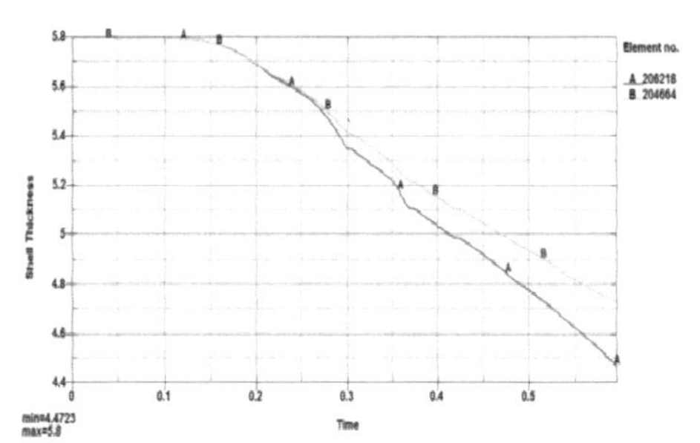

Figure 5. Change in the thickness of characteristic elements at the end of the semi-finished product of the 1st transition, mm.

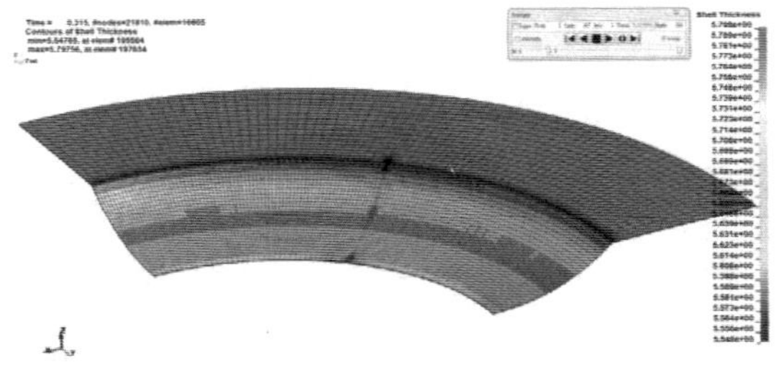

Figure 6. Change in the thickness of the semi-finished product 2nd transition, mm.

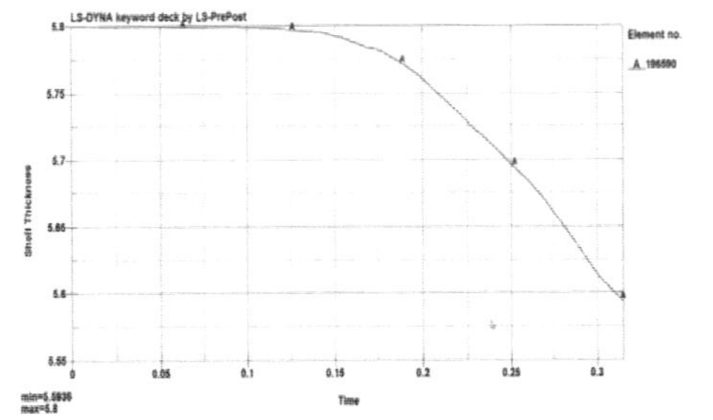

Figure 7. Change in the thickness of the characteristic element at the end of the semi-finished product of the 2nd transition, mm.

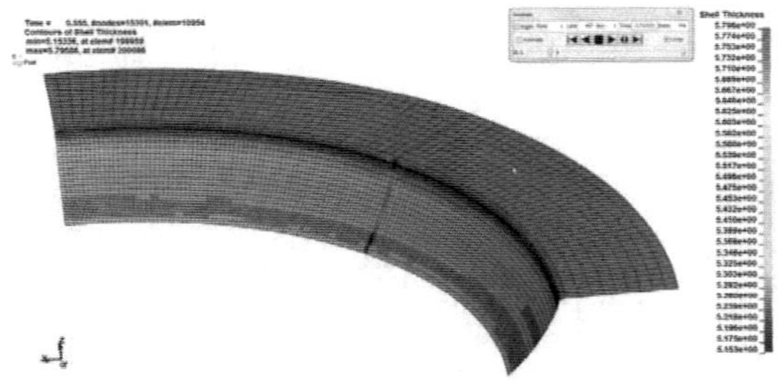

Figure 8. Change in the thickness of the characteristic element at the end of the semi-finished product of the 2nd transition, mm.

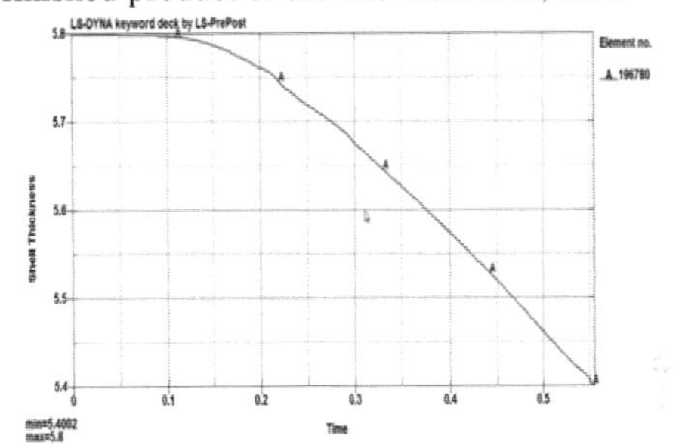

Figure 9. Change in the thickness of a characteristic element at the end of the semi-finished product of the 3rd transition, mm.

Formation of thin-walled and especially thin-walled bottoms is accompanied by intense stretching. In the considered method, the thickness difference is permissible and does not exceed the minimum permissible coefficient (>10%).

The main disadvantage of the proposed option is the use of large-scale die tooling. The technology is implemented rather laboriously, but when using the flanging process by the wrapping method, the process is simplified. The proposed option in all respects fits the Schuler press.

4. Conclusion

The proposed method for the manufacture of equal-thickness parts such as bottoms using a technology based on the flanging process (tightening method) and flanging of the central part of the bottom is effective and fairly simple to implement and involves the manufacture of parts from two parts by welding.

The central part, limited by the diameter D_{nn} and the spherical part, limited by the diameters D_{kk} and D_{nn}. In the latter case, it is possible to use strapping equipment to obtain petals, which represent the constituent elements of the spherical part (6 pieces).

The developed technique for calculating the shape and wall thickness of the part is confirmed by mathematical modeling and analytical values.

References

[1] Demyanenko E G 2013 *Proceedings of the SSC RAS* **4-1** 187 - 191.

[2] Demyanenko E G and Popov I P 2013 *Forging and stamping production* **7** 63-66.

[3] Demyanenko E G, Tlustenko S F and Popov I P 2006 *Bulletin of Samara State Aerospace University named after academician S.P. Koroleva* **2 (1)** 282-286.

[4] Grechnikov F V, Surudin S V, Erisov Y A, Kuzin A O and Bobrovskiy I N 2018 *IOP Conference Series: Materials Science and Engineering* **286 Issue 1.**

[5] Mikheev V A and Surudin S V 2016 *Key Engineering Materials* **684** 21-28.

[6] Khaimovich I N, Ramzaev V M and Chumak V G 2016 *CEUR Workshop Proceedings* **1638** 864-872.

[7] Khaimovich I N and Khaimovich A I 2017 *Key Engineering Materials* **746** 269-274.

[8] Khaimovich I N and Khaimovich A I 2018 *IOP Conference Series: Materials Science and Engineering,* **302(1)** 012062.

Research of technological properties of casting polyurethane compounds

R A Alekhina, Y A Lopatina, A S Sviridov

Federal State Budgetary Scientific Institution "Federal Scientific Agricultural Engineering Center VIM", 1st Institutsky proezd 5, Moscow, Russia

E-mail: rioraya9@gmail.com

Abstract. The technological properties of three marks of casting polyurethane compounds Advaform 80, Silagerm 6080, Polyflex 80A were considered in the article. Those products were intended for manufacturing piston pump diaphragm. The effect of preparation of the mixture, expressed in the presence of vacuum processing, on the quality of the product was investigated. The viscosity of these marks of polyurethane compounds were determined with a rotary viscometer. It is shown that the time of vacuum processing is an important parameter of quality of castings, which in turn depends on the pot time of the polyurethane. It was also noted that the low viscosity of the compound does not always guarantee the best casting quality in processes of manually pouring, and the pot time of the compound has a much greater effect on the quality. As the optimal material for producing diaphragm was chosen Advaform 80 due to its long pot time, which made it possible to carry out high-quality vacuum processing with the formation of a minimum amount of bubbles.

1. Introduction

Polymer materials and their composites, including polyurethane, are promising materials. They are increasingly used in various industries therefore the study of their properties is currently an urgent task [1-4]. Polyurethane has high mechanical characteristics, increased wear resistance and resistance to abrasion; it can be used for manufacturing such parts as bearings, couplings, bushings, flanges, elbows, gravity sectors [5-7]. Another example is the diaphragm, which is the main working part in the diaphragm piston pump. For the manufacture of diaphragms can be used resin casting method in which a liquid polymer compound is poured into a mold and cured [8-10].

The resin casting in silicone molds is one of the most demanded and economically reasonable technologies for single and small batch production of polymer parts. Casting in elastic silicone molds allows to quickly produce polyurethane parts with similar physical, mechanical and chemical properties, as well as the obtained by injection molding. This method has low cost of equipment and tools and allows to produce casting parts with different hardness range. One of the most important points in casting polyurethanes into elastic silicone molds is a correctly selected technological mode, which provides the highest strength of the product. This study is aimed at revealing the dependence of viscosity and hardness of different marks of polyurethanes on the conditions of casting samples and as a result directed to the choice of optimal technological mode for producing diaphragms.

The main material for the manufacturing of the diaphragm is a casting polyurethane compound. It does not require complex and expensive injection molds like thermoplastics and elastomers. The casting polyurethanes are flowable compositions which are poured into molds and cured in them. They have excellent physical and mechanical characteristics and a wide range of hardness (from 30 A to 90 D Shore).

2. Materials and methods

For the study, three marks of polyurethane compounds were used: PolyFlex 80A, Advaform 80 and Silagerm 6080. The main criteria for choosing a material were hardness in the range from 75 to 90 units on the Shore A

Content from this work may be used under the terms of the Creative Commons Attribution 3.0 licence. Any further distribution of this work must maintain attribution to the author(s) and the title of the work, journal citation and DOI.
Published under licence by IOP Publishing Ltd

scale and a pot life at a temperature of 25 °C for at least 10 minutes. The characteristics of these materials are presented in Table 1.

Table 1. Characteristics of the investigated polyurethane two-component compounds

№	Index	PolyFlex 80A	Advaform 80	Silagerm 6080
1	Shore hardness, scale A	80	80	85
2	Component ratio A/B, wt/wt	100 / 200	100 / 250	100 / 200
3	Density at 25 °C, kg/m^3	Polyol -1140 Isocyanate – 1100	Polyol -1000 Isocyanate – 1000	Polyol -1040 Isocyanate – 1000
4	Pot life at 23 °C, min.	10-15 (depends on the mass)	25-35 (depends on the mass)	20-30 (depends on the mass)
5	Curing time 23 °C, h	12-24 (depends on the mass)	2-4 (depends on the mass)	8-10 (depends on the mass)

The preparation of polyurethane compounds before casting samples consists of several main stages. Firstly, both components (A and B) should be thoroughly mixed until homogeneous condition within 3-5 minutes (the recommended rotation speed is from 100 to 200 rpm when using a mixing device), manual mixing is also possible, which was carried out in this work. Secondly, the release agent based on silicone or wax should be applied to the mold before casting polyurethane in order to prevent damage of the mold and smooth removal of the finished product without excessive physical impact on the unformed material. The release agent must be applied to all parts and surfaces on which contact with polyurethane will take place. A multipurpose silicone grease Si-M was used as a release agent in this study.

In this work we investigated the effect of preparing the mixture, expressed in the presence of vacuum processing, on product quality. The vacuum processing allows to eliminate dissolved air in the material that entering the compound during the mixing of the components. Vacuum processing was performed using the degassing system shown in Figure 1a. The samples were cast into a soft silicone mold shown in Figure 1b.

a)

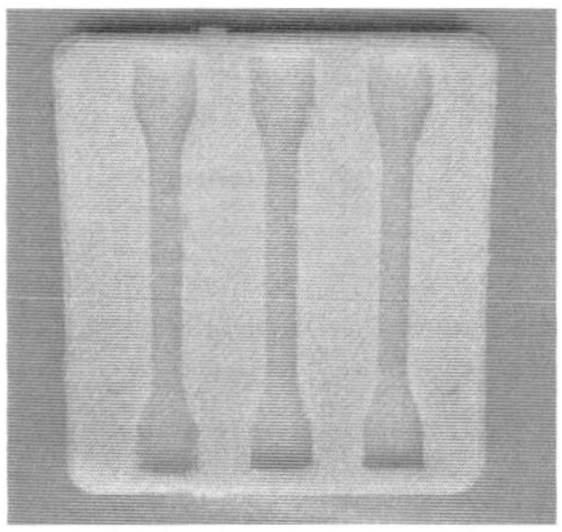
b)

Figure 1. Vacuum degassing system (a) and silicone mold for casting samples (b).

The study of the technological properties of polyurethane compounds was carried out using a ROTAVISC lo-vi Advanced viscometer. The measurement accuracy of the ROTAVISC rotary viscometer for both Newtonian and non-Newtonian fluids is +/- 1% of the measuring range. Reproducibility of measurements +/- is about 0.2%. The ROTAVISC viscometer has a stepless speed adjustment.

The principle of operation of a rotary viscometer is that it has two coaxial cylinders with a measured fluid between them, one of which moves by an electric motor with a constant angular velocity, the other is hung on a torsion wire. After starting the drive cylinder from rest, the suspension cylinder rotates until a position is reached where the force generated by the torsion wire is simply balanced by the viscous force transmitted through the fluid. When measuring viscosity torque is measured at a given angular velocity or vice versa.

Because of the curing time of the testing polyurethanes is different from each other, it is necessary to measure values within 15-25% of the polymerization period for the purity of the experiment to determine the viscosity. So for Silagerm 6080 this value will be 3.25-4.25 minutes, for PolyFlex 80A - 3.75-4.75 minutes and for Advaform Mark 80 - 5.5-6 minutes. The rotor speed during the tests was 1.6 rpm.

3. Results and discussion

As a result of viscosity measurements, graphs of the dependence of the viscosity of polyurethanes Silagerm 6080, PolyFlex 80A and Advaform Mark 80 on the test time were obtained. A typical view of the graph is shown in Figure 2.

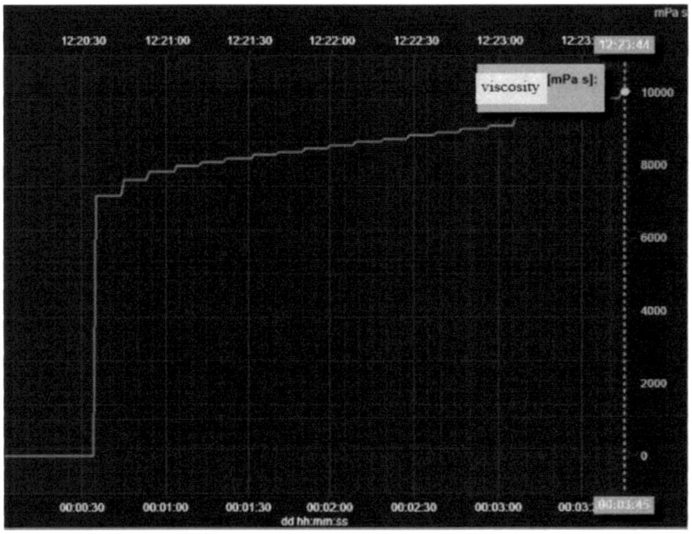

Figure 2. The viscosity-test diagram of polyurethane compound Silagerm 6080.

Table 2 shows the results of the measured viscosity for all grades of polyurethanes. Figure 3 shows the average values of the viscosity of polyurethane compounds. The diagram shows that the highest viscosity is possessed by the polyurethane compound Advaform 80, which is almost two times lower than the viscosity of the Silagerm 6080 compound and almost ten times lower than the viscosity of the PolyFlex 80A compound.

Table 2. Results of measuring the viscosity of polyurethane compounds

Mark	Average viscosity, Pa·s	Test period, min	Fluctuations between the minimum and maximum viscosity, Pa·s
PolyFlex 80A	2.04	5.00	0.1
Advaform Mark 80	18.75	5.45	0
Silagerm 6080	10.05	3.45	2.8

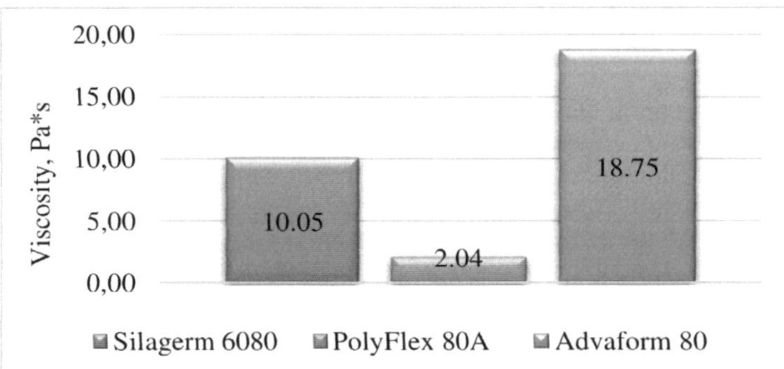

Figure 3. The average values of the viscosity of polyurethanes.

Figure 4a shows the appearance of the samples obtained without a degassing operation and with a degassing operation. It can be seen that the preliminary degassing operation significantly improves the quality of the casting. Figure 3b shows the appearance of castings from various grades of polyurethanes with preliminary degassing of the mixture.

Without vacuum processing With vacuum processing

a)

Advaform 80 Silagerm 6080 Polyflex80A

b)

Figure 4. Appearance of the samples: a) depending on the presence of the preliminary vacuum processing of the mixture; b) depending on the type of polyurethane.

It can be concluded from the obtained experimental data that the most suitable material for casting is PolyFlex 80A, since it has the lowest viscosity. However, it should be noted that due to the short pot life, the PolyFlex 80A is not possible to properly vacuumize, which affects the formation of bubbles in the samples, as shown in Figure 3b. The advantages of this material include the transparency of the material, which allows to visually estimate the presence of bubbles in castings. The Silagerm 6080 has a pot life in comparison with PolyFlex 80A approximately 1.5-2 times higher, which made it possible to vacuumize it more thoroughly. The material is not transparent and it is possible to visually distinguish the presence of bubbles. The Silagerm 6080 in comparison with PolyFlex 80A proved to be more technological and stable. The obvious disadvantage of both polyurethane compounds is a pungent odor that requires the use of personal respiratory protection. Samples from polyurethane Advaform 80, subjected to preliminary degassing, have practically no bubbles, and without degassing showed an appearance similar to the samples made in the same modes from two previously considered brands of polyurethane compounds: they contained a lot of bubbles. Based on the discussed above results and the visual analysis of the samples, it can be concluded that the time of vacuum processing has the effect on the absence of bubbles.

4. Conclusions

The PolyFlex 80A has the lowest viscosity (2 Pa·s), however, its short pot life (about 15 minutes) does not allow to vacuumize samples well enough. The best material is Advaform 80 because this material is well vacuumized with a minimal number of bubbles quality despite the relatively high viscosity of 18.5 Pa·s. The polyurethane compound Advaform 80 was chosen for further research and the manufacture of the diaphragm due to its long pot life and the best quality of the resulting castings.

References

[1] Sviridov A, Lopatina Iu, Kurganova Iu 2019 3D-printed polyether ether ketone samples mechanical properties estimation *IOP Conf. Series: Materials Science and Engineering* **589** 012021

[2] Mirzaev M A, Smirnov I G, Dyshekov A I, Slavkina V E, Shereujev M A, Aliev S A 2019 Design of the solar panel frame of the spacecraft by 3D printing with composite materials. *IOP Conference Series: Materials Science and Engineering* **747** 012009

[3] Lopatina Yu A and Slavkina V E 2020 Development of a high strength polymeric composite material using 3D-printing and vacuum impregnation technology. *IOP Conference Series: Materials Science and*

Engineering **934** 012019
[4] Tertyshnaya, Yu V, Levina N S, Popov A A, Moskovskii M N, Izmailov A Yu 2019 Hydrolytic destruction of agrofiber made of natural polymers *Fiber chemistry* **2(51)** 117-120
[5] Wang S, Yuan Y, Tan P, Zhu H, Luo K 2020 Experimental study on mechanical properties of casting high-capacity polyurethane elastomeric bearings *Construction and Building Materials* **265** 120725
[6] Xiang N, Zhang X, Zheng M, Ge Y, Wang T, Liu H, Maharaj C, Dear J P, Yan Y 2020 Microstructure and tensile properties of injection molded thermoplastic polyurethane with different melt temperatures *Journal of Applied Polymer Science* **137(29)** 48891
[7] Quintana C, Rull N, Deniro P J, Frontini P, Rueda F, Sergeeva O Yu 2020 Process simulation-based design, mold construction and mechanical performance evaluation of an insert-injection molded thermoplastic polyurethane *International polymer processing* **35(4)** 385-397
[8] Leal A, Frizera A, Marques C 2020 A fiber Bragg gratings pair embedded in a polyurethane diaphragm: Towards a temperature-insensitive pressure sensor *Optics and laser technology* **131** 106440
[9] Galineau J, Guiffard B, Severyat L, Lallart M, Gyomar D 2013 Study and modeling of an electrostrictive polyurethane diaphragm loaded with conductive carbon black **V. 139** 117-124
[10] Leal-Junior A G, Frizera A, Marques C 2020 Thermal and mechanical analyses of fiber bragg gratings-embedded polymer diaphragms *IEEE Photonics Technology Letters* **32(11)** 623-626

Fractal model of structure-properties effect of low carbon steel

V M Volchuk[1,3], O V Uzlov[1], O V Puchikov[2], S V Ivantsov[1] and I A Tiutieriev[1]

[1] Department of Materials and Materials Processing, State Higher Education Institution "Prydniprovska State Academy of Civil Engineering and Architecture", 24-a, Chernyshevskoho Str., Dnipro, 49600, Ukraine
[2] Department of Constructional Steels, Institute of Ferrous Metallurgy of Academy of Sciences of Ukraine, 1-a, Starodubov sqr., Dnipro, 49005, Ukraine

[3]Email: volchuky@gmail.com

Abstract. Fractal theory has been used for predicting of microstructure (ferrite, pearlite and non-metallic inclusions) effect on mechanical properties (KCV^{-60}, $\sigma_{0.2}$, σ_B, δ) of S355J2 steel. The fractal model which has been obtained may be considered as additive database. The database can be complemented by different treatment modes and microstructure. It has been stated that good response of structure-properties has been obtained for strength characteristics ($\sigma_{0.2}$, σ_B) of S355J2 steel and fractal size of lamellar pearlite. Also, correspondence of elongation to fractal size of ferrite has been obtained. As it has been determined the better correspondence of fractal size of non-metallic inclusions defined to impact toughness of S355J2 steel. The results have been confirmed by effect of low carbon steel microstructure on mechanical properties of the steel. The histogram of microstructure and non-metallic inclusions effect on mechanical properties of S355J2 steel has been plotted.

1. Introduction

Structure of materials forms at open systems with exchange of energy [1]. Therefore, structure elements forms at nonequilibrium conditions and have complicated geometrical shape. Thus, Euclid geometry is not always let the define real structure with required accuracy. But such complicated objects as steel structure at different size scale levels can be easily defined by Mandelbrot's fractal geometry [2], [3]. For example, at work [4] fractal modelling has been used for evaluation of non-metallic inclusions in steel. At works [5] and [6] fractal modelling of structure and properties of materials has been made. For example, strength/elongation ratio for S235 steel has been searched by means of fractal modelling [7].

At present work fractals theory will be used for structure-properties ratio modelling and impact trend definition for S355J2 steel.

2. Experimental

The chemical composition of low alloyed S355J2 steel are shown at Table 1.

Table 1. Chemical composition of investigated steels (weight %).

C	Mn	Si	S	P	Cr	Ni	Cu
≤0.12	1.3-1.7	0.5-0.8	≤0.04	≤0.035	≤0.30	≤0.30	≤0.30

The steel has been rolled to thick sheets and heat treated. Specimens were cut off at as-rolled condition and after normalization at 900^{0}C and 920^{0}C. Heat treatment let increase the elongation through some decrease of the strength. Mechanical properties of the S355J2 steel are shown at Table 2/

Content from this work may be used under the terms of the Creative Commons Attribution 3.0 licence. Any further distribution of this work must maintain attribution to the author(s) and the title of the work, journal citation and DOI.
Published under licence by IOP Publishing Ltd

Table 2. Treatment and properties of S355J2 steel.

#	Treatment	YTS, $\sigma_{0.2}$, MPa	UTS σ_B, MPa	Elongation, δ, %	Impact toughness $KCV^{-60°C}$, J/cm^2
1	As rolled	390	533	29	15, 20, 28
2	900°C 1.5 min/mm	360	516	35	337, 331, 332
3	920°C 1.5 min/mm	346	504	35	281, 235, 245

Fine ferrite-pearlite structure has been obtained as a result of heat treatment (Fig. 1). Structure has been investigated by Neophot-2 light microscope. It should be mentioned that investigated pearlite have had fine-lamellar structure.

Treatment 1 *Treatment 2* *Treatment 3*

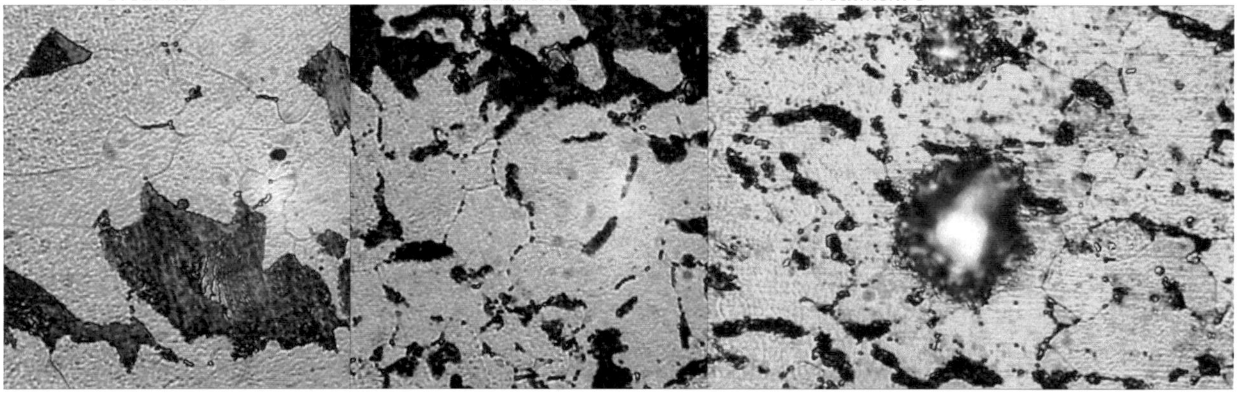

Figure 1. Ferrite-pearlite structure and non-metallic inclusions of S355J2 steel, ×1000.

Ferrite grain size has been measured according to standard (GOST 5639). Steel after hot rolling (Treatment 1) has 6-7 grade. Steel after 900°C normalization (Treatment 2) had 9-8 grade, and after 920°C normalization (Treatment 3) had 9-8 grade, also.

Non-metallic inclusions have been detected in structure of the S355J2 steel (Fig. 2). The inclusions have been identified as nondeformed silicates and brittle silicates. Such kind of non-metallic inclusions have strong effect on mechanical properties of the steel and especially on impact toughness [8].

Treatment 1 *Treatment 2* *Treatment 3*

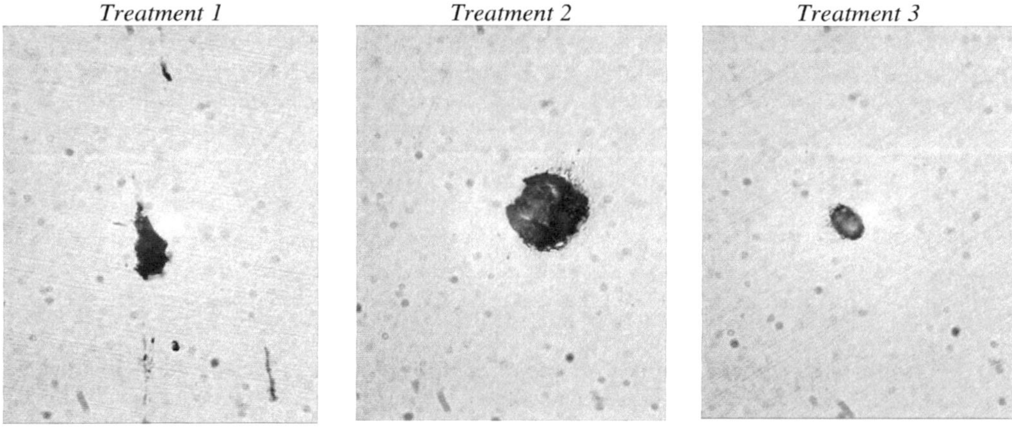

Figure 2. Non-metallic inclusions at structure of S355J2 steel, ×1000.

It should be noted that non-metallic inclusions are quasi-uniformly distributed at the structure of the steel.

3. Discussion

According to grade evaluation there is poor correlation of structure and non-metallic inclusions to mechanical properties (Table 3). Thus, fractal dimensions of the structure and properties have been used. For example, at work [9] strong correlation of multifractal characteristics of ferrite, pearlite, upper and lower bainite, Widmanstatten ferrite and properties of S235 steel has been established. Correlation coefficients were from 0.65 to 0.89. These and other results prove the response of fractal dimension to material properties.

For definition the fractal dimension of the investigated ferrite-pearlite structure a proprietary method has been used [10]. The method is based on convergence of grid [11] and point [12] numerical dimensions of the investigated object. The example of convergence fractal dimensions search for pores size at exclay-concrete is shown at Fig. 3 [10].

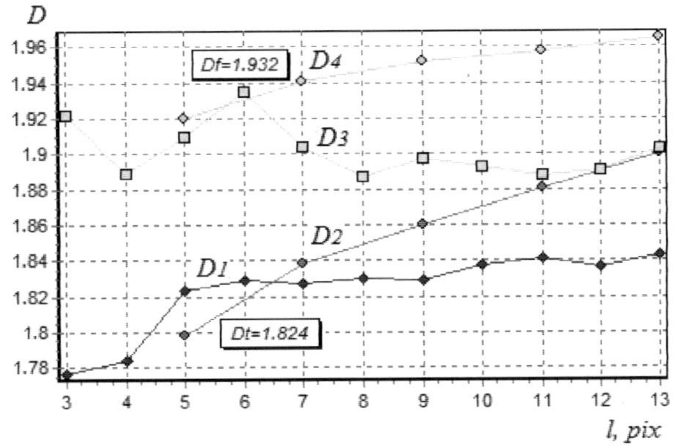

Figure 3. Correlation of fractal dimension to grid size.

The best correlation of pore size Dt=1.824 to grid size l = 6 pix has been obtained at the six iteration step, calculated by $D1$ grid and $D2$ point methods. Background correlation is described by $D3$ grid and $D4$ point methods. The best correlation for background is Df=1.932 and also has been obtained at the six step of iteration.

The fractal dimensions of pearlite D_p, ferrite D_F and non-metallic inclusions D_{inc} have been calculated. Fractal dimensions of the structure have been compared to properties in order to find the best response (Fig.4).

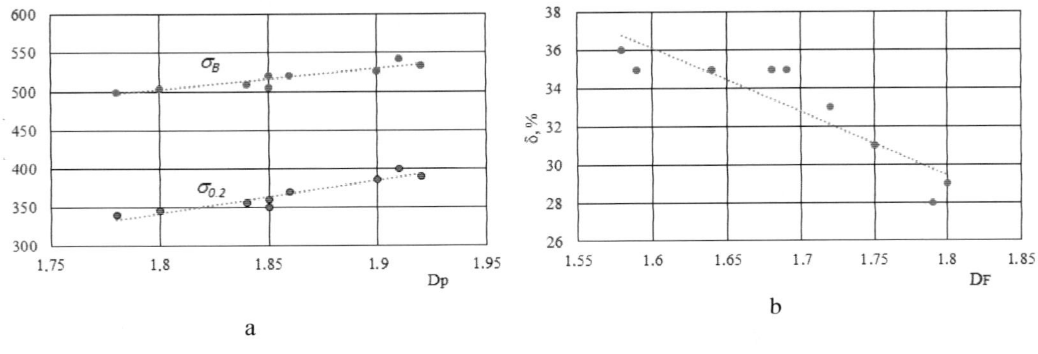

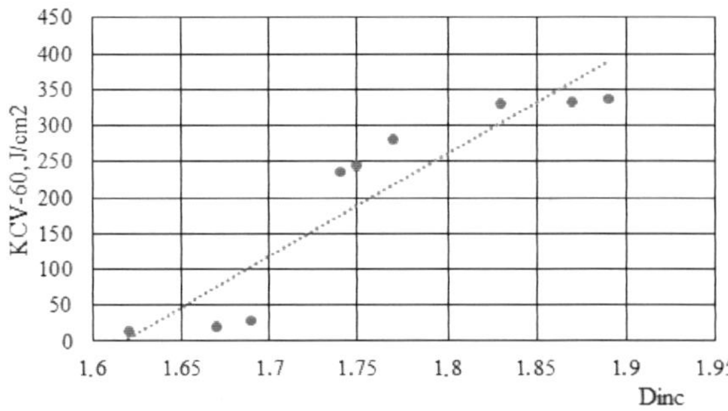

c

Figure 4. Response of the fractal dimensions of the structure to properties of the S355J2 steel.

Analysis of the Fig.4 and models (1÷4) prove the possibility of the predicting of the S355J2 steel properties by fractal method. The good response of elongation to fractal dimension of the ferrite has been found (Fig. 4 b). Also good response of the steel strength to fractal dimension of the pearlite should be noticed (Fig. 4 a). It can be explained by higher strength of the pearlite compared to ferrite due to higher (0.8%) carbon content. Response of the non-metallic inclusions to impact toughness can be explained by peculiarities of the fractal modelling.

$\sigma_{0.2} = 419.05D_p - 411.81,$ $R^2 = 0.89$ (Fig. 3 a) (1)
$\sigma_B = 272.71D_p + 11.445,$ $R^2 = 0.83$ (Fig. 3 a) (2)
$\delta = -33.398D_F + 89.555,$ $R^2 = 0.81$ (Fig. 3 b) (3)
$KCV^{-60} = 1420.9D_{inc} - 2296.5,$ $R^2 = 0.85$ (Fig. 3 c) (4)

The impact trend (structure to properties) evaluation (Fig. 5) has been calculated as a result of the analysis the pair correlation coefficients R^2 of the models (1÷4).

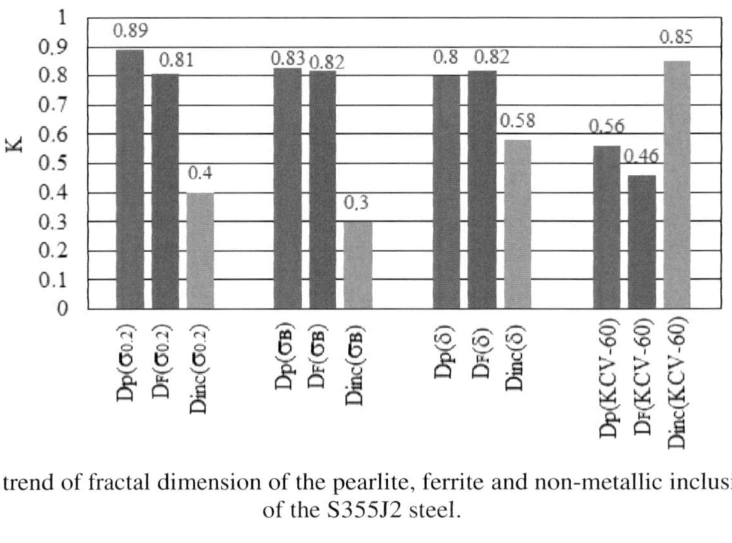

Figure 5. Impact trend of fractal dimension of the pearlite, ferrite and non-metallic inclusions to properties of the S355J2 steel.

Such approach to structure-properties correlation response let decrease the quantity of the mechanical tests due to effective use of the fractal model for mechanical properties predicting. The model is an additive and can be can be complemented by new experimental data.

4. Conclusions

It has been established that lamellar pearlite fractal dimension has good response to strength of the S355J2 steel. Also, fractal dimension of the ferrite has good response to elongation. The fractal dimension of the non-metallic inclusions has better response to impact toughness of the S355J2 steel. Fractal models let the evaluate an effect of the structure components to properties of the steel. The fractal approach let eliminate the models with poor response and complement the database for steel properties predicting.

References

[1] Berns H and Theisen W 2008 *Ferrous materials: Steel and Cast Iron* (Berlin Heidelberg: Springer)
[2] Mandelbrot B B 1982 *The Fractal Geometry of Nature* (New York–San Francisco: Freeman)
[3] Feder J. 1988 *Fractals* (New York: Springer Science Business Media).
[4] Guo, Luo-fang, Li, H. and Wang, Yajun 2012 Fractal dimension of nonmetallic inclusions and relationship with its morphology in steel *J. Iron Steel Res. Int.* **24** 33–38
[5] Hornbogen E 1986 Fine, Coarse and Fractal microstructures *Prakt. Metallographie* **23** 258-267
[6] Li J M, Lu L, Lai M O and Ralph 2003 *Image-Based Fractal Description of Microstructures* (Kluwer Academic Publishers: Springer Science & Business Media, LLC)
[7] Volchuk V, Ivantsov S, Tiutieriev I and Fortyhin, A 2020 Search for the Evaluation of 'strength-plasticity' Relation in Constructional Steel *KEM* **864**, 211–216 https://doi.org/10.4028/www.scientific.net/kem.864.211
[8] Arreola-Herrera R, Cruz A, Rivera-Salinas, J, Romero-Serrano J and Sánchez-Alvarado R 2018 The effect of non-metallic inclusions on the mechanical properties of 32 CDV 13 steel and their mechanical stress analysis by numerical simulation *Theor. Appl. Fract. Mech.* **94** 134–146
[9] Bolshakov V I and Volchuk V M 2011 Materials Science Aspects of Using of Wavelet-Multifractal Approach to an Evaluation of Structure and Properties of Low-Carbon Low-Alloyed Steels *Metallofiz. Noveishie Tekhnol*. **33(3)**, 347–360
[10] Kroviakov S, Volchuk, V, and Zavoloka M 2020 Fractal Model of the Influence of Expanded Clay Concrete Macrostructure on its Strength *KEM* **864**, 43–52 https://doi.org/10.4028/www.scientific.net/kem.864.43
[11] Hausdorff F 1919 Dimension und äußeres Maß *Math. Ann.* **79** 157–179
[12] Crownover R M 1995 *Introduction to Fractals and Chaos* (Boston, London: Jones and Bartlett Publishers, Inc.)

Thermal imaging method for quantitative and qualitative assessment of the acceptable heat losses in configurations of structures in the diagnosis of buildings

S V Fedosov[1], V N Fedoseev[2] and S A Loginova[2]

[1]National Research Moscow State Civil Engineering University, 26 Yaroslavskoye av., Moscow, 129337, Russia,
[2] Ivanovo State Politechnical University, 21, Sheremetevsky avenue, Ivanovo, 153000, Russia

E-mail: sl79066171227@yandex.ru

Abstract. The article discusses the possibility of using a hardware thermometric diagnostic method based on a thermal imaging device in energy-intensive low-rise industries. The proposed optoelectronic device, a thermal imager, can operate in real time, store and process information, and transmit the received data to the consumer. The solution, the creation of a model - an algorithm with software, can be used for the purpose of obtaining numerical values of heat losses and the accompanying characteristics of differentiated temperature states with the surface heat transfer coefficient.

1. Introduction

Modern building technologies, implying airtight and vapor-tight enclosing structures, now more and more need to diagnose the temperature state of premises [1, 2]. Today, in construction technologies, the hardware thermometric diagnostic method based on a thermal imaging device is objectively and subjectively in demand. Such an optoelectronic device as a thermal imager can operate in real time with the ability to save information recording and transfer the received data, and taking into account new developments (engineering and circuit solutions), it is possible to obtain numerical estimates of heat loss, infiltration, stratification and humidity, that is, to obtain the main parameters of comfortable room microclimate.

Thermal costs can be significantly reduced thanks to thermal imaging, using computed tomography and obtaining a picture of the temperature spectrum [3].

Various colors, obtained for convenience on the thermogram, show the change in the intensity of infrared rays - an approximate idea of the supercooled parts of the structure. As a result, we have on hand a reliable thermographic report with a picture of the problems identified in color, namely: to see the picture of construction defects, defects in windows, joints, ceilings, to diagnose the presence of "cold bridges".

2. Body

As a kind of spectrum of electromagnetic waves, thermal or infrared radiation (IR radiation) occurs in solids, liquids and gases due to vibrations of atoms in a crystal lattice or rotational-vibrational motion of molecules [4]. The range of this spectrum (IR) 0.75 μm to 1000 μm lies between visible light and radio waves. When radiation falls on a body, the following picture takes place, shown in Figure 1: interaction of radiation with heat: 1) absorption (coefficient); 2) reflection (coefficient); 3) transmission (coefficient).

Content from this work may be used under the terms of the Creative Commons Attribution 3.0 licence. Any further distribution of this work must maintain attribution to the author(s) and the title of the work, journal citation and DOI.
Published under licence by IOP Publishing Ltd

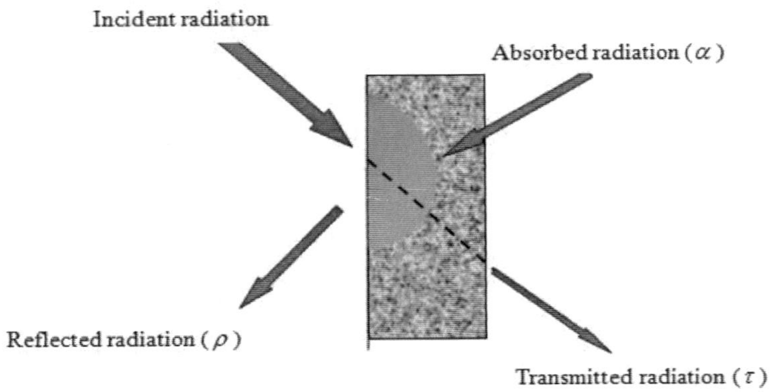

Figure 1. Interaction of radiation with heat

The output electrical signal of the IR thermal imager is proportional to the total absorbed flux. It is used in modern thermal imagers to determine true temperatures.

Physical radiation receivers generate an electrical signal proportional to the power of optical radiation absorbed by their sensitive areas [5, 6]. The principle of operation of thermal detectors is based on the bolometric effect, i.e. on the change in the electrical resistance of the sensitive element when heated by absorbed IR radiation. A thermal imager registers thermal radiation from objects using the bolometer sensor. The signal from it is converted by the processor into a thermogram - taking heat from the object, i.e. the picture that we see on the monitor. The warmest places are red, and the coldest places are blue.

3. Methodical part

Our proposed solution is the creation of an electronic - digital model - an algorithm with software for obtaining numerical values of heat losses.

To obtain a quantitative assessment of these heat losses, the existing thermal imaging survey, taking into account some thermograms, is generally not enough. We know that heat loss from the surface of a building to the environment is determined by the formula:

$$Q = \alpha \cdot (t_c - t_в) \cdot S, \quad (1)$$

where:

α - surface heat transfer coefficient, $W/m^2\ {}^oC$;

$\boldsymbol{t}_c$ - average surface temperature, $\circ\ C$;

$t_В$- ambient temperature, $\circ\ C$;

S - surface area, m^2.

Determination of the numerical or percentage value of heat loss is associated with the determination of all quantities included in the equation and is performed during the operating mode, including both thermal imaging and, if necessary, contact measurements. As an example, we use a thermometer (TH90 from AMTAST) with the function of measuring the temperature and humidity of the ambient air, integrated with the thermal imager system (Fig. 2).

Figure 2. Thermometer (TH 90) with the function of measuring the temperature and humidity of the ambient air

Thermometer type TH 90 - Specifications:

- Indoor temperature measurement range: from - 50 °C to + 70 °C.
- Indoor relative humidity measurement range from 10% to 99%.

For a comprehensive determination of indicators, additional functions of the thermal imager model are used: a method of stocking images (hard disk, SD memory card, USB, flash card), a method of image processing.

The implementation of the diagnostic algorithm (approach) is most successful and economically feasible with an optoelectronic device - a thermal imager with the following parameters. The UTi 160B model, which has a temperature measurement range from -20 °C to 300 °C, with a resolution of 160 × 120 pixels, with video output to an NTSC and PaL television system, has a USB serial port. (made in China).

Consider the possibility of determining the type of defect using the matrix sensitive area of the UTi160B thermal imager. The thermal imaging image on the screen is a temperature matrix, which creates advanced software so that the user can use the extensive capabilities of digital data processing in order to obtain geometric dimensions - an area S (defect). Consider an example of fixing a defect (area) with a thermal camera (Fig. 3) in a thermal enclosure in the form of a flat figure, digitally represented in the form of a matrix$K$, whose elements $k_{i,j}$ - represent line by line the coordinates of the centers of gravity of the points of the defect image in the coordinate system XOY.

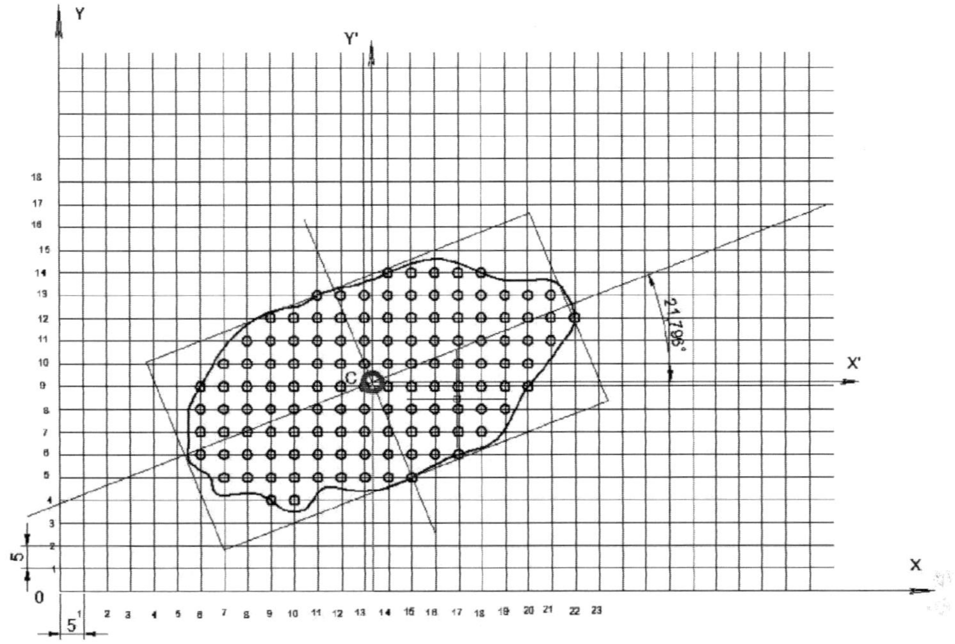

Figure 3. A flat figure representing a defect fixed by a heat chamber

There were 123 points with two coordinates in total. In addition, let, programmatically define a function that describes the contour of a flat figure of a defect $L(x,y) = 0$, which can be represented in any form. In the simplest case, it can be presented in tabular form.

$$L_{XY} = \begin{vmatrix} 6 & 7 & 8 & 9 \\ 7 & 5 & 10 & 0 \\ 8 & 5 & 11 & 0 \\ 9 & 4 & 12 & 0 \\ 10 & 4 & 12 & 0 \\ 11 & 5 & 13 & 0 \\ 12 & 5 & 13 & 0 \\ 13 & 5 & 13 & 0 \\ 14 & 5 & 14 & 0 \\ 15 & 5 & 14 & 0 \\ 16 & 6 & 14 & 0 \\ 17 & 6 & 14 & 0 \\ 18 & 7 & 14 & 0 \\ 19 & 8 & 13 & 0 \\ 20 & 9 & 13 & 0 \\ 21 & 11 & 13 & 0 \\ 22 & 12 & 12 & 0 \end{vmatrix} \quad (2)$$

Next, we determine the position of the center of gravity of this figure, using (Fig. 3) and the matrix K and taking the areas of the points of the figure as a unit. As a result, we get: the area of the figure is 123 units, the coordinates of the center of gravity$X_C = 13.341$;$Y_C = 9.187$.

$$\bar{S} = \sum_{i=1}^{123} \Delta S_i = 123 \quad (3)$$

$$X_C = \frac{1}{S} \cdot \sum_{i=1}^{123} (\Delta S_i \cdot A_{i,2}) = 13.341 \quad (4)$$

$$Y_C = \frac{1}{S}\cdot \sum_{i=1}^{123}(\Delta S_i \cdot A_{i,3}) = 9.187 \quad (5)$$

We connect a new coordinate system $X'CY'$with the center of gravity of a plane figure and calculate the axial and centrifugal moments of inertia relative to the new axes, taking $\delta_S = 1$. As a result, we get:

$$J_{CY} := \delta_S \cdot \sum_{i=1}^{123}(C_{i,2})^2 = 2.08 \times 10^3 \quad (6)$$

$$J_{CX} := \delta_S \cdot \sum_{i=1}^{123}(C_{i,3})^2 = 860.699 \quad (7)$$

$$J_{CXY} := \delta_S \cdot \sum_{i=1}^{123}(C_{i,2} \cdot C_{i,3})^2 = 580.146 \quad (8)$$

Using the obtained expressions, we find the parameters of the ellipse of inertia: the direction of the main axes of inertia and the value of the main central moments of inertia.

$$J\,\frac{J_{CX}+J_{CY}}{2}\,\frac{1}{2}\sqrt{(J_{CX}-J_{CY})^2+4\cdot(J_{CXY})^2}^{\,3}{}_{Cmax} \quad (9)$$

$$\bar{J}\,\frac{J_{CX}+J_{CY}}{2}\,\frac{1}{2}\sqrt{(J_{CX}-J_{CY})^2+4\cdot(J_{CXY})^2}{}_{min} \quad (10)$$

$$R_{ct} := \frac{-2\cdot J_{CXY}}{J_{CX}-J_{CY}} = 0.952 \qquad \frac{atan\cdot(0.952)}{2\cdot deg} = 21.796 \quad (11)$$

From here we find the equations of the main central axes of the ellipse of inertia:

y_{max}- equation of the major axis of the ellipse of inertia;

y_{min}- equation of the minor axis of the ellipse of inertia.

Solving these equations, together with the function $L(x, y) = 0$, we obtain the values of the coordinates of the points of intersection of the axes of the ellipse with the contour line of the flat figure characterizing the defect of the thermal enclosure, from which we find the dimensions of this figure - length -17.6; width - 8.9 units. Thus, the problem of determining the dimensions of the defect figure is solved.

In general, the work shows the possibility of determining the geometric parameters and the simplest types of defects in thermal fencing of buildings and structures. As a result, in order to obtain a quantitative assessment of any section of heat loss by thermal imaging, taking into account the obtained area of the defect S, temperature $t_{в}$ and t_{c} the tabular heat transfer coefficient α , we determine the amount of heat loss taking into account all quantities included in the equation.

4. Conclusions

Thermal imaging control in building structures has become widespread, especially at the stage of operation during the acceptance of buildings [7]. Modern diagnostic thermal imagers with extended functionality provide an opportunity to obtain in the control mode the temperature range using thermograms and then diagnostic characteristics in numerical or percentage form, comparing them with the existing standards. Such widespread use of thermal imaging, low-energy laser and other optoelectronic systems, closed on a computer, makes it possible to evaluate the image of an object with a certain logic of calculus, where the analysis of an optical image of the surface state of an object serves as the basis for characterizing its qualities.

References

[1] Smirnov N V, Dudin - Barkovsky I V 1965 Course in Probability Theory and Mathematical Statistics for Technical Applications (Moscow: Nauka) p 501

[2] Fritsch V 1984 Application of microprocessors in control systems (Moscow: Mir) p 464

[3] Larioshina I A, Vavilov V P 2012 Thermal imaging as a tool for energy audit *Energetic, No 8, pp 38 - 39*

[4] Cowden D 1961 Statistical Methods of Quality Control (Moscow: Fizmatgiz)

[5] Daniel W B, James V A 2001 Polarization sensitive QWIP thermal imager *Infrared Physics & Technologyб, Volume 42, pp 323 - 328*

[6] Peric D, Livada B, Peric M, Vujic S 2019 Thermal Imager Range: Predictions, Expectations, and Reality *Sensors, No 15, p 3313*

[7] Nikolsky E K, Eriskina T O, Belyakova M S 2014 Remote Methods of Monitoring of fires and their consequences *Geoinformation Sciences and Environmental Development: New Approaches, Methods, Technologies. Collection of articles of the II International conference. Southern Scientific Centre of RAS, pp 7-14*

Topological optimization of a part taking into account technological constraints

D V Lozhkin and P V Maksimov

Perm National Research Polytechnic University, 29 Komsomol ave., Perm, Russia, 614990

e-mail: denvmm@yandex.ru, pvmperm@mail.ru

Annotation. In this work, the topological optimization of an aluminum alloy bracket with specified mass restrictions and maximum von Mises stress intensities arising under the influence of load is carried out. The part was redesigned for the additive manufacturing method without forming a support structure, for this, restrictions on the overhang angle between the surfaces of the part and the build platform were used, all surfaces are considered self-supporting. Comparison of strength and stiffness characteristics of parts for bracket options, including both the original one, created by the designer, and those obtained after optimization, taking into account and without taking into account restrictions on the method of technological manufacture of the product, is performed.

1. Introduction

In this paper, the topological optimization of the bracket used in the aircraft engine industry is carried out. The topological optimization algorithm is based on the SIMP (Solid Isotropic Material with Penalization) method. The main idea of the method is the introduction of an additional characteristic - the field of the virtual density of the material of the part being optimized, the physical and mechanical properties of the material at a point or some area are taken to be dependent on the virtual density. The SIMP algorithm is implemented in conjunction with the finite element method. With a predetermined final mass of the product, the result of the algorithm operation is the optimal distribution of the material in the discretized area, at which the structure has maximum rigidity, taking into account the specified additional constraints. The mathematical formulation of the method is detailed in the work of M P Bendsoe and N Kikuchi [1].

2. Material and Limitations

The bracket is made of aluminum alloy AlSi10Mg. The material properties are shown in table 1. It is necessary to redesign the original bracket, reducing its weight, while the stress-strain state that occurs in the product during its operation should not lead to the destruction of the part. To ensure the strength of the bracket in the new version, when performing optimization, an additional restriction on the stresses arising in the bracket is used - the stress intensity according to von Mises, taking into account the safety factor, should not exceed 185 MPa at the design load.

Table 1. Material properties.

Yield Strength, MPa	Young's modulus, GPa	Density, g/cm^3
275	76	2.67

Topological optimization is impractical to carry out on the original geometry proposed by the designer, because the solver will be seriously limited in the choice of possible options for the mass distribution and will not be able to obtain the maximum possible rigidity of the structure. To increase the possible options for the final topology, it is necessary to expand the computational domain, for which the initial volume of the original geometry increases, while the enlarged computational domain should not intersect with other parts located in the assembly next to the redesigned bracket as part of the finished product. The initial geometry of the bracket (hereinafter referred to as Initial) and the extended computational domain are shown in figure 1.

Content from this work may be used under the terms of the Creative Commons Attribution 3.0 licence. Any further distribution of this work must maintain attribution to the author(s) and the title of the work, journal citation and DOI.
Published under licence by IOP Publishing Ltd

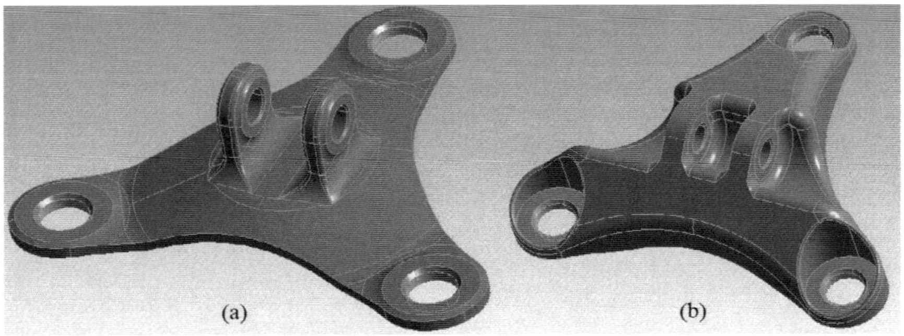

Figure 1. Computational domain: (a) initial volume, (b) extended volume

The bracket is bolted to the base at three locations (green areas in figure 1a). The load on the bracket is transmitted through the central fasteners. For the bracket, the work considers three independent loading options (table 2), each of which is characterized by a certain direction and magnitude of the resulting force.

Table 2. Force. Loading options

No.	Px, N	Py, N	Pz, N
1	0	800	1000
2	700	500	700
3	1500	0	0

After the completion performing of the topological optimization algorithm, a topology is obtained that must satisfy all three specified options for loading the bracket, as well as the specified restrictions on the mass of the part and the magnitude of the stress intensity. Due to the discrete finite element representation of the design area, automated or manual refinement of the optimization result is required by smoothing, repairing and refining the STL model in any CAD system (for example, by using the SpaceClaim system integrated in ANSYS), after which it becomes possible to perform verification strength calculations using the final topology.

Note that after topological optimization, complex-shaped products with many curved surfaces are obtained, the manufacture of which is complicated by the classical subtractive methods, additive technologies are often used, in particular, the selective laser melting (SLM) technology, which assumes the sequential layer-by-layer formation of the part ("bottom-up") by fusion powder particles using a laser. The presence of surfaces hanging over the build platform requires the creation of a structure of supports that provide support for such surfaces at the stage of their formation, in addition, the presence of support structure provides better heat removal from the formed hot part, which ensures a better quality of the additive process, and also support structure act as an additional rigid frame , which does not allow the grown part to come off the build platform or collapse under the influence of significant internal stresses arising in the process of additive formation of the part. The disadvantage of such support structure is the increased consumption of powder spent on their manufacture, a longer process of manufacturing the part, the need to separate the supports from the final part after the completion of the technological process, as well as the need to perform additional operations for machining the points of attachment of support structure to the part, finishing the surfaces, taking into account the requirements for accuracy of geometry, roughness, etc. An urgent technological task is to manufacture a part without a support structure, however, in this case, the part must be redesigned for additive technology in such a way that its surfaces, for a given orientation of the placement of the part on the build platform, would be self-supporting, that is, the overhang angle between such surfaces and the build platform does not go out beyond predetermined limits. Further in the work we will assume that the surface is self-supporting and the creation of support structure for it is not required if the overhang angle between this surface and the build platform is more than 45°.

An example of a technique that allows you to take into account the restriction on the overhang angle is presented in [2], in which all the surfaces of the part are initially divided into self-supporting surfaces and not. Then an indicator is introduced that characterizes the proportion of self-supporting surfaces in the total number of surfaces of the part, after which the restriction on the value of this indicator is used in the formulation of the topological optimization problem in the form of an additional constraint. The mathematical formulation of the problem used in this paper is presented in the equations (1)-(6):

$$\min c(\rho) = \mathrm{U}^{\mathrm{T}}\mathrm{K}\mathrm{U} = \sum_{e=1}^{N} E_e(\rho_e)\mathrm{u}_e^{\mathrm{T}}\mathrm{k}_0\mathrm{u}_e \quad (1)$$

$$E_e(\rho) = E_{\min} + (E_0 - E_{\min}) \cdot \bar{\rho}_e^{p}(\rho) \quad (2)$$

$$0 \le \rho \le 1 \quad (3)$$

$$\mathrm{K} \cdot \mathrm{U} = \mathrm{F} \quad (4)$$

$$V(\rho) \le V_0 \quad (5)$$

$$\tilde{\Phi}(\rho) \ge \Phi_0 \quad (6)$$

where equation (1) is the condition for minimization of the compliance function, written using the well-known matrix formulations of the finite element method, equation (2) is an introduction to the virtual density optimization algorithm in the form of the Young's modulus law, equation (3) Is the range of variation of the virtual density parameter, equation (4) is the constraint for the optimization problem, the fulfillment of which ensures the satisfaction of the equilibrium conditions when solving the solid mechanics problem, (5) is the constraint on the volume of the body after optimization, (6) is the condition on the proportion of self-supporting surfaces during optimization.

The procedure for topological optimization of the bracket, taking into account various requirements and constraints, was carried out further using the ANSYS package, Topology Optimization module.

3. Results and Discussion

Figure 2 shows the smoothed geometry of the bracket (hereinafter referred to as Topology 1), obtained after performing the topological optimization procedure, taking into account the restrictions on the final volume of the part and the maximum value of the von Mises stress intensity.

Figure 2. Bracket model: (a) top view, (b) side view

Figure 3 shows the fields of maximum displacements and stress intensity arising in the bracket under the most unfavorable loading case.

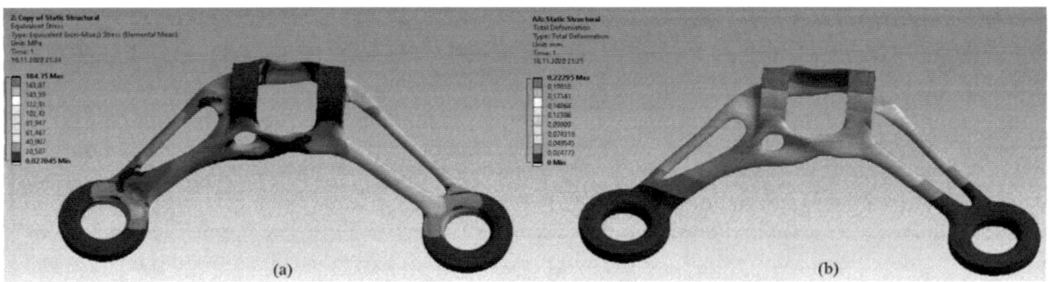

Figure 3. Results: (a) stress, (b) displacement

The resulting result shows that for given loading options and constraints on stresses and mass, it is sufficient to fix the bracket in only two places. It should be noted that the resulting geometric model has a significant number of surfaces for which the formation of support structure is required with the additive method of manufacturing a new part. Figure 4 shows the new geometry of the bracket (hereinafter – Topology 2), obtained by optimization taking into account the conditions for the formation of a topology containing mainly self-supporting surfaces.

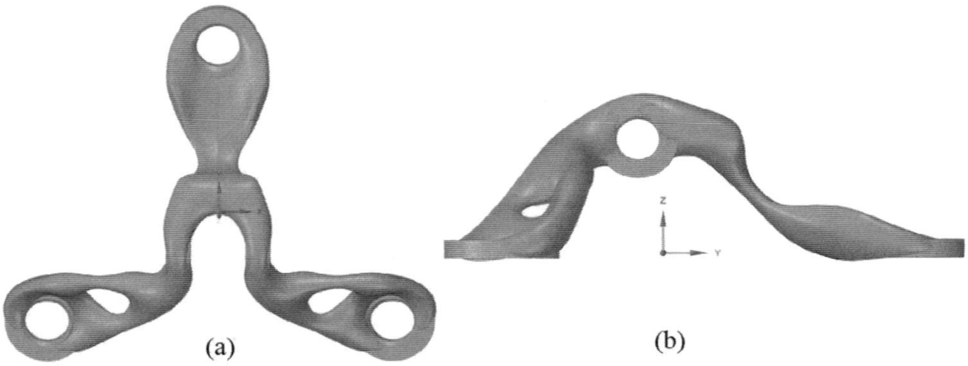

Figure 4. Bracket model: (a) top view, (b) side view

Figure 5 shows the fields of maximum displacements and stress intensity arising in the bracket under the most unfavorable loading case.

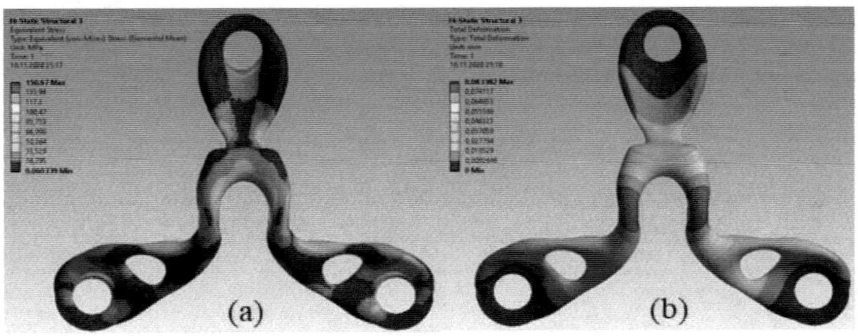

Figure 5. Results: (a) stress, (b) displacement

As a result of the optimization, a part (Topology 2) is obtained that can be manufactured mainly without creating a support system. Due to the difference in manufacturing conditions, the final geometry of the bracket, in contrast to Topology 1, remains with three legs.

Let's compare all three parts – Initial, Topology 1 and Topology 2. The results of estimated calculations of the strength and stiffness of the three brackets are presented in table 3.

Table 3. Results.

Geometry	Mass, kg	Max stress, MPa	Displacement, mm
Initial	$4.367 \cdot 10^{-2}$	180.65	0.220
Topology 1	$1.578 \cdot 10^{-2}$	184.35	0.223
Topology 2	$4.282 \cdot 10^{-2}$	150.67	0.080

4. Conclusion

The results obtained demonstrate that topological optimization without constraints on the overhang angle of the surfaces of the part to the build platform turned out to be 64% lighter than the original model, without losing rigidity, which is provided by the SIMP topological optimization algorithm itself, in the frame of which, for a given volume of material, topology is searched with the least pliability. The topology with self-supporting surfaces is close in mass to the original bracket, but with a stiffness of 2.75 times the original. During the design of the Topology 2 part, the task was not set to minimize the weight of the bracket as much as possible (table 3 shows that for Topology 2 the stress intensity limit has not yet been reached, there is potential for further weight reduction). The results obtained clearly show that when redesigning products using the SIMP topological optimization method, the designer has the opportunity to choose whether to ensure the greatest reduction in the weight of the product, or, at the same weight, due to the new topology, provide greater rigidity of the part in cases where it is necessary.

Funding
The study was carried out with financial support from the Ministry of Science and Higher Education of the Russian Federation under the state order **FSNM-2020-0028**.

References

[1] Bendsoe M P, Kikuchi N 1988 *Computer Methods in Applied Mechanics and Engineering* **71** 197-224
[2] Garaigordobil A et al 2018 *Structural and Multidisciplinary* **58** 2003-18

IOP Conf. Series: Materials Science and Engineering 1100 (2021) 012037 doi:10.1088/1757-899X/1100/1/012037

Experimental verification of the of contamination reduction of silicon during electron beam melting due to the use of a gas-dynamic window

A Kravtsov[1] and G Chikvaidze[2]

[1]KEPP EU AS, Riga, Latvia
[2]Institute of Solid State Physics University of Latvia, Riga, Latvia

E- mail: doc@keppeu.lv

Metallurgical process of producing initial rods for FZ silicon single crystal application was described before. Purity of pulled ingots was satisfied, but some impurities level was over required level. Concentrations of these impurities such Fe; Cr; Cu; Al; O2 been decreased by gas dynamic window and electric trap in laboratory scale equipment. Present article described results of experiments on industry scale equipment with application gas dynamic window for purifying indicated impurities in initial rods and results received in FZ silicon single crystals, grown from that rods, for concentrations of Aluminum and Oxygen.

1. Introduction

Monocrystalline silicon used for the manufacture of power semiconductor devices and detectors is produced by crucible-free zone melting (FZ). The reason for this is the purity of the single crystals, which cannot be achieved in any other process. The development of the method in terms of increasing the diameter of single crystals lags behind the method of growing from a crucible, primarily due to the absence of corresponding initial rods. Today their maximum diameter does not exceed 180 mm. In our opinion, a promising process is the production of initial rods for FZ not -using the conventional Siemens process, which is already used in industry today, but by growing the rods from silicon melt. We develop the most promising process of growing the silicon rods from a melt located in its own skull using electron beam heating (Skull Melting with Electron Beam Heating - SMEBH). A number of attempts have been made to apply silicon skull melting earlier. However, this did not succeed, since the authors, for example, [1,2], faced the difficulty in increasing the diameter of the growing crystal over 40 mm. In the process of growing from a crucible using the electron beam heating, we obtained a rod suitable for the FZ process, namely 300 mm in diameter and 80 kg in weight [3]. The concentration of most of the controlled impurities in the resulting rod is comparable to the purity of the reference sample made from silicon obtained by the FZ process. However, in the silicon residues after the process, a whole spectrum of metallic impurities was observed, indicating the introduction of contaminants into the melt during the growth (see Table 2, line 1, process 28) and the increased oxygen content. In the case of electron beam melting of metals, this amount of contamination can be neglected, but for semiconductors this level of impurity concentration is too high. In [3], assumptions were made about the sources of the impurity background formation of the process and it was proposed to control concentration of impurities not in the grown crystal, where, due to the small distribution coefficients of metallic impurities, their content is not significant, but in the silicon residue from the process, enriched with introduced impurities. In our opinion, the most interesting is the control of the aluminum content. The distribution coefficients of these metal impurities, except for aluminum, when drawn from the melt are in order of 10^{-5} and they are practically absent in the rod and even more in the single crystal. In addition, to determine the concentration of aluminum in single crystals, there is a method for controlling its content by Low-Temperature FTIR spectroscopy, with sensitivity up to ppta units.

Content from this work may be used under the terms of the Creative Commons Attribution 3.0 licence. Any further distribution of this work must maintain attribution to the author(s) and the title of the work, journal citation and DOI.
Published under licence by IOP Publishing Ltd

2. Experimental

The aim of the work is experimental verification of aluminum and oxygen content in silicon during its processing when growing rods in a SMEBH in the installation described in [3], and in silicon single crystals obtained from these rods using the FZ process. To carry out the SMBEH process, a cold crucible is used with a heat insulator placed inside, in which silicon is placed [3]. In this process, silicon is heated by two cold-cathode electron-beam heaters (EBH), using a mixture of gases consisting of hydrogen and less than 1% oxygen. Under the influence of the applied voltage, in the gas-discharge chamber of an electron-beam heater with a cold cathode, plasma ions bombard the cathode and initiate the creation of electrons directed to the treated surface [4]. Work [5] describes the use of devices for reducing contamination of the melt in laboratory equipment and shows that the most difficult impurity to remove is aluminum.

The silicon rod was grown in the SMEBH process. The EBH of installations were equipped with gas dynamic windows (GDW) (Fig. 1). The results of using the GDW are described in [5]. In this case, the crucible contained a skull, as illustrated in Fig. 1. Silicon doped with boron to a concentration of ~ 0.3-0.5 ppma was used as a raw material. The aluminum content in the initial silicon, the residues of molten silicon after the process, and in the rod was determined by ICP MS, and the oxygen content by FTIR spectroscopy at room temperature.

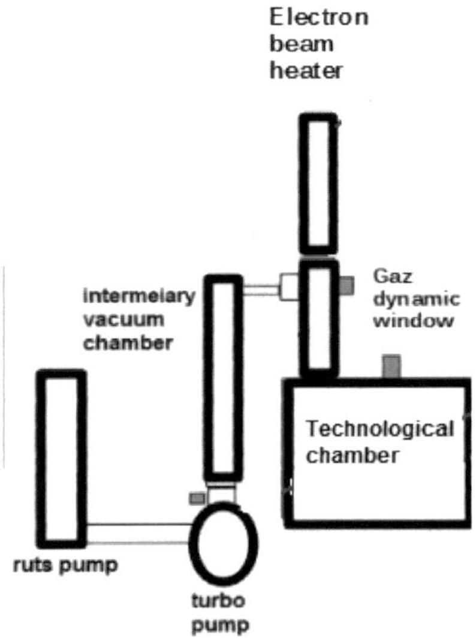

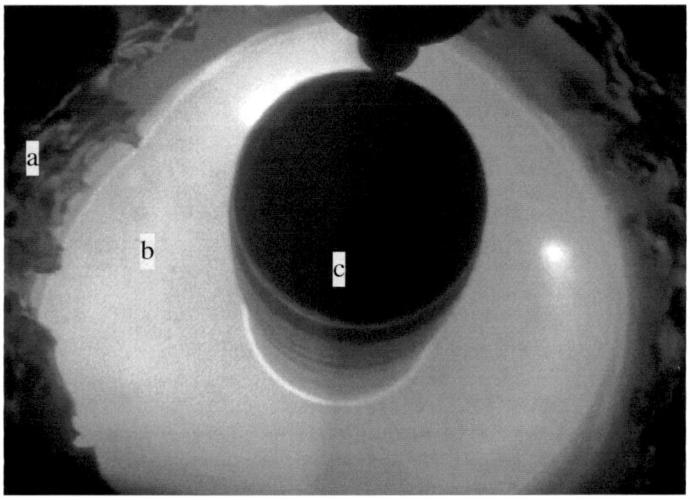

Figure 1. Connection diagrams of the gas dynamic window and the process of growing a rod
a- silicon skull. b- silicon melt. c- silicon rod

A single crystal with a diameter of 65 mm was grown from a part of the obtained rod by the FZ method in an argon atmosphere. The single crystal was cut as shown in Fig. 2. In samples (1, 3, 5, 7), the content of aluminum and oxygen was controlled at a temperature of 7K by FTIR spectroscopy method.

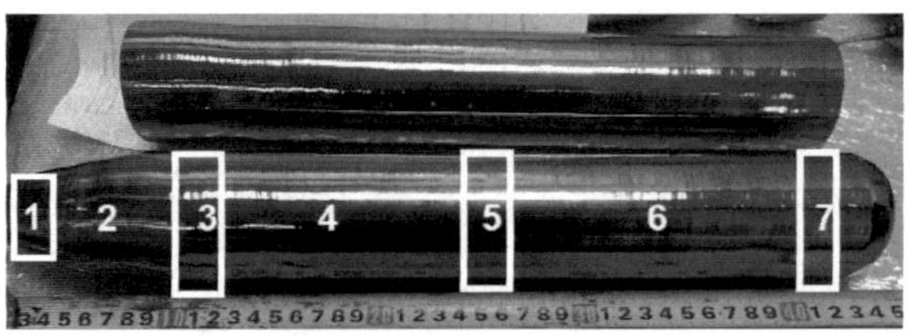

Fig 2. Scheme of sampling from the FZ single crystal of silicon.

The measurement results are shown in Table 1. The same table contains data on process # 28 described in [3] and measurements of the aluminum concentration in a single crystal grown from a similar rod [6].

Table 1. Experimental and comparative data on the processes of obtaining rods in SMEBH and control FZ of single crystals produced from these rods

Impurities	concentration of impurities			k ef of impurities in SMEBH	concentration of impurities in grown single crystal				k ef of impurities in FZ process
	raw materials	residues of melt	archeved rod		Sample 1	Sample 3	Sample 5	Sample 7	
Aluminium, ppba	40	4400	240	0.0545	14.0	20.4	21.2	21.2	0.0883
Oxygen, ppba	160	250	400	1.6	60.0	60.0	60.0	60.0	0.15
Aluminium, ppba [3,6]		24000 [3]	996 [3]	0.0415	34 [6]	94[6]	96 [6]		0.0964

3. Discussion of the results

Table 1 shows the concentration of aluminum and oxygen in the entire process of manufacturing a FZ crystal of a silicon single crystal using the technology with growing an initial rod in SMEBH. From the presented data it can be seen that that in the process of melting and growing, the loaded silicon is doped with aluminum and oxygen. The concentration of aluminum in the rod decreases. In the FZ process, after deep crystallization purification at the beginning of the process (sample 1), the melt was saturated with aluminum due to displacement, after which a stable concentration of impurities along the length of the single crystal was established. The oxygen content in the rod after SMEBH is higher than in the residual of melt. The oxygen concentration in the FZ process dropped by a factor of ~ 7 with respect to the original rod.

Obviously, the high-energy particles formed in the EBH plasma and the electrons directed to the melt bombard all surfaces of the gas-discharge chamber and the beam guide. In these collisions, metal atoms are knocked out of the surfaces, of which structural elements are composed, both in the form of neutral atoms and anions. This mixture, getting into the process chamber, creates a partial pressure of metal and oxygen vapors above the melt, which leads to alloying of the melt with the mentioned impurities. Verification of sources of pollution with the use of laboratory equipment and their assessment are given in [5]. From the introduction of contaminants from the gas generating the plasma, GDW was used. In addition, a "trap" was used [5] in the form of an electrode with a positive potential to reduce contamination from materials located inside the process chamber. The authors of [5] found that the “trap” does not affect the concentration of aluminum in the melt, and GDW removes the hydrogen-oxygen gas mixture saturated with oxygen and vapors of iron, chromium, aluminum and copper. For a more complete assessment of the results we obtained, Table 2 provides a comparative information on the effect of GDW on contamination of the melt residues (row 1 process 28 without GDW and row 2: processes 80_81 with the use of GDW). The purifying factor shown in row 3 of the table: it is the ratio of the data in row 1 to the data in row 2. The values of the concentrations of the studied impurities in the rod grown in process 28, measured by the ICP MS method, are given in row 4. By the ratio of these lines 1 and 4 we can calculate the effective distribution coefficient of impurities during growth in the SMEBH, and in the 5th row is showed the equilibrium distribution coefficients of these impurities during silicon crystallization are given for reference.

Table 2. Comparative data obtained in [3] and [5].

Process data	**Al**	**Cu**	**Fe**	**Cr**	**O_2**
#28, without GDW, ppba [3]	20	57	190	7	6420
#80_81_85, with GDW, ppba,[5]	7.83	7.15	34.4	2.5	250
Purifying factor	3	8	6	3	26

Rod of 28 SMEBH, ppba [3]	0.987	0.0467	0.152	0.0198	8519
k_{ef}	0.0493	0.00082	0.0008	0.0029	1.327
k_0 [7]	0.002	0.0004	0.000008	0.000011	1.25-1.4

It can be seen from the presented data that the effective distribution coefficients of all impurities (except copper) are in satisfactory agreement with the values of their equilibrium distribution coefficients. The use of GDW increases the cleaning efficiency for aluminum 3 times according to Table 2, and more than 5 times according to Table 1. In our opinion, taking into account the concentration of controlled impurities, the data show satisfactory agreement. The value of the effective distribution coefficient of aluminum in the SMEBH is - (4.15-5.45) $x10^{-2}$, and in the FZ process (8.83-9.64) $x10^{-2}$.

The indicated values are estimated, since the process of silicon doping with aluminum occurs during the entire growing process.

Conclusions

1. Experimentally, using an installation for growing large-diameter rods from a skull with electron-beam heating, silicon rods were obtained, from which single crystals were grown by the FZ method. During the entire technological process, the concentrations of aluminum and oxygen impurities were determined.
2. The concentration of aluminum in silicon rods obtained using a Gas-Dynamic Window, decreased 3 times according to [5], and five times according to the results of our process compared to the process 28 described in [1]. In single crystals grown from the corresponding rods, the aluminum concentration also differs by a factor of 5. Thus, the experimental results obtained in the laboratory setup [5] are in satisfactory agreement with the data of our industrial process.
3. The oxygen content in the remnants of the melt in our process remained at the same level as in the experimental processes [5].
4. It was found that the Gas-Dynamic Window provides practically the same results when carrying out the processes of electron beam melting of silicon in laboratory and industrial installations.

Acknowledgement

The research leading to these results has been supported by the European Regional Development Fund project "Competence Centre of Mechanical Engineering", contract ID Nr. Nr.1.2.1.1/18/A/008 signed between Competence Centre of Mechanical Engineering and Central Finance and Contracting Agency, Research No. 2.1 "Process research and optimized development of electron beam heater for production of high purity materials".

References

[1] An. Kravtsov. “Developing of silicon growth techniques from melt with surface heating”, IOP Conf. Series Materials Science and Engineering 355 (2018) 012037
[2] Charles W Hanks, Charles D A Hunt “Method for growing crystals”, Patent US 3494804 A Continuation-impart of application Ser. No. 659,175,Aug. 8, 1967. This application July 15, 1968, Ser.No. 744,874, Int. Cl. B013 17/06; US. Cl. 148—1.6 11
[3] T.F. Ciszek “Growth of 40 mm diameter single crystals by a pedestal technique using electron beam heating”, "Journal of crystal growth" No. 12 (1972), page 281-287.
A. Krauze, J.Virbulis, An. Kravtsov. “Modeling electron beam parameters and plasma interface position in an anode plasma electron gun with hydrogen atmosphere”, IOP Conf. Series Materials Science and Engineering 355 (2018) 012008
[4] Kravtsov, J. Virbulis, A. Krauze “Reduction of impurity sources in Si crystal growth system
[5] with electron gun beam heating” FM&NT 2020, will be printed in IOP Conf. Series Materials Science and Engineering.
A. Kravtsov. “Contamination of Silicon during Electron Beam Melting”, in 32 European Photovoltaic Solar Energy Conference and Exhibition, Munich, Germany, 2016,
[6] E. Falkevich; E. Pulner; I. Chervony; L. Shvartsman; V. Yarkin; I. Sally "Technology of semiconductor silicon" Moscow Metallurgy, 1992, 408 pages.

Application of two types of embedded fiber-optic sensors for process-induced strain measurement in cement mixture

N Kosheleva, G Serovaev and G Gusev

Institute of Continuous Media Mechanics UB RAS, Perm, Russia

E-mail: kosheleva.n@icmm.ru, serovaev@icmm.ru, gusev.g@icmm.ru

Abstract. Two types of fiber-optic sensors applied to process-induced strain measurements of cement mixture are compared in this study. Point fiber-optic strain sensors based on the Bragg grating and distributed fiber-optic strain sensors were embedded into cement sample at the manufacturing stage. Embedded optical fibers did not have any additional protective coating except standard one (polyimide or acrylate). Both systems remain their integrity and operability after casting and hardening of the studied material and show the good agreement in strain measurements at the period taken into consideration.

1. Introduction

Structural health monitoring of various engineering structures is a complex and important task. Timely maintenance and safe operation of existing structures, as well as increased requirements for monitoring of the state of new buildings, have led to a growing demand for cost-effective and reliable measurement and control systems. Currently, a large number of different sensors (accelerometers, extensometers, etc.) are actively used to monitor the structures state. In particular, to assess the mechanical state of structures made of concrete and cement compositions, it is promising to use fiber-optic sensors (FOSs) embedded into the material [1–4]. The possibility of embedding the FOSs into the material allows to assess the strain state not only during the operation of the structure, but also during manufacturing stage. The use of optical fiber as a sensing element offers numerous advantages over traditional sensors. Fiber-optics sensors can be divided into two types: point FOSs and distributed FOS. FOSs can be multiplexed and immune to electromagnetic interference; are dielectrics, and also have high reliability and relatively small dimensions [5–7]. There is a big amount of studies, when various types of protective structures are used in the area of the fiber Bragg grating (FBG) during optical fiber embedment. Metallic, silicone, composite and other protective coatings are of great practical importance [8] since an optical fiber itself, even with a polymer coating (polyimide, acrylate), is prone to rupture during pouring of the material. One of the good solutions is to protect the optical fiber in the cable sheaths and use of additional protective coatings in the area of the sensitive element (FBG) [9,10]. However, the cost of using protective coatings is the incomplete strain transfer from the host material (concrete, cement) to the optical fiber and therefore the inconsistency of the mechanical behavior of the material in the grating area and in the host material in which the optical fiber is embedded. To restore the real picture of the mechanical state, it is necessary to use the recalculation of the values obtained with the FOSs [11]. The correction factors must be established in laboratory conditions and require knowledge of the strain distribution of the host material at the points, where the embedded sensors are located.

The possibility of using point and distributed fiber-optic sensors to monitor the mechanical state of a cement sample during the formation of the material was investigated in this paper. The optical fiber used for embedding has no additional protective coatings other than polyimide or acrylate. Despite the fact that this variant of optical fiber embedment has significant limitations for practical use in construction, the use of an unprotected optical fiber in laboratory conditions makes it possible to obtain a more reliable picture of the mechanical state of cement samples at the early stages of formation.

Content from this work may be used under the terms of the Creative Commons Attribution 3.0 licence. Any further distribution of this work must maintain attribution to the author(s) and the title of the work, journal citation and DOI.
Published under licence by IOP Publishing Ltd

2. Sample description and sensors used in the experiment

In the course of the work, the registration of internal strains arising in the cement mixture, poured into a cylindrical mold was carried out with the help of two types of fiber-optic sensors (FOSs). The strain along the length of the sample was measured at the stage of material formation and is limited in this work to first 72 hours after material casting. The cement mixture was poured into a polypropylene mold, in which the sample was subsequently formed and hardened. Before casting, two types of FOSs were placed in the empty mold: point FOSs based on the fiber Bragg gratings (FBG) and distributed FOS. Optical fiber with Bragg gratings was suspended in the center of the mold, passing through the free top surface and bottom surface of the cylindrical polypropylene mold. The distributed fiber-optic sensor was placed at a distance of 37 mm from the FBG sensors. After placing the fiber-optic sensors, layer-by-layer (with compaction and vibration) filling of the mold with a cement mixture to the level of the free surface was performed so that both lines of fiber-optic sensors remained immersed in the cement mixture. The vertical positioning and centering of the optical fibers were performed by attaching the weight. Figure 1 shows a scheme of a sample with embedded fiber-optic sensors based on FBG (s1–s6) and distributed fiber-optic sensor which utilizes the optical backscatter reflectometry (OBR) for measurements.

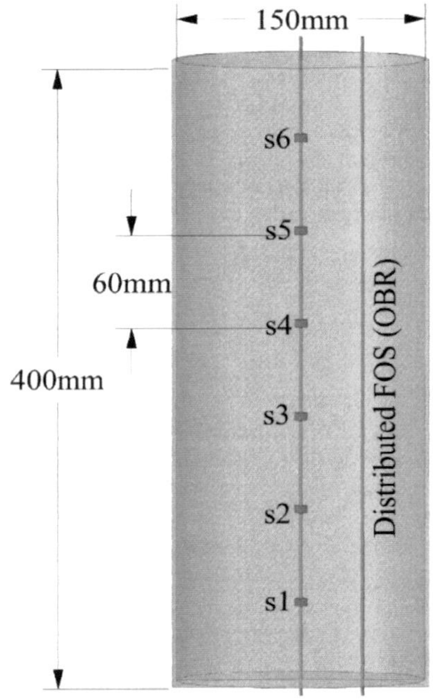

Figure 1. The scheme of fiber-optic sensors location in the sample.

The first fiber-optic line located inside the cement sample contains six FBGs and was designed in such a way that the FBGs were evenly distributed over the entire length of the sample. FBG represents the periodically changing refractive index of the fiber core over a certain length of an optical fiber [12]. Such a change in the refractive index profile makes it possible to filter the introduced broadband optical signal, reflecting only its narrow part with a central wavelength λ_B, which depends on the effective refractive index n of the optical fiber core and the grating period Λ [13]:

$$\lambda_B = 2n\Lambda \tag{1}$$

Under external action on the Bragg grating (strain or temperature change), the central wavelength of the reflected optical spectrum shifts. Such relationship between the shift of the Bragg wavelength, the longitudinal strain component along the fiber and temperature change can be written in the form [14]:

$$\frac{\Delta\lambda}{\lambda_B} = k\varepsilon_3 + S_T\Delta T \tag{2}$$

where ε_3 – longitudinal strain component along the fiber, $\Delta\lambda = \lambda - \lambda_B$ – central wavelength shifts in the current λ and initial λ_B moments of time, k and S_T – strain and temperature sensitivities, ΔT – temperature change In case, when the optical fiber in the region of the Bragg grating is in, or close to uniaxial stress state, the strain sensitivity k can be calculated by the given expression:

$$k = 1 - \frac{n^2}{2}\left(p_{12} - \nu(p_{11} + p_{12})\right) \quad (3)$$

where p_{11}, p_{12} – strain optic coefficients. For silica glass fiber $p_{11} = 0.113$, $p_{12} = 0.252$, n =1.458, ν = 0.17. Thus, under the uniaxial stress state of the Bragg grating, the coefficient k~0.78 [15].

In contrast to a point FOSs based on FBG, where only a small region of the optical fiber is sensitive to strain and temperature change, in a distributed FOS the entire length of the optical fiber is used as a sensitive element. Thus, no preliminary action on the optical fiber is required in order to change the refractive index. The distributed strain measurement method used in this work is based on measuring Rayleigh backscatter as a function of length by optical frequency-domain reflectometry (OFDR) technique [16]. The refractive index of any optical fiber along its length undergoes minor changes due to the small imperfections of the material. For different optical fibers, the change in refractive index along the length will be different, but will remain from measurement to measurement (in the absence of external influences) for a single optical fiber. Thus, a reference profile of the reflected signal can be formed for an optical fiber under test. Under an external influence on an optical fiber (strain, temperature), the signal shifts in the frequency domain. To calculate the strain or temperature change, the signal shift relative to the reference signal is estimated on a certain window and this shift is multiplied by the coefficient of strain or temperature sensitivity of the optical fiber. In this way distributed FOS is similar to weak Bragg grating so the relation (2) is valid. Distributed FOSs based on Rayleigh backscatter allows to obtain the distribution of the measured value with a very high resolution, however, it is associated with complex calculations and, as a consequence, the use of expensive equipment. In this work, an OBR 4600 backscatter reflectometer by Luna Innovations was used to carry out distributed measurements. Another limitations of distributed sensing are the short maximum length and low sampling frequencies compared to point sensors based on Bragg gratings.

In this study fiber-optic sensors which were embedded in the sample didn't have any additional protective coating, apart from the polyimide coating with thickness 12 μm for the optical fiber with Bragg gratings and acrylate coating with thickness 60 μm for optical fiber used for distributed sensing. After pouring the sample, measurements made by point FOSs were recorded with a frequency of 1 measurement per second and a distributed FOS with a frequency of 1 measurement per minute. The length of the used Bragg gratings was 5 mm. Each measurement point recorded with a distributed FOS was calculated on a 5 mm base.

3. Results and Discussion

The Figure 2 shows the strain values for a cement sample as red lines, obtained using point sensors based on FBG, within 72 hours from the moment of sample pouring. Also the strain values in a cylindrical sample, recorded using a distributed fiber-optic sensor are shown on the figure. These readings are displayed as a surface in the same time period. This figure clearly shows the difference between the two measuring systems based on fiber optics. Whereas FBG point sensors allow to obtain strain values only at certain points, the distributed fiber-optic sensor allows data to be obtained along the entire length of the sample.

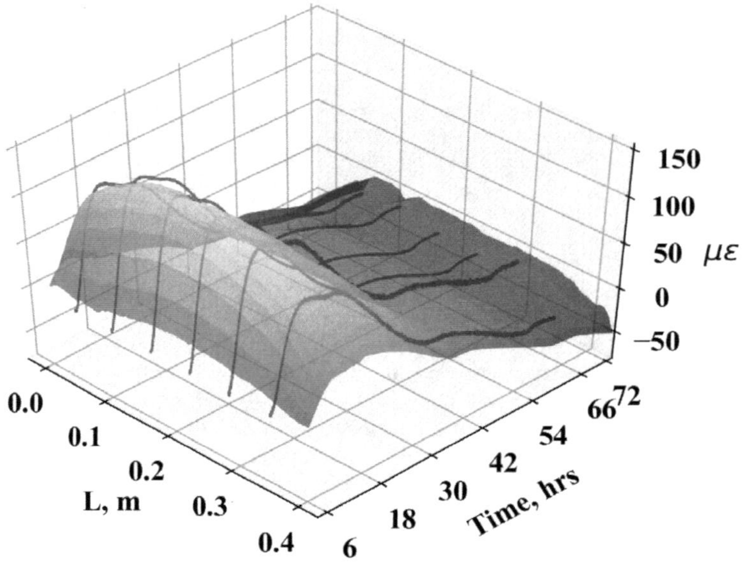

Figure 2. The strain values recorded by FBG and distributed fiber-optic sensor.

To compare the readings obtained from two different fiber-optic measuring systems, a graph (Figure 3) is given of the averaged values for all FBG sensors and the values obtained by distributed fiber-optic sensor averaged over the same fiber-optic sections where the FBG point sensors are located. It can be noted that in this case the difference between the averaged readings of the two types of sensors does not exceed 4% in the area of maximum strain and 6 με in the absolute value during the whole measurement period.

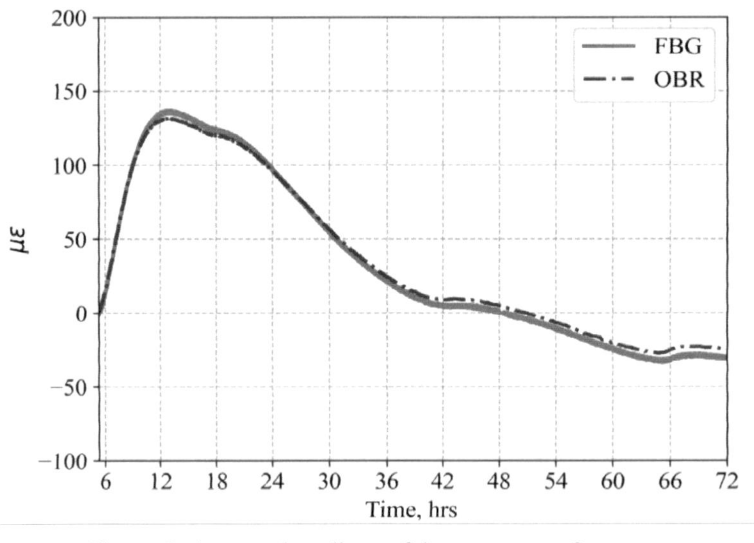

Figure 3. Averaged readings of the two types of sensors.

Figure 4 shows the strain distribution along the sample length, obtained at the time when the strains in the sample reached their maximum value. For clarity, an additional diagram of the sample is presented, where the position of the point sensors corresponds to the position of the red dots on the graph.

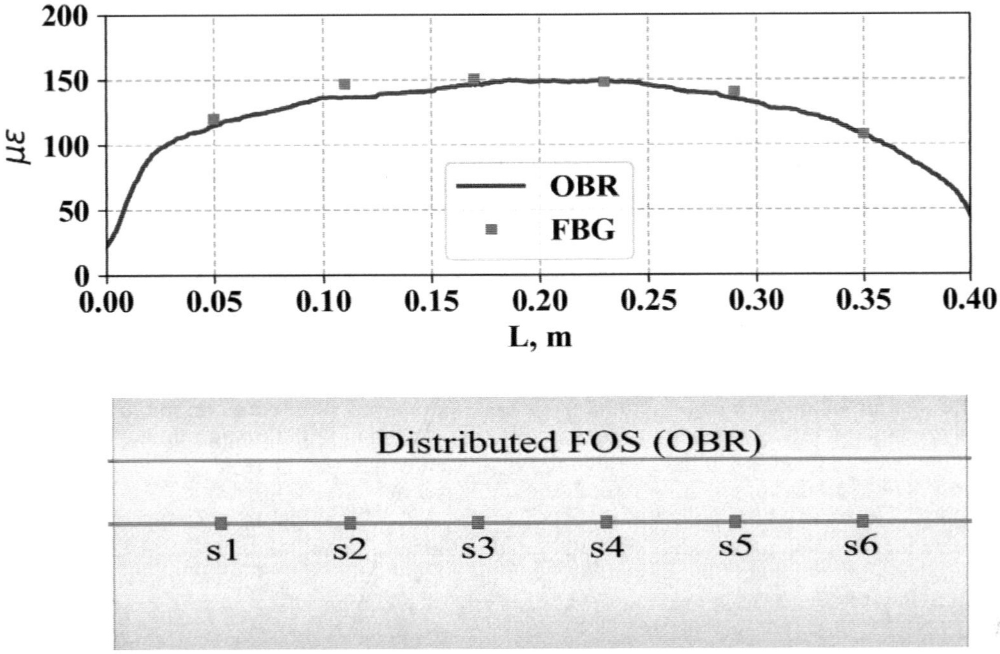

Figure 4. Strain distribution along the sample length.

The strain values obtained from point sensors correlates well with the strain values obtained from distributed fiber-optic strain sensors. Application of embedded fiber-optic sensors gives great opportunity to measure strain distribution inside studied structure with resolution which strongly depends on the type of sensing system.

4. Conclusions

The paper presents a study of the possibility of using two types of fiber-optic sensors: point, based on the Bragg grating, and distributed, based on measuring Rayleigh backscatter, for investigating processes and monitoring the mechanical state of a cement sample. For better understanding the process-induced strain formation process, the embedded fiber-optic strain sensors collected data from the moment the cement sample was poured. Using two systems, the history of process-induced strains arisen in the process of sample manufacture was recorded and stored. The work shows the results of measurements for the first 72 hours after sample pouring.

Both optical measuring systems have shown their good performance, despite the absence of additional protective coatings, apart from the standard ones. Analysis of the obtained data shows a good agreement of the measurement results of the two optical systems. It can be noted that the combined use of both systems can significantly improve the quality of the obtained data. However, the special features of each measuring system should be taken into account when using these measuring systems separately.

Acknowledgments

This work was supported by the grant of the President of Russian Federation for support of young Russian scientists and leading scientific schools (MK-2401.2019.1).

References

[1] Huo L, Cheng H, Kong Q and Chen X 2019 Bond-Slip Monitoring of Concrete Structures Using Smart Sensors—A Review *Sensors* **19** 1231
[2] Slowik V, Schlattner E and Klink T 2004 Experimental investigation into early age shrinkage of cement paste by using fibre Bragg gratings *Cem. Concr. Compos.* **26** 473–9
[3] Barrias A, Casas J R and Villalba S 2019 SHM of Reinforced Concrete Elements by Rayleigh

Backscattering DOFS *Front. Built Environ.* **5**

[4] Scott R H, Chikermane S, Vidakovic M, McKinley B, Sun T, Banerji P and Grattan K T V 2019 Development of low cost packaged fibre optic sensors for use in reinforced concrete structures *Measurement* **135** 617–24

[5] Leung C K . 2001 Fiber optic sensors in concrete: the future? *NDT E Int.* **34** 85–94

[6] Grattan K T V and Sun T 2000 Fiber optic sensor technology: an overview *Sensors Actuators A Phys.* **82** 40–61

[7] Barrias A, Casas J and Villalba S 2016 A Review of Distributed Optical Fiber Sensors for Civil Engineering Applications *Sensors* **16** 748

[8] Scott R, Vidakovic M, Chikermane S, McKinley B, Sun T, Banerji P and Grattan K 2019 Encapsulation of Fiber Optic Sensors in 3D Printed Packages for Use in Civil Engineering Applications: A Preliminary Study *Sensors* **19** 1689

[9] Leng J S, Winter D, Barnes R A, Mays G C and Fernando G F 2006 Structural health monitoring of concrete cylinders using protected fibre optic sensors *Smart Mater. Struct.* **15** 302–8

[10] Majumder M, Gangopadhyay T K, Chakraborty A K, Dasgupta K and Bhattacharya D K 2008 Fibre Bragg gratings in structural health monitoring—Present status and applications *Sensors Actuators A Phys.* **147** 150–64

[11] Huang Y, Bao Y, Chen G and Zhou Z 2019 A constrained cylinder model of strain transfer for packaged fiber Bragg grating sensors embedded in inelastic medium *Struct. Control Heal. Monit.* **26** e2335

[12] Othonos A 1997 Fiber Bragg gratings *Rev. Sci. Instrum.* **68** 4309–41

[13] Mao J, Xu F, Gao Q, Liu S, Jin W and Xu Y 2016 A Monitoring Method Based on FBG for Concrete Corrosion Cracking *Sensors* **16** 1093

[14] Lammens N, Luyckx G, Degrieck J and De Waele W 2011 Experimental determination of the multi-axial strain transfer from CFRP-laminates to embedded Bragg sensor *Smart Struct. Mater. 5th ECCOMAS Themat. Conf. Proc.* 482–5

[15] Kablov E N, Sivakov D V., Gulyaev I N, Sorokin K V., Fedotov M Y, Dianov E M, Vasil'ev S A and Medvedkov O I 2011 Application of optical fiber as strain gauges in polymer composite materials *Polym. Sci. Ser. D* **4** 246–51

[16] Güemes A, Fernández-López A and Soller B 2010 Optical Fiber Distributed Sensing - Physical Principles and Applications *Struct. Heal. Monit. An Int. J.* **9** 233–45

Formation of a composite structure in Ti-Cr alloys using reversible hydrogen alloying

S V Skvorcova[1], G T Zaynetdinova[1], M B Afonina[1] and A Ya Anackaya

[1] Moscow Aviation Institute (National Research University), 4 Volokolamskoe shosse, Moscow, 125993, Russia

E-mail: GZaynetdinova@gmail.com

Abstract. The aim of this work is to study the regularities of the influence of hydrogen doping on the course of phase and structural transformations in Ti-Cr alloys. The influence of the amount of hydrogen on the preferred sites of intermetallic separation was studied. The influence of hydrogen on the lines of variable solubility of the state diagram Ti-Cr is studied. The possibility of controlling the process of structure formation in titanium alloys with β-eutectoid-forming stabilizers by additional alloying with hydrogen and obtaining such types of structures that are not typical for them with traditional methods of technological influence is shown.

1. Introduction

The development of science and technology is associated with a constant increase of material properties requirements. One of the main trends in this direction is the creation of new composite materials, both polymer- and metal-based. Reinforcing components of very hard compounds such as TiB, TiC, TiN are usually used to create metal-matrix composite materials. Selfpropagating high-temperature synthesis, reactive hot pressing, spark plasma sintering, etc. can be used for this cause. One of the main disadvantages of these methods is their high cost and the impossibility of obtaining large-sized workpieces. In addition, the materials obtained have rather low plasticity values so far. Thus, the development of a new method for creating a composite structure in industrial titanium-based alloys, based on reversible hydrogen alloying, corresponds to the general world trend and is very relevant. Reversible hydrogen alloying is one of the most effective ways to transform the structure in titanium alloys. The unique feature of hydrogen as an alloying element lies in the fact that due to its high sorption capacity, diffusion mobility and strong influence on the processes of phase transformations and structure formation, it allows reversible alloying without changing the material aggregate state. Hydrogen increases the stability of the β-phase, reduces the critical cooling rates and the characteristic temperatures of martensitic transformation. This makes it possible to obtain a large number of hydrogenstabilized metastable phases under "mild" temperature-rate cooling conditions. As a result of the uneven distribution of hydrogen between the α- and β-phases and a significant difference in the diffusion coefficients of hydrogen and the main alloying elements between the α- and β-phases, a redistribution of the main alloying components occurs. It creates conditions for the phase transformations occurrence and the formation of a structure unusual for alloys in equilibrium condition. Thus, changing the thermo-kinetic and concentration parameters of the thermal hydrogen treatment process significantly expandsthe ability to control the structural state of titanium alloys, and, accordingly, the level of their properties. To solve the fundamental problem of creation of new composite materials based on high-alloyed industrial titanium alloys, it is necessary to conduct a study of added hydrogen content influence on the phase composition and structure formation, the preferred locations for the precipitation of intermetallic phases and their morphology as well. It is necessary to study structure transformations in the process of hydrogen desorption and to establish the temperature-time parameters of structure stability. It is vital to establish the dependence of the physical and mechanical properties of the composite material on the formed metastable structure. Formation of the unusual composite structure containing intermetallic phases in materials based on high-alloyed industrial titanium alloys will provide a good combination of strength and

Content from this work may be used under the terms of the Creative Commons Attribution 3.0 licence. Any further distribution of this work must maintain attribution to the author(s) and the title of the work, journal citation and DOI.
Published under licence by IOP Publishing Ltd

plasticity, and will also improve their thermal stability. In general, it is expected to have a higher level of properties on commercial alloys compared to other titanium-based composite materials.

2. Results and discussion

In this work, the nature of the interaction of hydrogen with titanium alloys alloyed mainly with β-eutectoid stabilizers was studied. The research was carried out on bars made of experimental Ti-Cr alloys (Ti-1%Cr; Ti-2%Cr; Ti-4%Cr; Ti-5%Cr; Ti-6Cr; Ti-8%Cr; Ti-12%Cr; Ti-15%Cr; Ti-20%Cr).

If hydrogen alloying of alloys containing β- isomorphic stabilizers causes the appearance of intermediate hydrides, then, as was shown in [2], in titanium alloys with β-eutectoid-forming elements, it stabilizes titanium intermetallides. Chromium was considered one of the most promising alloying additives, so the study of phase and structural transformations occurring in Ti-Cr alloys was devoted to many works, summarized in the monograph [9]. Chromium has a high solubility in β - titanium at elevated temperatures, and in α-phase its solubility does not exceed 0.5%. At a temperature of 670°C in the region of titanium-rich alloys, the β- phase undergoes eutectoid decay, which proceeds very slowly and is not realized at normal cooling rates. Therefore, the structure of alloys in the annealed state is usually represented by α- and β-phases. Intermetallide $TiCr_2$ of stoicheometric composition, containing 68.4%Cr, can have two modifications: at temperatures below 1150°C it has a FCC lattice with 24 atoms in the unit cell, and above 1150°C it has a hexagonal lattice. Alloying titanium with chromium allows you to get high strength both due to the solid-solution hardening of the β - phase, and due to the structure obtained as a result of strengthening heat treatment. However, during long-term isothermal exposures at elevated temperatures, eutectoid decomposition occurs and $TiCr_2$ intermetallide is released mainly along the grain boundaries, which leads to a sharp loss of plasticity, so the chromium content in industrial alloys was subsequently limited [3].

Additional alloying of alloys with up to 1.4% hydrogen was carried out by thermal diffusion in the Siverts unit at a temperature of 700°C.

After saturation of Ti-Cr alloy samples with hydrogen, they were heated to a single-phase β- state and cooled in water after 2 hours of exposure. The selection of the heating temperature for quenching was based on the Ti-Cr state diagram [3], taking into account that hydrogen lowers the temperature of the polymorphic β→α- transformation. The formation of the phase composition and structure in hardened titanium alloys with a chromium content from 1 to 8% with additional alloying with hydrogen up to 0.4% follows general regularities [1]. Suppression of the β→ω- transformation in Ti-6%Cr and Ti-8%Cr alloys and Expansion of the β-phase existence region are observed. In alloys with 1, 2, and 4%Cr, after quenching, martensite α′ and δ-hydride TiH_2 are formed in the structure, and in Ti-5%Cr alloy with an increase in the hydrogen content from 0.15 to 0.4%, the two-phase (α′+TiH_2)- structure is replaced by a three-phase structure – (α′+TiH_2+β).

The most interesting behavior under the action of hydrogen is that of alloys with a chromium content of more than 12%. Before hydrogenation, ti-12% Cr, Ti-15% Cr, and Ti-20% Cr alloys were in the single-phase β-state. After hydrogenation annealing, the β-solid solution was stratified into hydrogen-rich and hydrogen-depleted micro-volumes and the intermetallic compound TiCr2 was isolated from the latter, which was confirmed by the data of micro-x-ray spectral analysis (figure 1).

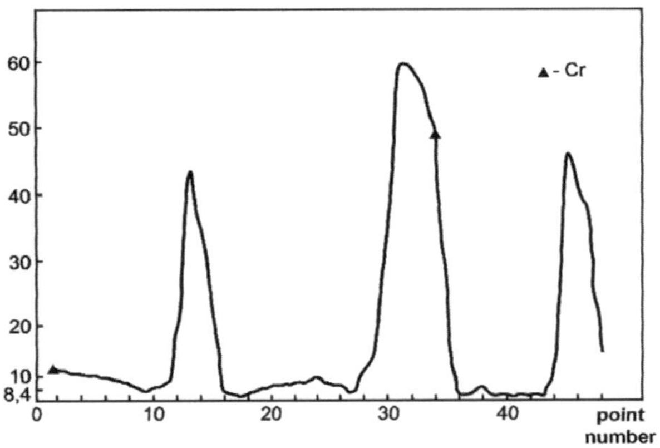

Figure 1. Results of micro-x-ray spectral analysis of a sample made of Ti-12%Cr alloy alloyed with 0.4% hydrogen.

In order to determine whether hydrogen initiates the release of an intermetallic compound, samples of the same compositions with an initial hydrogen content of 0.006% were subjected to isothermal treatment at 700°C, i.e. at the temperature at which hydrogenation annealing was performed. In alloys containing 12 and 15%Cr, the formation of an intermetallic compound was not detected, and in the ti-20% Cr alloy, during isothermal aging, the intermetallic compound was released mainly along the grain boundaries (figure 2a), as in the case of the Ti-20% Cr alloy, additionally alloyed with 0.15% hydrogen (figure 2b).

At the next stage of the work, the temperature-concentration region of existence of a two-phase (β+TiCr2) structure was determined by the test quenching method. In a Ti-20% Cr alloy with an initial hydrogen content (0.006%), the transition temperature β+TiCr2/β corresponds to 840°C, after quenching, the single-phase β-state is fixed in the structure.

At the same time, for Ti-20%Cr alloy, additionally alloyed with 0.15% hydrogen, quenching with 840°C does not cause visible changes in the structure, and it is possible to obtain a single-phase β- state only after heating to 940°C. It was found that with an increase in the hydrogen content in the alloy, the amount of intermetallide increases and the preferred locations of its release change.

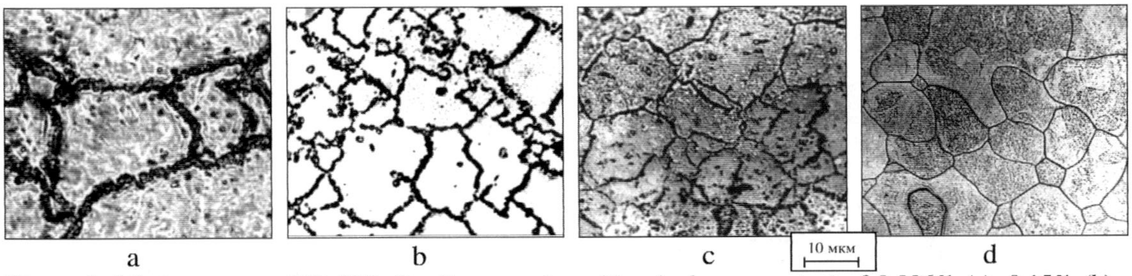

Figure 2. Microstructure of Ti-20% Cr alloy samples with a hydrogen content of 0.006% (a); 0.15% (b), 0.4% (c) and 0.8% (d) after quenching at various temperatures: a, b-700°C; c-840°C; d-900°C

When up to 0.14% of hydrogen is introduced, its release occurs as separate fairly large particles, mainly located along the boundaries of the initial β- grains (figure 2a and 3a). Additional alloying of 0.4%H reduces the size of particles released both along the body of the β- grain and along its boundary (figure 2c and 3b). With a further increase in the amount of hydrogen introduced to 0,8%, the size of intermetallic particles decreases even more, and the preferred place for its release becomes the body of the β- grain (figure 2d and 3c).

In addition, it should be noted that with an increase in the amount of hydrogen in the alloy, the temperature range of existence of the intermetallic compound also expands. If the temperature of the β+$TiCr_2$/β-transition (A_{cm}) in an alloy of Ti-20% Cr with 0.15% H corresponds to 940°C, then at a hydrogen content of 0.4% it increases to 1000°C, and at 0.8%H it becomes higher than 1100°C.

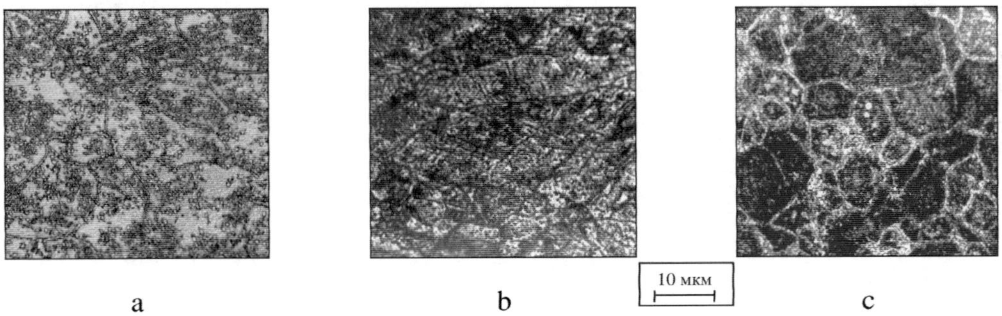

Figure 3. Microstructure of Ti-12%Cr alloy samples after hydrogenation at 700°C and isothermal exposure for 10 hours.

Studies on Ti-Cr alloys have shown that additional alloying with hydrogen leads to a shift of the line of variable chromium solubility in the β-phase in the state diagram towards a lower chromium content and a significant expansion of the two-phase (β+$TiCr_2$)-region. Based on the conducted studies, a fragment of the diagram was constructed showing the change in the position of the line of variable solubility of $TiCr_2$ in the β-phase depending on the hydrogen content (figure 4). In addition, the diagram shows dotted lines corresponding to the Ti-Cr equilibrium diagram and the α/β-transition line (Ac_3) for alloys with a low chromium content, additionally alloyed with 0.4% hydrogen.

Thus, studies have shown that additional hydrogen alloying of titanium alloys containing up to 5% chromium lowers the temperature of the α/β-transformation and expands the area of existence of the β-phase, while in alloys with a chromium content of more than 6%, there is a significant narrowing of the area of existence of the β- phase and expansion of the two-phase (β+$TiCr_2$)-region.

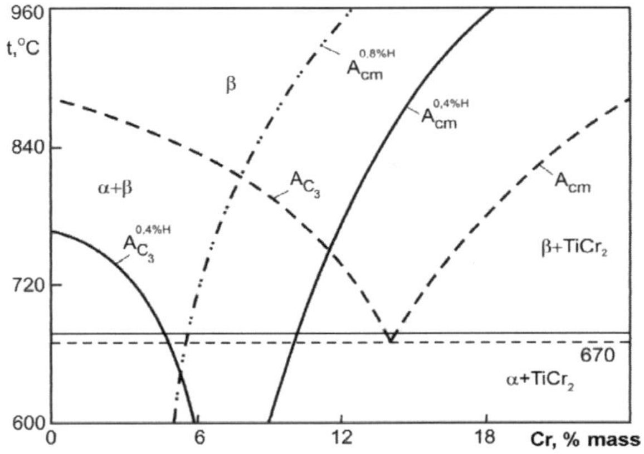

Figure 4. Diagram showing the change in the line of variable solubility of $TiCr_2$ intermetallide in the β- phase (Acm) depending on the hydrogen content in Ti-Cr alloys (the dashed line corresponds to the equilibrium diagram of the Ti-Cr state [3])

The study of structure formation in titanium alloys with a high content of β- eutectoid-forming stabilizers under the action of hydrogen is of interest for further research.

3. Conclusions

The possibility of controlling the process of structure formation in titanium alloys with β-eutectoid-forming stabilizers by additional alloying with hydrogen and obtaining such types of structures that are not typical for them with traditional methods of technological influence is Shown.

It has been established that additional hydrogen alloying in Ti-Cr alloys containing up to 5% Cr lowers the α/β-transformation temperature and increases the stability of the β- phase with respect to the α-phase, and in Ti-Cr alloys with a chromium content greater than 6% reduces the stability of the β - phase with respect to the $TiCr_2$ intermetallic compound.

It was found that the preferred locations for intermetallic separation depend on the hydrogen content in the alloy.

References

[1] Ilyin A A 1994 Mechanism and kinetics of phase and structural transformations in titanium alloys (Moscow: Science) p 304.
[2] Makkvillan M K 1967 Phase transformationes in titanium alloys (Moscow: Metallurgy) p 75
[3] Cvikker U 1979 Titanium and titanium alloys (Moscow: Metallurgy) p 512

IOP Conf. Series: Materials Science and Engineering 1100 (2021) 012040 doi:10.1088/1757-899X/1100/1/012040

Metal matrix composites based on valence-unstable systems

E S Clementyev[1], D A Serebrennikov[1] and V V Sikolenko[1,2]

[1]REC "Functional Nanomaterials", Immanuel Kant Baltic Federal University, 14 Nevskogo str, Kaliningrad, 236016, Russia
[2]Joint Institute for Nuclear Research, 6 Joliot-Curie str, Dubna, 141980, Russia

Email: dimafania@mail.ru

Abstract. The work is focused on the thermal expansion of composites based on valence-unstable systems. The main motivation was to create invar-like materials of a new type and to develop approaches to the quantitative description of coefficients of thermal expansion (CTE) of both the composite itself and its key component - a material with a negative contribution to thermal expansion due to valence fluctuations. The experimental data on the linear CTE of Al/SmB6 composite was performed by capacitive dilatometry in the temperature range 5-220K. It turned out that this composite shows invar properties up to 60K. Several models were used to describe thermal expansion in Al/SmB6. A quantitative approach is proposed for describing the negative magnetic contribution and the lattice contribution to the CTE of systems based on intermediate-valence samarium hexaboride.

1. Introduction

The terminology accepted in solid state physics reserves the names "invar" and "invar alloys" for transition metal compounds, usually based on iron and nickel [1-3]. These compounds exhibit close to zero CTE values over a wide temperature range due to the competition of the magnetic and lattice contributions to thermal expansion [4]. Invars have found many application niches, but there are still many open questions in the physics of this class of materials. Compounds that do not contain magnetic constituents based on d-elements, but at the same time demonstrate CTE values less than $|2|x10^{-6}$ K^{-1} in a given temperature range are called invar-like, and the related physical mechanisms may be different in origin. One of the ways to obtain an invar-like material is to synthesize a composite of two components with negative and positive contributions to the CTE and their compensation in a fairly wide temperature range. Reviews with a detailed description of such composites, which are characterized by invar-like properties in terms of their thermal expansion are given in [5]. The choice of material with a positive CTE is quite wide, while there are very few materials with a negative CTE. Most of invar-like composites described in the literature are based on the use of complex oxides such as: ZrW_2O_8, $Sc_2W_3O_{12}$, $Y_2W_3O_{12}$ and others [6-8]. In these systems the negative thermal expansion is driven by transversal thermal motion of oxygen atoms leading to apparent decrease in cation-anion distances [9]. Meanwhile, systems with a pure electronic mechanism of negative thermal expansion are known. The electronic degrees of freedom associated with this anomalous contribution to thermal expansion rely on phenomena such as strong electronic correlations and valence instabilities. These phenomena are observed in some compounds of f-elements like samarium, ytterbium, and europium. Such compounds are classified in condensed matter physics as systems with strongly correlated electrons (SCES). Systems with intermediate valence (IV) represent a special class of SCES. Valence instability in IV systems leads to strong fluctuations between two valence configurations. In the case of samarium ions, fluctuations occur between electron configurations with Sm2+ and Sm3+ ions, which have very different radii. A strong dependence of the samarium valence on temperature, coupled with a strong difference in the size of the configurations, leads to a negative contribution to the CTE, since a smaller configuration (namely, Sm3+) is populated as the temperature increases.

Content from this work may be used under the terms of the Creative Commons Attribution 3.0 licence. Any further distribution of this work must maintain attribution to the author(s) and the title of the work, journal citation and DOI.
Published under licence by IOP Publishing Ltd

Table 1. Negative thermal expansion values of selected IV systems

Materials	NTE temperature range, K	Minimum CTE value, $x10^{-6}$ K^{-1}	Reference
$Sm_{0.8}B_6$	5-90	-4	10
SmB_6	10-130	-4.2	10
$Sm_{0.9}La_{0.1}B_6$	55-156	-3.1	10
$Sm_{0.78}La_{0.22}B_6$	20-200	-3.1	11
$Sm_{0.5}La_{0.5}B_6$	10-160	-1.3	11

SmB_6 is one of the most interesting objects in the SCES domain. Samarium hexaboride, which belongs to the class of IV systems, shows numerous anomalies of physical properties and continues to attract the attention of both theorists and experimenters. SmB6 and systems based on it show a significant negative contribution to CTE. Table 1 shows the quantitative parameters of several SmB6-based systems, as well as the temperature range of negative CTE values and the amplitude of this anomalous contribution.

As we can see, these systems based on valence-unstable samarium look promising for creating a composite with invar-like properties in the temperature range below 100 K. The main goal of the present work was to create and investigate invar-like material of a new type – a composite based on a metal and an IV system. Additional important goals were to develop model approaches to the description of the CTE of the composite itself, as well as the electronic and lattice contributions to the CTE for a key component - a system with negative thermal expansion.

2. Experimental results

A sample with a component ratio of 25% SmB6 and 75% Al was prepared by hot pressing. All details concerning the initial components, the synthesis process, diagnostics, and primary characterization of the sample (optical and electron microscopy, x-ray micro-tomography, and elemental composition analysis) will be described in a separate publication. In this paper in the chapter related to sample characterization we show the results of a study of the composite by x-ray diffraction (XRD) technique.

The composites were characterized by XRD using synchrotron-generated X-Ray beam with E = 15.2 keV (λ = 0.826 Å) at the STM beamline of Kurchatov synchrotron radiation source KSRS (Figure 1). XRD pattern is fully defined by peaks attributed to Al and SmB6 thereby indicating that no chemical reactions occurred to form novel phases.

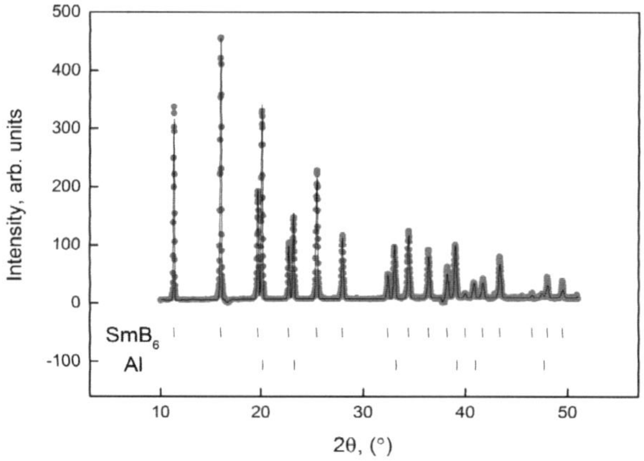

Figure 1. XRD pattern of Al/SmB_6 composite at room temperature. Symbols correspond to the experimental data. Line is a result of Rietveld refinement using the FullProf software package.

The key experiment of this work and the feasibility study of creating composite invar-like materials based on samarium hexaboride was the measurement of relative elongation on a dilatometer. The dilatometric method allows calculating the value of such a thermodynamic parameter as CTE. The measurements of Al/SmB6 composite were carried out by capacitive dilatometry using a measuring cell/option installed in the PPMS (Quantum Design) system (for more details see [12]) in the temperature range 5-210 K.

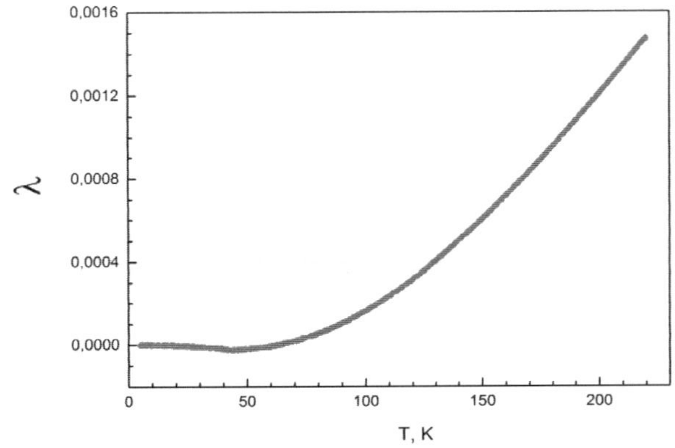

Figure 2. Relative thermal expansion of Al/SmB_6 composite.

Figure 2. shows relative expansion (elongation) (dL/L) as a function of temperature for the composite sample under study. The contribution of the cell material to the relative elongation was estimated by separate measurements and was extracted from the data. CTE of Al/SmB6 was calculated based on the measurements of relative elongation. The temperature dependence of CTE is depicted in Figure 3. In should be noted that to prevent oscillations of the CTE, the temperature dependence of relative elongation was preliminarily smoothed using the spline function.

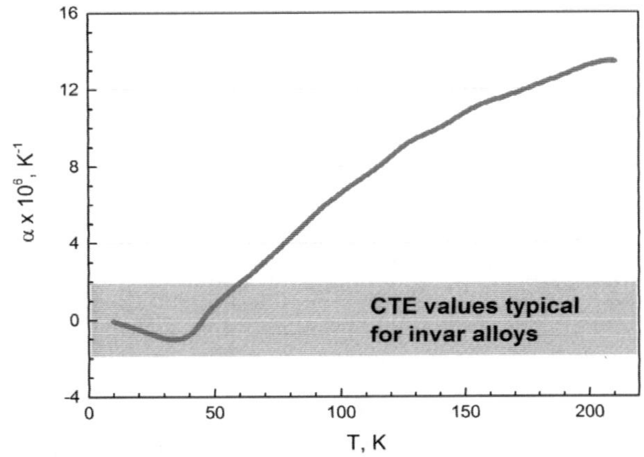

Figure 3. Linear coefficient of thermal expansion of Al/SmB_6 composite.

3. Discussion

The results of fitting of the X-ray diffraction pattern showed the compliance of the sample with the claimed composition and proved the absence of parasitic phases in the accuracy of X-ray phase analysis. Measurements of the thermal expansion of the sample under study allow us to conclude that as a result of compensation of the positive contribution of aluminum by the negative contribution to the CTE from the intermediate-closed component of the composite, the sample is indeed an invar-like material up to a temperature of 60 K.

Several models have been developed aiming to quantitatively describe the thermal expansion of composites. Many of them are listed in [13]. The simplest approach is the rule of mixtures (ROM) (1):

$$\alpha_c = \alpha_f v_f + \alpha_m v_m \quad (1)$$

where the subscripts *c*, *f* and *m* denote the composite, filler and matrix respectively; α – CTE; υ – volume fraction. More sophisticated treatments are based on widely used Turner (2) and Kerner (3) models:

$$\alpha_c = (\alpha_f \upsilon_f K_f + \alpha_m \upsilon_m K_m)/(\upsilon_f K_f + \upsilon_m K_m) \tag{2}$$

$$\alpha_c = \alpha_m \upsilon_m + \alpha_f \upsilon_f + \upsilon_m \upsilon_f (\alpha_f - \alpha_m) \frac{K_f - K_m}{\upsilon_m K_m + \upsilon_f K_f + 3K_m K_f / 4G_m} \tag{3}$$

where K – bulk modulus; G – shear modulus.

In the following part we analyze the dilatometric experimental data based on the thermo-elastic models given by Eqs. (1-3). In our calculations we used literature values for thermal expansion and elastic properties of Al and SmB_6 [10,14]. A comparison of experimental and computed data is depicted in Figure 4.

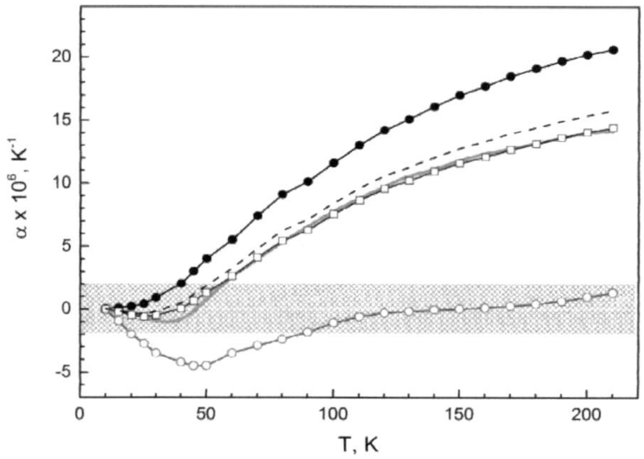

Figure 4. Plot of experimental CTE values (orange solid line) along with curves representing CTE values calculated using thermo-elastic models for the Al/SmB_6. Black filled circles – CTE of Al; red open circles – CTE of SmB_6; dashed line – results of the ROM model assuming SmB_6-25 vol.%; blue open squares – results of the ROM, Terner and Kerner models assuming various SmB_6 volume fraction (30%, 25% and 29%, respectively). The shaded area corresponds to the linear CTE values typical for invar alloys.

Our calculations have shown that in the case of SmB6, two model approaches can be applied to simplify the problem. These approaches rely on phenomenological models with a relatively small number of parameters. To calculate the lattice contribution to CTE, we can use the superatom model, which was previously successfully used to describe the dispersion of phonons, the density of phonon states, and the specific heat of higher borides, including hexaborides [15]. At the same time negative electronic contribution to CTE caused by intermediate valence of samarium ions can be described using the Weiss phenomenological model (the invar model) [16], which was previously successfully used to describe CTE in classical invar alloys, as well as in intermediate-valence systems based on plutonium [17]. In the domain of SCES, plutonium is considered as a 5f electron homologue of the 4f element samarium due to a similar structure of the partially filled electron shells and related anomalous properties.

Figure 5 shows the Weiss model fit of the electronic contribution to the CTE of samarium hexaboride. The anomalous electronic part of CTE was estimated by subtracting the lattice contribution calculated within the superatom model from the experimental data obtained by XRD.

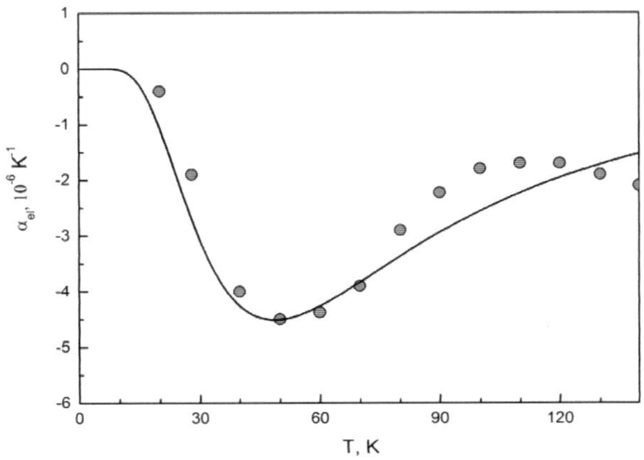

Figure 5. Electronic contribution to the CTE of SmB6. Dots represents experimental data; line is a result of Weiss model.

4. Conclusion

The synthesis of a new type of composite and its study by several experimental techniques methods demonstrate that intermediate-valence systems can be successfully used to create invar-like materials. In particular, the negative anomalous electronic contribution to CTE caused by valence-unstable samarium ions allowed to fully compensate for the positive lattice contribution to CTE in the composite Al/SmB6-25% at temperatures up to 60K. The Turner model was proved to perform very well to reproduce the thermal expansion in Al/SmB6. So this model is probably well-suited for the new type of metal matrix composites based on intermediate-valence systems. Two model approaches have been successfully applied to describe both the electronic contribution (phenomenological invar Weiss model) and the lattice contribution (phenomenological superatom model) to the CTE in SmB6.

Acknowledgments

This work was financially supported by the Russian Foundation for Basic Research, grant 19-02-00953-a. We are grateful to Dr. A.V. Aborkin for his help with the composite synthesis, Dr. A.L. Freydman for dilatometry experiment, and to Dr. A.L. Trigub and Dr. N.A. Kolyshkin for their help with the XRD experiment.

References

[1] Wasserman E F, Acet M in *Magnetism and Structure in Functional Materials* (eds Planes A, Manosa L, Saxena A) Springer, 2005, 177-196
[2] Shiga M 1996 *Current Opinion in Solid State and Materials Science* **1** 340-348
[3] Wasserman E F *in Ferromagnetic Materials* Vol. 5 (eds Buschow K H J &Wohlfarth E), North-Holland, Amsterdam, 1990, 237-322
[4] van Schilfgaarde M, Abrikosov I A, Johansson B 1999 *Letters to Nature* **400** 46-49
[5] Sidhu S S, Kumar S, Batish A 2016 *Critical Reviews in Solid State and Materials Sciences* **41** 132-157
[6] Holzer H, Dunand D C 1999 *J. of Materials Research* **14** 780-789
[7] Gao S, Zhao N, Liu Q. et al. 2019 *J. of Alloys and Compounds* **779** 108-114
[8] Das Sa, Das Si, Das K 2014 *Ceramics International* **40** 6465-6472
[9] Sleight A W 1998 *Annu. Rev. Mater. Sci.* **28** 29-43
[10] Nefedova E V , Alekseev P A, Klement'ev E Sm et al. 1999 *JETP* **88** 565-573
[11] Nefedova E V , Alekseev P A, Lazukov V N, Sadikov I P 2003 *JETP* **96** 1113-1121
[12] Freidman A L , Popkov S I, Semenov S V, Turchin P P 2018 *Technical Physical Letters* **44** 123-125
[13] Hsien C L, Tuan W H 2007 *Materials Science and Engineering A* **460-461** 453-458
[14] Nakamura S, Goto T, Kunii S, et al 1994 *J. Phys. Soc. Jpn.* **63** 623-636
[15] Serebrennikov D A, Clementyev E S, Alekseev P A 2016 *JETP* **3** 452-460
[16] Weiss R J 1963 *Proc. Phys. Soc.* **82**, 281
[17] Lawson A C, Roberts J A, Martinez B, et al. 2006 *Phil. Mag.* **86** 2713;

Mechanochemical method for recycling construction industry waste

M A Abramov [1], E G Stepanov [2], V B Dobrokhotov[1] and E S Egorov[1]

[1] Yaroslavl State Technical University, 150023, Russia, Yaroslavl, Moskovsky prospect, 88
[2] Rybinsk State Aviation Technical University, 152934, Russia, g. Rybinsk, Yaroslavskaya oblast, ul. Pushkina, 53

E-mail: abramovma@ystu.ru

Abstract The article deals with the use of metallurgical slag, which served as insulation in the floors of residential buildings, as a secondary technogenic raw materials. The principal possibility of creating components from it for reaction powder concrete using the method of Mechanochemistry has been shown.

1. Introduction

Russian Federal law No. 384 establishes the basic principles according to which construction waste, as well as waste from the dismantling of buildings and structures, must be safely disposed of. Construction waste in Russia is mainly stored in landfills, which negatively affects the ecology of the environment, and also leads to unjustified losses of raw materials [1]. On the other hand, major repairs and renovation of residential buildings lead to a significant increase in the amount of waste generated.

In the 60s of the 20th century, the insulation of attic floors of residential buildings was carried out by backfilling with slag of metallurgical industries. This material was included in the construction normative documents for construction heat engineering with a density of 400-800 kg / m3 and a thermal conductivity coefficient of 0.11-0.18 W/(m^2*°C).To date, the slag has crumpled, including as a result of regular humidification, and has ceased to function as a heat-insulating material.

Standard technologies for building reconstruction involve replacing the slag with its disposal at the landfill.

We have conducted research on application of backfill slag insulation and its use as a technogenic raw material for creating a functional product [2,12].

2. Result and discussion

The object of the study was backfill slag insulation, removed during the overhaul of the roof of a residential building in Yaroslavl. The characteristics of the investigated slag are shown in table 1 and figure 1. The true density of the initial slag was 3.8 g / cm^3

Table 1. Fractional composition of the initial slag

Specifications	Particle size, mkm / fraction of particles, % of mass.								Specific surface area, m2 / g
	>10	5-10	2.5-5	25-2.5	0.63-1.25	0.315-0.63	0.14-0.315	<0.14	
Fractional composition	9.5	14.5	5.5	15.5	13	12	19.5	10.5	8.57

Content from this work may be used under the terms of the Creative Commons Attribution 3.0 licence. Any further distribution of this work must maintain attribution to the author(s) and the title of the work, journal citation and DOI.
Published under licence by IOP Publishing Ltd

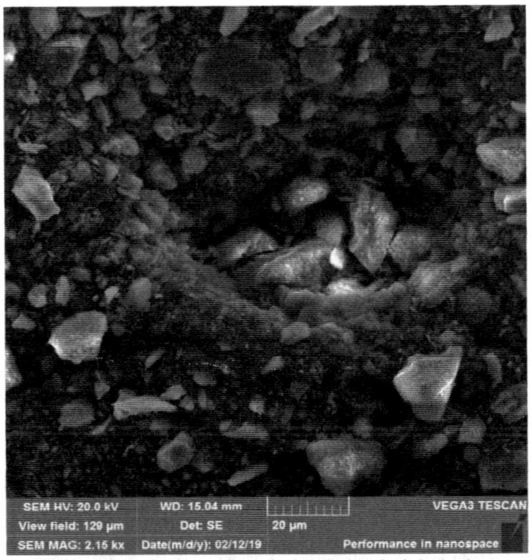

Figure 1. Micrography of the initial slag

Thus, waste is a polyfractional powder of 3-rd class of danger.

The derivatogram of the slag sample shown in figure 2 shows the presence of sorbed water, as well as organic and carbonate impurities with an annealing peak of about 500 °C.

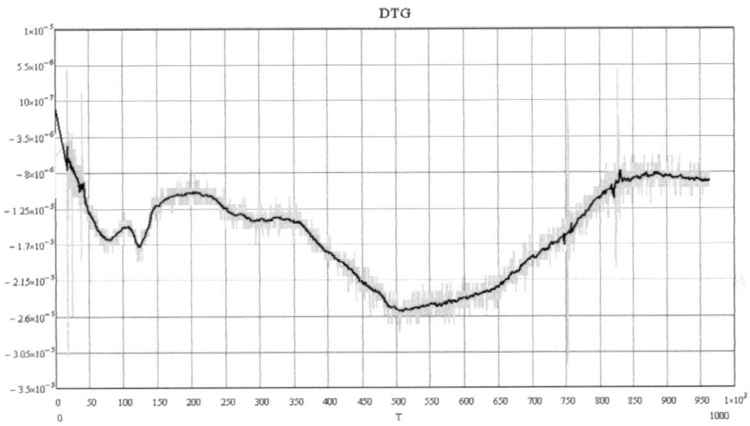

Figure 2. DTG curves of the source slag sample

Recycling of construction waste usually involves their grinding. Vibro-jaw crushers [3] and planetary ball mills [4] are used as grinders.

We proposed to use method of mechanochemistry and disintegrator technology for directed regulation of the properties of the recycled slag, and a universal disintegrator-activator (UDA) was chosen as the device that performs such a deliberate change. Disintegrator technology allows a wide range of physical and chemical properties of raw materials-particle size, reactivity, rheological characteristics [5].

It is known [11] that as a result of mechanical activation (table. 2) the transition of crystalline bodies to an amorphous state is possible. The aim of the study was to evaluate the possibility of obtaining a fine inorganic powder from the slag, which can be reactively active due to the amorphization of components.

IOP Conf. Series: Materials Science and Engineering 1100 (2021) 012041 doi:10.1088/1757-899X/1100/1/012041

Table 2 Influence of ultrafine grinding on the structure and properties of solids [11]

Grinding	1. The increase of surface
Mechanical activation	2. Increased surface activity
	3. Changes in the structure of the lattice a) violations in the structure of the lattice b) transformation in the structure of the lattice C) transition of crystalline bodies to an amorphous state
Chemical-mechanical activation	4. Chemical change
	5. Formation of chemical compounds
	6. Destruction of organic macromolecules

In the inertial separation mode, the slag is crushed into a fine powder with particles less than 20 microns (table 2). The size of the specific surface determined by the BET method suggests that various compounds can be grafted onto the slag surface [9] to modify the resulting powder and transfer the required properties to it.

Table 3 Characteristics of powdered slag

Mode of grinding	Particle size, mkm / fraction of particles, % of mass.						Specific surface area, m2 / g
	>63	40-63	30-40	20-30	10-20	<10	
Inertial separation	-	-	-	-	29.9	70.1	7.3

The study of the x-ray image of a sample of mechanically activated slag in the peak region1. 3750, 2.2873, 2.7117 by their broadening allows us to confirm the previously proposed hypothesis about the amorphization of some compounds of mechanically activated slag.

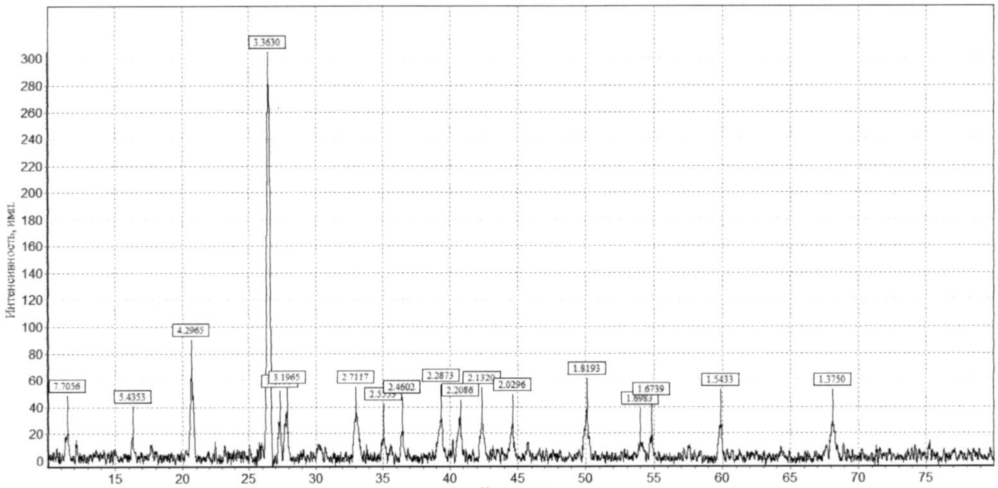

Figure 3. The radiograph of the slag after mechanochemistry activation

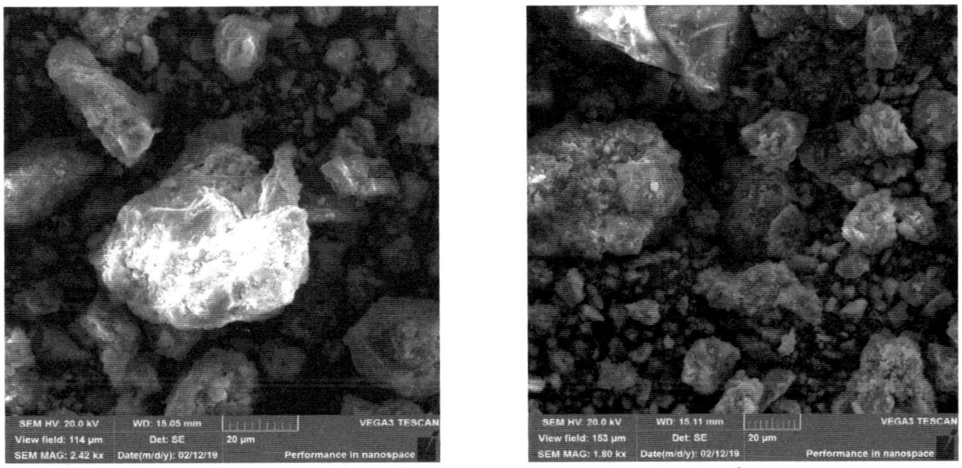

a b

Figure 4 a) Micrography of the concrete without additives; b) Micrography of the concrete with powdered slag

The slag obtained as a result of mechanochemical grinding was introduced as a filler in the composition of fine-grained low-grade concrete. The mobility of the mixtures was the same. Cone draft was 15 cm.

Table 4 Strength characteristics of concrete samples

Mechanochemistry	Sample	water-cement ratio	Bending strength, MPa	Design compressive strength, MPa	Actual compressive strength, MPa
	without additives	0.76	2.24	12.3	12.3
Disintegration	10% slag	0.72	3.34	13.4	16.1
	10% slag + plasticizer	0.6	4.58	17.5	23.8

The strength indicators of the tested samples not only do not decrease, but also there is a slight increase in them. This can be explained by the fact that due to the introduction of a fine component, the pores amount and, accordingly, defects of the concrete structure are reduced.

3. Conclusion

Mechanochemistry and disintegrator technology methods make it possible to have a regulating effect on the properties of the resulting materials. In addition, the principal possibility of using secondary technogenic raw materials (metallurgical slag) has been shown to create reactive components of concrete.

References

[1] P. P. Oleynik, S. P. Oleynik, Organization of the construction waste processing system, Moscow: MGSU, 2009. -251 p

[2] Abramov M. A., Stepanov E. G., Yablonsky O. P. Application of Mechanochemistry methods for quality management of the process of processing deactivated catalysts. RGATU Bulletin № 1 (22).- 2012.- P. 250-255.

[3] L. A. Weisberg, I. D. Ustinov Technology of concrete utilization // Construction and building materials No. 8 - 2003.- S. 11-13

[4] Fedorkin S. I., Makarova E. S., Elkina I. I., Kogay E. A. Mechanochemical activation – an effective direction for recycling secondary raw materials in the production of building materials // Economics of construction and environmental management #1 (66) - 2018. Pp. 154-161

[5] Stepanov E. G., Kotelnikov G. R. Disintegrator technology of preparation and utilization of heterogeneous catalysts// Yaroslavl: yagtu, 2005.

[6] Official website of JSC "Skokovo" [Electronic resource] / / URL:http://www.oao-skokovo.ru/index.php?s=6 (accessed 2.02.2020).
[7] Yu. I. Rebrin. Quality management Tutorial. Taganrog: TRTU Publishing house, 2004.
[8] [O. P. Yablonsky, V. A. Ivanova Fundamentals of standardization, Metrology, certification, Rostov-on-don: "Phoenix", 2010. - 475 p.
[9] Fedyukin V. K., Durnev V. D., Lebedev V. G. Methods of assessment and quality management of industrial products, Moscow: "Filin". 2000. 328 p.
[10] Stepanov E. G., Abramov M. A., Shatulsky A. A. Preparation of highly dispersed powders for composite materials and casting molds by methods of disintegration technology.// Vestnik rgata. - 2011. - no. 1. - p. 49-58.
[11] N. G. Krasnobay, L. P. Leiderman, A. F. Kozhevnikov Production of iron oxide pigments for construction/ Construction materials, 2001, no. 8, p. 19.
[12] L. A. Kazitsyna, L. B. Kupletskaya Application of UV, IR and NMR spectroscopy in organic chemistry / M. Higher school, 1971, 264 p.
[13] V. I. MOLCHANOV, T. S. YUSUPOV PHYSICAL and CHEMICAL PROPERTIES of FINELY DISPERSED MINERALS Moscow "NEDRA" 1981 Physical and chemical properties of finely dispersed minerals . - M: Nedra, 1981 160 p
[14] Kochetkov S.P., Bryl S.V., Smirnov N.N., Rukhlina N.I., Rukhlin G.V. Conditioning techniques technogenic raw materials used for binders // Ekologiya & Stroitelstvo. – 2017. – № 2. – P. 16–24.

Composites based on polypropylene and Georgian minerals

J Aneli, D Gventsadze, L Shamanauri, A M Chkonia and Z V Samkharadze

R.Dvali Institute of Machine Mechanics, Mindeli Str,10, Tilisi 0186 Georgia

E-mail:jimaneli@yahoo.com

Abstract. Polymer composites on the basis of secondary polypropylene and high dispersive powders of minerals wide spread in Georgia (andesite, quartz sand and a slam) were obtained and investigated their some physical-mechanical, thermal and hydrophobic properties. It is shown that the ultimate strength, and thermal stability extremely depend on the filler concentration and at definite significant of the concentration have maximums. For the composites with binary fillers (quartz sand/slam, quartz sand /andesite and slam/ andesite) revealed synergistic effect of anomaly increasing of the ultimate strength at definite proportion of fillers. It is established that modification of the mineral fillers with tetraethoxisilane enhances technical characteristics of the composites.

1. Introduction

Environment ecological protection and utilization of the industrial wastes present today very important and actual problems. From the scientific- technical literature it is known that if the development of the composites based on secondary thermoplastic materials, in which the different dispersive or natural and artificial fiber fillers are used, about 40 % of the primary ores can be spared. In the secondary polymer composites the industrial technological wastes as trimming, injection molding heads technical tare, films, bottles and so on are used. The content of such wastes was varied in the range 10 - 40 %. In our case the high pressure polyethylene industrial wastes as binder are used.

Polypropylene (PP) now is one of more spread polymer (after polyethylene), which is due to many positive exploitation properties and low cost [1-4].

2. Experimental

2.1 Methods of obtaining of research objects

The high dispersive powders obtained in result of fine disintegration of domestic polypropylene bags were used as binder of composites. The fine dispersed minerals as andesite from village Bakuriani, slam from village Okami and quartz sand from the village Sachkhere were used as fillers for composites. Below the short characteristics of the named fillers are presented:

- *andesite* (word basis -American mountains Ands) dark red color dense, but sometimes is porous material. Andesite is used as a building and acid proof material;
- *Quartz sand* includes the quartz particles, contenting silicon oxide near70-85% and rest are iron, calcium and magnesium oxides. Besides of the sand includes 5% clay and dust particles.
- *slam* is red color micro-porous volcanic generated mineral with high specific surface. In Georgia this mineral is used as warm-isolated material . The slam belongs to basalt type porous variety. It is the glass with alumina-silicate (contents 75-80%). Density 2630 kg/m^3.

At the initial stage the mixing of composite ingredients in the propeller mill during 2-3 min. In result of mixing of the polymers and different fillers the homogeneous powder was obtained, which after preliminary drying at 50-

Content from this work may be used under the terms of the Creative Commons Attribution 3.0 licence. Any further distribution of this work must maintain attribution to the author(s) and the title of the work, journal citation and DOI.
Published under licence by IOP Publishing Ltd

70°C underwent to the pressing in the standard press-forms (cylindrical and rectangular). The samples were obtained after pressing at 8-10 MPa and temperature 140-150°C during 10-15 min.

2.2 Methods of composite testing

The samples were tested on the strengthening at compression, bending and impact viscosity. The water absorption was defined separately. The specific weight was defined with use of standard method. Mechanical parameters were defined on the Germany device of type Dinstant. The temperature stability was defined on the apparatus of Vicat.

3. Results and discussion

First of all it was necessary determination of the dependence of properties of the research objects on the filler concentration. With this aim there were obtained the composites with following concentrations of fillers 20, 30,40 50 and 60wt%.

On the Fig.1 it is shown the dependence of the ultimate strength (at compression) on the filler type and content. The corresponding curves are characterized with maximums at different concentrations of fillers, however the filler concentrations, to which are correspond the maximums little differ one from another. It takes place near 40wt%. This fact indicates on the optimal content of the fillers, when swelling of filler particles realized maximally. At more high concentrations of fillers the mechanical strengthening of composites decreases because of starting of formation of filler particle associates in the polymer matrix. Among the fillers the andesite is one, containing of which enhances the strength of composite higher than for other fillers. The selected mechanical properties of composites with andesite appear at their testing on mechanical strengthening (at bending and impact viscosity (Fig-s 6 and 7).

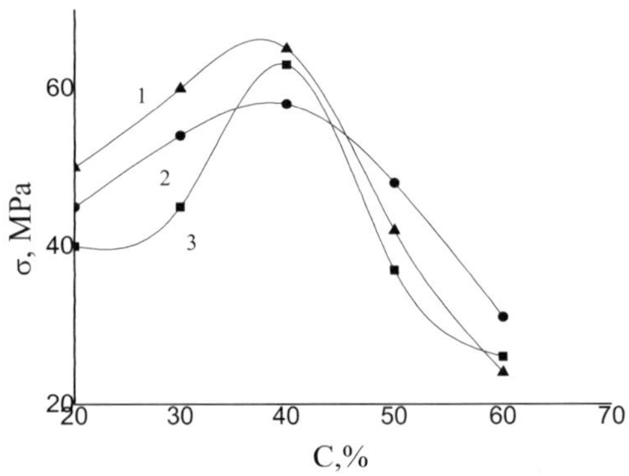

Figure1. Dependence of ultimate strength (at compression) of composites containing andesite (1), slam (2) and quartz sand (3) on the filler concentration

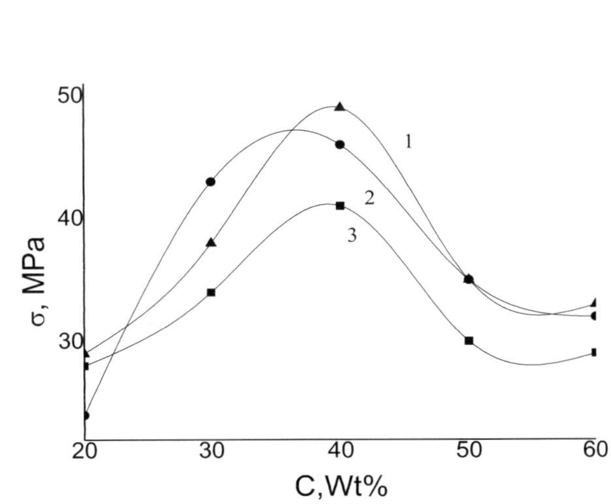

Figure 2. Dependence of ultimate strength (at bending) of composites containing andesite (1), slam (2) and quartz sand (3) on the filler concentration

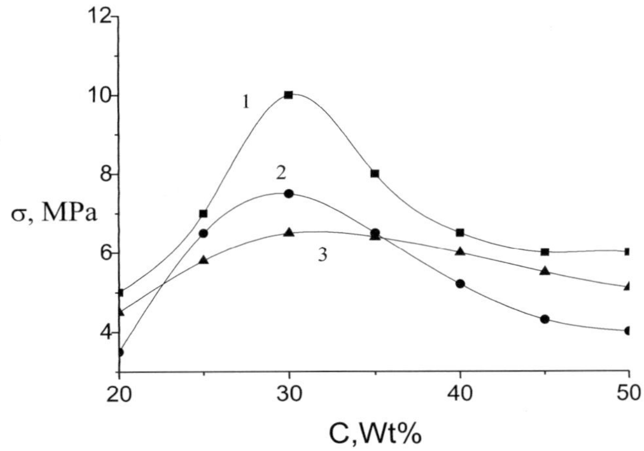

Figure 3. Dependence of impact viscosity of composites based on PP containing andesite (1), slam (2) and quartz sand (3) on the filler concentration

The Dependences shown on the Figs 1-3 lead to conclusion that the differences between corresponding curves character for different fillers are determined by different level of interaction intensity between polymer matrix and filler particles. First of all the essential effect on the mechanical characteristics has a micro structure of the filler particles. It is clear that the higher is a sum surface of filler particles the higher is the inter-phase square and consequently -the adhesive forces between polymer matrix and filler particles. Besides of the existence of active chemical groups on the surface of filler particles additively can enhance composite because of formation of covalent bonds between phases. On the basis of these supposes it can be concluded, that mineral andesite particle surfaces are characterized with relatively inhomogeneous structure on surface of which some active chemical groups are displaced.

Experiments provided on the composites contenting the binary filler particles show that in this case technical characteristics of the composites can be proved by selection of type and proportion of two different type fillers.

This opinion is confirmed by numerical data of the Table 1. Comparison of these data, obtained for composites containing 40wt% of sum concentration of binary fillers with analogical one obtained for the PP composites containing one of these fillers gives full right to express the opinion presented above.

Table 1. The characteristics of composites based on PP and binary fillers

Filler (proportion of the fillers in composite)	Impact viscosity, kJ/M^2	Ultimate strength at binding, MPa	Ultimate strength at compression, MPa	Thermal stability, ^{0}C (by Vica)
Slam	7.0	41.5	62.3	160
Andesite	6.0	46.8	65.1	165
Quartz sand	5.1	45.5	60.4	180
Slam/andesite (2/3)	8.2	50.5	70.0	185
Slam /quartz sand (1/1)	7.7	56.6	69.5	201
quartz sand / andesite (2/3)	7.8	54.2	80.1	196

Comparison of the thermal stability of the composites less structural changes undergoes the composite containing slam, which definitely show on the good thermal stability of this filler and the composite filled by them (Fig.4). Analogically can be estimated the effect of quartz sand. Besides of high physical-mechanical characteristics of composites with andesite at high temperatures the pores near the particles surfaces are expanding relatively more intensively and in result of exit of macromolecule fragments diffused in these pores (by other words - degradation of the physical bonds between phases) the thermal stability of this composite decreases.

The experiments on investigation of hydrophobic properties of the composites show that water absorption by composites is rather low (no more than 1.5%). Therefore we decided not to include the data of this parameter to the work.

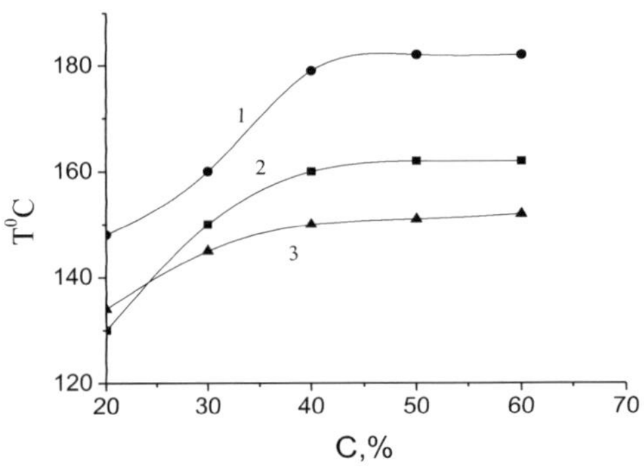

Figure 4. Dependence of the softening temperature of the composites based on PP containing quartz sand (1), slam (2) and andesite (3) on the filler concentration

4. Enhancing of composite properties by use of modifier

With an aim of improvement of the exploitation properties of the obtained composites the structural modifier of type tetraetoxisilane (TEOS) was used. This modifier was introduced to the composite with amount 5wt% solved in the toluene for obtaining of homogeneous solution, which was evaporated after definite time.

Table 2. Technical characteristics of composites containing of modified by TEOS fillers (in brackets corresponding characteristics for analogues with unmodified fillers)

#	Filler	Specific weight, Kg/m^3	Impact viscosity, kJ/M^2	Ultimate strength at binding, MPa	Ultimate strength at compression, MPa	Thermal stability, ^{0}C (by Vica)
1	Slam	1082 (1174)	7.5(7.0)	63.9 (62.3)	43.8 (41.5)	170 (160)
2	Quartz sand	1115 (1242)	5.0 (5.0)	58.0 (60.4)	51.3 (45.5)	200 (180)
3	Andesite	1183 (1180)	6.0 (6.0)	53.7 (65.1)	30.5 (46.8)	160 (150)

On the Table 2 there are presented some technical characteristics of the composites with modified and unmodified by tetraetoxisilane fillers. Data of the Table 2 show that in general the effect of the modifier on mechanical characteristics of the composites is relatively weak, however the effect of modifier is manifested in thermal stability of composites. This parameter increases on 10-20°C. Here the protector role of silicone-organic compounds appears.

Conclusion

1. On the basis of industrial residues of the polypropylene it is obtained and studied the composite containing the fine-dispersed mineral fillers spread in Georgia as andesite from Bakuriani, Okami slam and quartz sand from Sachkhere.
2. Experimentally is established that composites physical-mechanical, thermal and hydrophobic properties essentially depend on the type and concentration of the fillers; at definite (for each fillers separately) concentrations of the fillers the materials with best properties are obtained.
3. For the composites contained binary fillers there are fond the optimal concentrations , which ensure the obtaining of composites with improved properties (the synergistic effect is appeared).
4. The composites of PP containing modified by tetraetoxisilane fillers differ from analogues with unmodified fillers with improved technical characteristics.

Acknowledgments

This research was supported by Sh.Rustaveli Georgian National Scientific Foundation. Grant CAYRUS -19 -2692.

References

[1] Maddah H.A. Polypropylene as promising plastic: a Review. American Journal of Polymer Science, 2016, 6(1)2152 -2163. Doi: 10.5923/j.ajps.20160601.01
[2] Shubra .K.T, Alam A.M., Kuaiyyum.. Properties of polypropylene composites: A.Review. Journal of Thermoplastic Composite materials, 2011, 26(3), 362-391. Doi: 10, 1177/0892705711428659.
[3] Nikmatin S., Syafiuddin A., A.Beng Hong Kueh. Physical, thermal and mechanical properties of polypropylene composites filled with rattannanaoparticles. Journal of of Applied Research and Technology. 2017, V.15. Issue 4, 386-395.Doi: 10.1016/J.Art.2017, 03.008.
[4] Prambauer M., Paulik C.Burgstaller. Composites, Part A: Applied Science and Manufacturing. 2016, 59-66. Doi.org/10.1016/J.Composites, 2016.05.016

Construction of primitive cells of the cubic syngony on the basis of the molecular acoustics method

S A Yurchenko

LLC Scientific Production Firm "Tehosnastka-RTD"

662925, 2 Novaya Str., Petropavlovka, Russian Federation

E-mail: tehrtd@gmail.com

Abstract: Topicality of the present article is connected with introduction of new verifying condition, which is required for estimation of structures of cores of the "elementary" lattices of the body-centred cell and face-centred cell (these lattices were proposed in accordance with the Bravais theory), on the basis of theoretical calculations of the ultrasonic velocities, which are directed along the edges of the selected cells and which are verified with the help of the experimental data that were obtained in accordance with the molecular analysis method. *The purpose of the present investigation was as follows*: to carry out evaluation of the core structures of the elementary and primitive cells of the body-centred and face-centred cell of the crystal lattice of the cubic syngony (for alpha iron and cuprum) in respect of their conformity with methods of the molecular acoustics. *The method of the present investigation* is based on comparison of values of the ultrasonic velocities (longitudinal velocity, transversal velocity, and velocity within the rod), which were calculated on the basis of the relevant theory and which are transmitted over edges of the primitive cell that is constructed (for alpha iron and cuprum), with the values of these velocities of the ultrasound, which were obtained experimentally with the help of the molecular acoustics method for these materials. Introduction of the additional criterion (requirement) in respect of selection of the least cell on the basis of the molecular analysis has made it possible to draw the conclusion that cores of the "elementary" cells of the body-centred and face-centred crystal lattices, which are determined in accordance with the Bravais theory and which are assumed as the basis of the analysis with the help of the analytic geometry method, are inconsistent with actually existing mechanism of transmission of the ultrasound over the interatomic bonds (edges of the cell). In addition, this conclusion has made it possible to reveal new type (that is, a new syngony), which is the basis of the core structure of these "elementary" cells. The theoretically calculated ultrasound velocities, which are directed along the edges of the primitive cell of the new type (for alpha iron and for cuprum), have been coincident with high accuracy with the values of the longitudinal and transversal sound velocity, as well as with the sound velocity within the rod, which were experimentally established with the help of the molecular acoustics method.

Keywords: interatomic distance, parameters of the crystal lattice, core of the cell, longitudinal velocity, transversal velocity, frequency of vibrations, monoatomic two-dimensional planes.

1. Introduction

Physical and mechanical parameters (PMP), such as velocity of the ultrasound in the relevant material or mechanical strength characteristics (Young's modulus of elasticity, modulus of transverse elasticity G, and other parameters) are usually determined in the course of relevant experiments. These characteristics are always connected with the structure of the relevant material, that is with the relevant parameters of the elementary cell [1-2]. However, ambiguity of the criteria, which are usually applied and which determine types of various crystal lattices (see Figure 1), desire for introduction of maximum quantity of straight angles, which simplify description of elementary cells, as well as different quantities of atoms, which are contained within the unit volume of "elementary" cells, violate relevant principle of unity in respect of

Content from this work may be used under the terms of the Creative Commons Attribution 3.0 licence. Any further distribution of this work must maintain attribution to the author(s) and the title of the work, journal citation and DOI.
Published under licence by IOP Publishing Ltd

selection of the cell structures in accordance with the Bravais theory. In addition, this theory does not envisage any possibilities of existence of the monoatomic two-dimensional materials [3], which were found during previous years.

The main attention in the present article was paid to the cubic syngony (see Figure 2), because of majority of materials, which are used in the metal goods industry, are always characterised by this kind of crystal lattices. It is well known that geometrical theory of development of the crystal structure was proposed by Bravais in 1849 [4], as well as that this theory is utilised (up to the present time) for development of mathematical models in the course of investigation of structure of various materials with the help of the following methods: method of deformation, method of the X-ray analysis of crystal structure, as well as with the help of the molecular acoustics method.

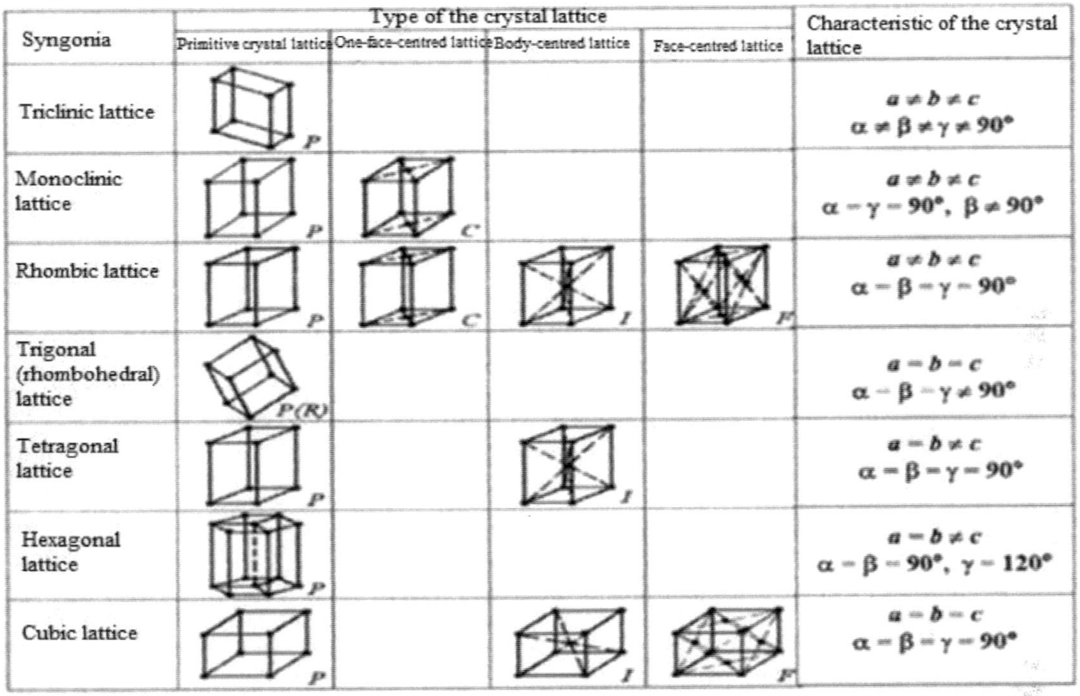

Figure 1. Types of the crystal lattice depending on the syngony

2. Materials and Methods

In the course of analysis of all the syngonies, which are presented in the Figure 1, in accordance with the Bravais theory, it is necessary to pay attention to the fact that volumes of the primitive cells (with the tetragonal, rhombic, rhombohedral, monoclinic, triclinic and cubic syngonies) are only created by means of interaction of the atoms, which are only situated in two neighbouring parallel planes. It is possible that, in the course of introduction of such determination Bravais supposed existence of certain monoatomic two-dimensional materials. However, Mr. Bravais did not introduce any determination of the monoatomic two-dimensional materials. Therefore, taking into consideration relevant classification of types of the crystal lattices, which are presented in the Figure 1, it is possible to demonstrate that there exist certain cores of crystal lattices, structure of cells of which is created in three atomic planes.

So, for example, crystal lattices in the cubic syngony are created both by two atomic planes α and β (see Figure 2a), and three atomic planes α, β, and γ (see Figure 2 b-c), which are revealed in the course of analysis of the "elementary" cells of the BCC-lattice (body-centred cubic lattice) and the FCC-lattice (face-centred cubic lattice) by means of assignment of the "primitive" cell, within which the cell volume is created at the expense of the interatomic interaction of atoms, which are situated in two neighbouring parallel planes (see Figure 3 a-b) [5].

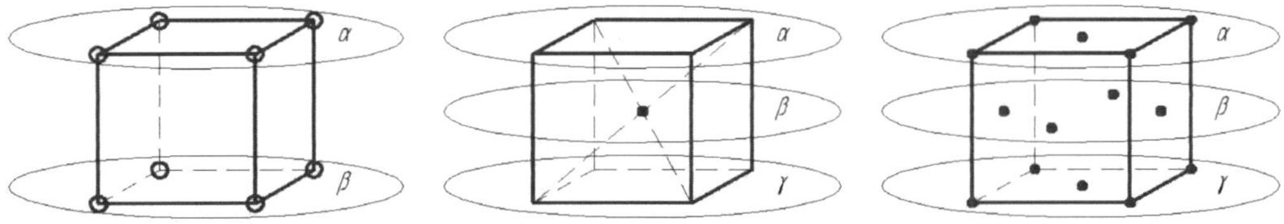

a b c

Figure 2. Structure of elementary cells of the cubic syngony

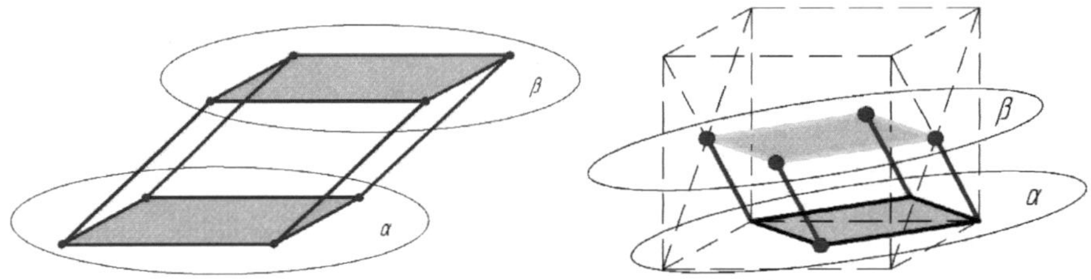

a b

Figure 3. Primitive cells of the BCC-lattice and FCC-lattice

Therefore, a primitive cell is usually presented as the minimum volume of the crystal. Translations of this crystal volume in three dimensions (parallel translations) make it possible to construct the three-dimensional crystal lattice of the entire volume of the relevant material. Such elementary cell is usually characterised by the fact that it must comply with the symmetry elements of the final volume, as well as with the properties of the relevant crystal [6].

In his theory, Mr. Bravais has formulated three rules for selection of elementary cells:

1. Symmetry of the elementary cell must comply with the symmetry of the relevant crystal.
2. The elementary cell must have maximum quantity of equal edges and equal angles.
3. Subject to fulfilment of these two rules, the elementary cell must have the least volume.

Let us introduce the additional condition in respect of selection of elementary cells of crystal lattices of various materials (because of the fact that elementary cell must comply with the physical and mechanical parameters of the crystal) with the help of the following statement:

4. Longitudinal velocity and transversal velocity of the ultrasound transmission within the material volume, which are determined in the course of relevant experiments, must comply with the theoretically calculated sound velocities, which propagate along the edges of the elementary cell that was constructed (that is, selected), thus confirming conditions of actual existence of such elementary cell. Therefore, in order to ensure strict assignment of the elementary (primitive) cell in this article, we will utilise the molecular acoustics method, which ensures possibility to compare velocities of transmission of the sound (ultrasonic) waves in the material [7], provided that these velocities were obtained in accordance with relevant experiments, as well as with the help of theoretical calculations.

Taking into consideration the fact that longitudinal velocity of the ultrasound propagates over the interatomic bonds within the monoatomic planes along the edge length, which is constructed (that is, selected) for the elementary (primitive) cell, as well as taking into consideration the fact that the wave length is limited by any maximum value, which (in accordance with results of article [8]) is approximately equal to double radius of atoms, then it is possible to present the wave length (λ) with the help of the following expression:

$\lambda \approx 2 \cdot R$ of the relevant atom, metres, (1)

Then, velocity of the longitudinal ultrasound wave within the plane will be presented with the help of the following equation:

$$\upsilon = \lambda \times \omega \text{ m/s}, \quad (2)$$

where λ denotes the interatomic distance, which is equal to the edge length within the plane, while ω denotes the natural frequency of vibration of atoms in the planes, which create volume of the elementary (primitive) cell of the relevant material.

We have to take into consideration the fact that natural frequency of vibration of atoms within the elementary cell is limited by certain maximum frequency ω_{max} [9] that has the same value for all the atoms, which create volume of the elementary (primitive) cell [10]. At the same time, it is assumed that lengths of edges within the cubic primitive cells are the same ($a = \lambda$). Therefore, in accordance with equation (2), values of the longitudinal and transversal velocities of ultrasound must have the same values (this fact can be confirmed in the course of investigation of the polonium, which has the same structure: the cubic crystal lattice).

However, it is known [11-12] that longitudinal and transversal sound velocities for the iron (BCC-lattice elementary cell) and copper (FCC-lattice elementary cell), which were determined in the course of relevant experiments with the help of the method of molecular acoustics, are different. It is only possible to explain this discrepancy taking into consideration the fact that all volumes of the primitive cells are created at the expense of the interatomic interactions of the atoms, which are situated in two neighbouring monoatomic parallel planes. At the same time, in the elementary crystal lattice, within which the intermediate plane β exists (see Figure 4), the prohibition occurs in respect of interatomic interactions of the atoms between planes α and γ. It is possible that in this case the ultrasound velocity can distribute over the interatomic bonds at the level of minimum resistance. Therefore, let us introduce relevant prohibition in respect of interatomic interaction of atoms through the intermediate plane in order to ensure further theoretical analysis.

Now let us calculate velocity of the ultrasonic wave for iron (Fe) and copper (Cu) on the basis of the introduced condition. From the very beginning, it is necessary to do the following: we have to construct the "elementary" cell of iron with the help of the atoms, which are situated in three parallel planes α, β, and γ (see Figure 4).

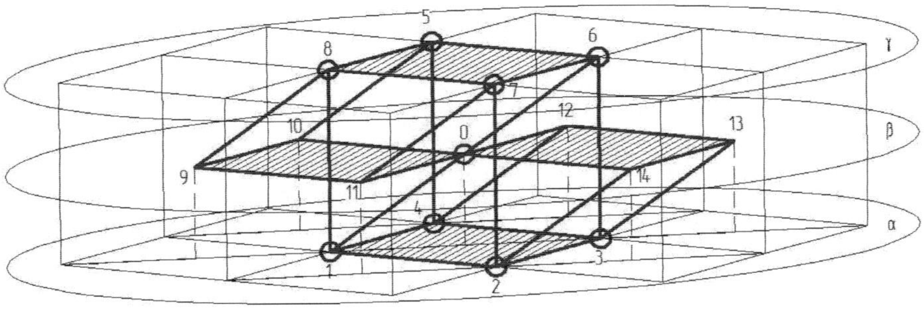

Figure 4. "Core" of the elementary body-centred cubic cell of the crystal lattice of iron. This core is created with the help of the atoms, which have sequential numbers from 0 to 8 and which are situated within the monoatomic two-dimensional parallel planes α, β, and γ.

Having assumed that energy of the ultrasound has already achieved the atom No. 0, let us analyse further transmission of energy over the bonds between the atoms, which create the "core" of the elementary cell of the BCC crystal lattice of iron (see Figure 4). This "core" is created with the help of two "primitive" cells, type of which is described in article [5]. The first cell is created with the help of the atoms, which have sequential numbers 1, 2, 3, and 4 and which are situated within the plane α, as well as with the help of the atoms, which have sequential numbers 0, 12, 13, and 14 and which are situated within the neighbouring parallel plane β. The second cell is created with the help of the atoms, which have sequential numbers 11, 10, 9, and 0 and which are situated within the plane β, as well as with the help of the atoms, which have sequential numbers 5, 6, 7, 8 and which are situated within the neighbouring parallel plane γ. In this case, these primitive cells comply with the relevant criteria for selection of cells. These criteria were proposed in accordance with the Bravais theory and they were described above already. Due to introduction of the restriction in respect of the interatomic interactions through a plane, there are no transmissions of the ultrasonic waves over the following edges of the core of the "elementary" cell: L1-8, L2-7, L3-6, and L4-5. The longitudinal velocity of ultrasound exists within the monoatomic two-dimensional planes α, β, and γ over the edges (interatomic bonds) of primitive cells of the following types: L1-2, L9-11, and L7-8. With the help of the experimentally established value of the longitudinal velocity of ultrasound in iron (in accordance with the data of article [13]), which is equal to 5.940 m/s, as well as with the help of the edge length L1-8, which is equal to 3×10^{-10} m (in accordance with article [10]; this fact is connected with introduction of the unified principle of the interatomic interaction for the cell volume), let us calculate frequency of vibration of atoms in accordance with the following formula:

$$\omega_{max} = \upsilon_{longitudinal} / \lambda, \quad (3)$$

$$\omega_{max} = 5{,}940\ (m/s) / 3 \times 10^{-10}\ (m) = 1.98 \times 10^{13}\ Hz. \quad (4)$$

3. Results and Discussion

In order to calculate transversal velocity of ultrasound within the core of the "elementary" cell of the BCC crystal lattice of iron, let us determine the length of the interatomic bond L 1-0 with the help of the length of the interatomic interaction L 1-2 of the atoms, which are situated within the plane α (length of the interatomic interaction in any of three monoatomic planes α, β or γ has the constant value: L 1-2 = L 9-11 = L 8-7). Due to the fact that the central atom (which has sequential number No. 0 and which is assumed as the atom, through which sound is transmitted) is situated in the middle of the space diagonal, then (with the help of the theorem of Pyphagoras) it is possible to calculate length of the interatomic interaction of the atoms, which are situated within the plane β, with the atoms, which are situated in the planes α and γ:

$$L\ 1\text{-}0 = L\ 0\text{-}6 = L\ 1\text{-}2 \times (\sqrt{3}) / 2, \quad (5)$$

$$L\ 1\text{-}0 = 3\times10^{-10} \times (\sqrt{3}) / 2 = 2.598 \times 10^{-10}\ \text{m}, \quad (6)$$

Time τ(in seconds) of transmission of the ultrasound wave over the length of the interatomic bond L 1-0 (taking into consideration the fact that frequency of vibration of atoms for the core of the elementary cell has the constant value) will be equal to:

$$\tau\ 1\text{-}0 = 1 / \omega_{max} = 1/ 1.98 \times10^{13}\ \text{Hz} = 0.505 \times10^{-13}\ \text{seconds}, \quad (7)$$

Theoretical value of velocity of the ultrasound transmission between the monoatomic planes α – β and β – γ will have the following value:

$$\upsilon\ 1\text{-}0 = L\ 1\text{-}0 / \tau\ 1\text{-}0 = 2.598 \times 10^{-10}\ \text{m} / 0.505 \times10^{-13}\ \text{s} = 5.144\times10^{3}\ \text{m/s}, \quad (8)$$

Then, it is possible to calculate the theoretical value of velocity of the transversal wave in the following manner:

$$\upsilon_{transversal} = L\ 7\text{-}2 / (\tau\ 1\text{-}0 + \tau\ 0\text{-}6), \quad (9)$$

where L 7-2 denotes the "fictitious" interatomic bond between planes α and γ.
τ 1-0 denotes duration of transmission of the ultrasound energy over the interatomic bond L 1-0;
τ 0-6 denotes duration of transmission of the ultrasound energy over the interatomic bond L 0-6.

Because of the fact that (in accordance with equation (5)) lengths of the interatomic bonds (that is, edges of cells) have the same length, then (respectively) durations of transmission of the ultrasound wave over the lengths of these interatomic bonds will be equal to each other:

$$\tau\ 1\text{-}0 = \tau\ 0\text{-}6 = 1 / \omega_{max}. \quad (10)$$

In addition, equation (9) can be presented in the following form:

$$\upsilon_{transversal} = L\ 7\text{-}2 / ((1/\omega) + (1/\omega)) = L\ 7\text{-}2 \times \omega_{max} / 2. \quad (11)$$

If we will substitute value of the "fictitious" length of the interatomic bond (in accordance with the condition, which we have introduced, this interatomic bond does not exist), which is equal to 3×10^{-10} m, and if we will substitute value of frequency of vibration of atoms in the planes, which create core of the elementary cell (this frequency is equal to 1.98×10^{13} Hz), into equation (11), then we will calculate the following theoretical value of transversal velocity of ultrasound:

$$\upsilon_{transversal} = 3\times10^{-10}\ \text{m} \times 1.98 \times10^{13}\ \text{Hz} / 2 = 2.973 \times10^{3}\ \text{m/s} = 2.940\ \text{m/s}, \quad (12)$$

It is well known that value of the transversal velocity of ultrasound, which was obtained in accordance with relevant experiments with the help of the molecular acoustics method, has the value which is equal to $\upsilon_{transversal}$ = 3,100 ± 100 m/s in accordance with the data of article [13]. In fact, the transversal velocity of ultrasound, which was calculated on the basis of new model of construction of the core of the "elementary" cell (taking into consideration the introduced restriction in respect of the interatomic interactions of atoms through the intermediate plane), has coincided with the lower boundary of the experimentally established value of transversal velocity of ultrasound in iron.

Theoretical value of the ultrasound velocity in the equation (8) $\upsilon 1\text{-}0 = 5.144\times10^{3}$ m/s. This value describes transmission of the ultrasound over the face of the primitive cell between two monoatomic parallel planes is in line with the ultrasound velocity within the rod of iron. The experimentally established value of the ultrasound velocity (which was calculated with the help of the method of molecular acoustics) within the iron rod has the following value: $\upsilon_{transversal}$ = 5.170 m/s [13]. Values of these velocities coincide with each other in accordance with the following percentage:

$$\% = [5.148/5.170] \times100\% = 99.574\ \%, \quad (13)$$

Coincidence of the ultrasound velocity along the edge of the primitive cell between planes and velocity of ultrasound within the rod can be caused by the fact that the rod has the strictly assigned ratio between the length and diameter.

Coincidence of the theoretically calculated values of the longitudinal velocity, transversal velocity and velocity of ultrasound within the rod was obtained due to taking into consideration the single frequency ω_{max} = Jdem of vibration of all atoms within the core of the "elementary" cell of iron in respect of all interatomic bonds (both within parallel atomic planes and within the interatomic bonds, which connect neighbouring planes). All theoretical calculations of the ultrasound velocity can be presented with the help of the single equation, which was proposed in article [10] and which is presented below:

$$\upsilon = L \times \omega \times (1/n), \quad (14)$$

where L denotes length of the interatomic distance, which is expressed through the length of the interatomic bond of atoms, which are situated within the plane.

ω denotes frequency of vibration of atoms within the core of the "elementary" (primitive) cell of the crystal lattice of iron.

n denotes quantity of the interatomic interactions, which are used for transmission of the ultrasound to different planes within the core of the elementary (primitive) cell of the crystal lattice.

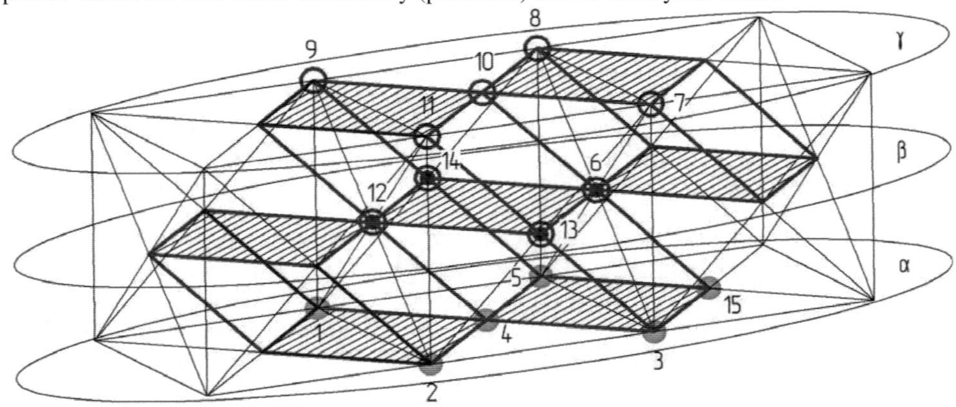

Figure 5. Core of the "elementary" of the face-centred cubic crystal lattice that is created with the help of the atoms, which are situated in three parallel planes α, β, and γ

Introduction of the new verifying condition in the course of selection of the elementary cell, as well transmission of the ultrasound wave over edges of the cell, which is constructed, have demonstrated that the "elementary" cell of the BCC crystal lattice (this "elementary" cell was proposed in accordance with the Bravais theory) does not exist, because of the interatomic bond between planes α and γ does not exist. In this case, relevant cell (Figure 3a), which is considered as the "primitive" cell and which is the basis of structure of the "elementary" cell, describes actual process of transmission of the ultrasound over edges of the present cell. Such "primitive" cell complies with conditions of construction, which are used for the rest types of primitive cells in accordance with the Bravais theory, in accordance with which the cell volume is only created with the help of the atoms, which are situated in two neighbouring parallel monoatomic planes α and β.

Let us perform similar theoretical analysis of structure of the "elementary" cell of copper, which has the FCC crystal lattice and which belongs to the cubic syngony as well. In order to do that, we will construct the core of the "elementary" cell of copper with the help of the atoms, which are situated in three parallel planes α, β, and γ (see Figure 5).

Atoms with subsequent numbers 1-14 (see Figure 5) are in correspondence with the entire core of the "elementary" face-centred cell, which is created with the help of atoms of four primitive cells [5] (see Figure 6).

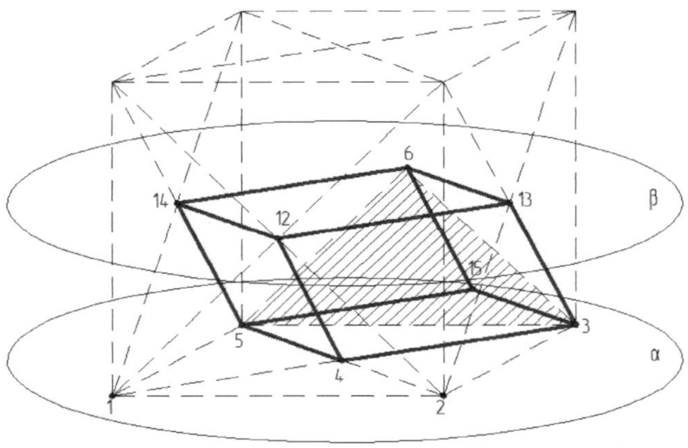

Figure 6. Primitive cell of the FCC crystal lattice is created with the help of the atoms 3, 4, 5, and 15, which are situated within the plane α, as well as with the help of the atoms 6, 13, 12, and 14, which are situated within the parallel plane β

Let us calculate frequency of vibration of atoms in the planes α, β, and γ with the help of values of the longitudinal velocity of ultrasound, which were determined experimentally with the help of the molecular acoustics method. Value of this longitudinal velocity is presented in article [13] as $\upsilon_{longitudinal}$ = 4.700 m/s. Let us assume (on the basis of condition of the unified principle of the interatomic interaction) that value of parameter of the edge of the "elementary" cell of copper is equal to L 1-2 = 3.65×10^{-10} m. This value was determined in the course of theoretical calculation, which is presented in article [10]. As concerns the core of the "elementary" cell of the FCC crystal lattice, the edge length of the primitive cell L 3-4 does not comply with the edge length L 1-2 (L 3-2) of the core of the elementary cell. This situation is observed within the primitive cell of the BCC crystal lattice. Because of the fact of introduction of the fourth point in respect of transmission of the ultrasound over edges of primitive cells, let us assume that longitudinal velocity of ultrasound will propagate along the edge length L 3-4. Let us present the edge length of the primitive cell through the edge length of the "elementary" cell L 1-2 as follows:

L 3-4 = L 1-2 × (√ 2/2) = 3.65×10^{-10} × 0.707 = 2.58 $\times10^{-10}$ m, (15)

Let us calculate value of frequency ω_{Cu} for the atoms, which create volume of the primitive cell:

$\omega_{Cu} = \upsilon_{longitudinal}$ Cu / L 3-4 = 4.700(m/s) / 2.58×10^{-10} m = 1.821×10^{13} Hz, (16)

We will calculate velocity of the ultrasound transmission along the length of the edge, which connects planes α and β of the primitive cell L 3-13, in the following manner:

υ 3-13 = L 3-13 × ω Cu, (17)

In order to perform this calculation, we will present length of the space edge of the primitive cell L 3-13 with the help of the length of the interatomic interaction of atoms within the core of the FCC crystal lattice L 3-2 = L 1-2 = 3.65×10^{-10} m:

L 3-13 = L 1-2 × (√2 /2) = 2.58 $\times10^{-10}$ m, (18)

If we will substitute value of length of the space edge of the primitive cell L 3-13 from equation (18) into equation (17), then we will calculate the following value of velocity of the ultrasound transmission:

υ 3-13 = 2.58 $\times10^{-10}$ m × 1.821×10^{13} Hz = 4.700 m/s, (19)

This theoretically calculated value of velocity of the ultrasound transmission along the edge, which connects atoms in two neighbouring parallel planes, has coincided with the value of longitudinal velocity of ultrasound within the FCC crystal lattice of copper. At the same time, the experimentally established value of the ultrasound velocity within the copper rod is equal to 3.700 m/s. Taking into consideration the fact that velocity of ultrasound within the iron rod has been complied with the theoretically calculated value of the ultrasound velocity along the edge (that is over the interatomic interaction of planes α and β of the primitive cell of the BCC crystal lattice), then it is possible to expect that mechanism of the ultrasound transmission velocity within the rod (in accordance with the principle of uniformity) within the primitive cell of the FCC crystal lattice of copper must be the same or similar. Therefore, it is not expedient to perform further analysis of such primitive cell (see Figure 6).

This is the reason why we will introduce the additional condition for selection of primitive cells: maximum length of the primitive cell edge must coincide with the maximum length of the interatomic interaction of atoms within the monoatomic plane. So, let us construct new primitive cell, which creates the core of the "elementary" cell of the FCC crystal lattice of copper (see Figure 7).

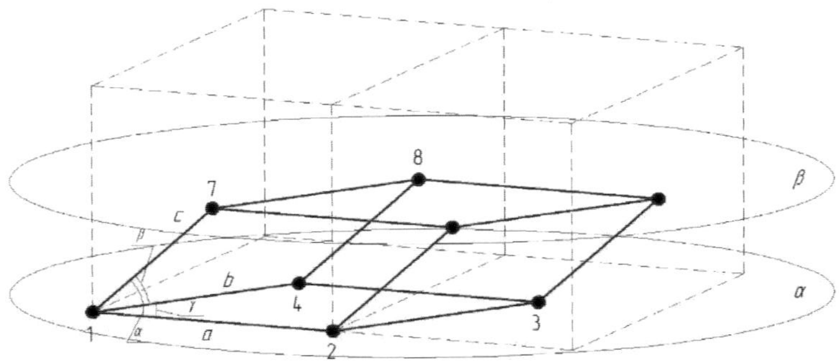

Figure 7. Structure of the primitive cell of the FCC crystal lattice that is created with the help of the atoms, which are situated in two parallel planes and which comply with the introduced condition in respect of the maximum length of the interatomic bond within the plane.

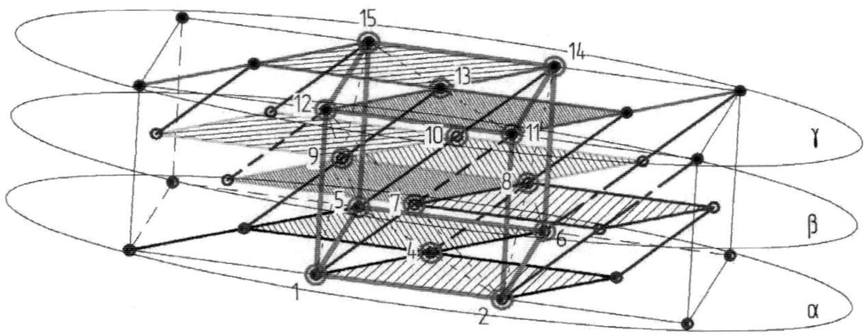

Figure 8. Core of the "elementary" cell of the FCC crystal lattice. This core is created with the help of four new primitive cells

Figure 7 presents structure of the primitive cell of the FCC crystal lattice. This structure is created with the help of the atoms that are situated in two parallel planes and this structure complies with the introduced condition in respect of the maximum length of the interatomic bond within the plane.

Let us perform construction of the core of the "elementary" cell of the FCC crystal lattice with the help of the atoms, which are situated in three planes α, β, and γ, taking into consideration structure of such new primitive cell (see Figure 8).

Now let us calculate frequency of vibration of atoms ωCu in the planes α, β, and γ for the volume of such new primitive cell. In accordance with the results of article [10], maximum edge length of the primitive cell is equal to L 1-2 = 3.65×10^{-10} m. This length complies with the velocity of ultrasound, which is equal to 4.700 m/s (see article [9]) and which was determined experimentally with the help of the method of molecular acoustics. Therefore, frequency of vibration of atoms will have the value, which is equal to:

$$\omega Cu = 4.700\ (m/s)/\ 3.65\times10^{-10}\ m = 1.287\times10^{13}\ Hz, \tag{20}$$

Taking into consideration the fact that value of frequency of vibration of atoms, which was determined in the equation (20) in the same manner for all atoms (both within the primitive cells of the FCC-lattice, and within the core of the "elementary" cell of the FCC crystal lattice of copper), let us calculate velocity of ultrasound along the edge that determines interatomic interaction of the atoms, which are situated in the planes α and β, L1-7 (see Figure 7):

$$\upsilon\ 1\text{-}7 = L1\text{-}7 \times \omega Cu, \tag{21}$$

Let us present value of the space edge length L1-7 with the help of the edge length L1-2 in the following manner:

$$L1\text{-}7 = L3\text{-}2 \times (\sqrt{2}/2) = 2.58\times10^{-10}\ m, \tag{22}$$

Therefore, if we will substitute this value into (21), then we will calculate value of the ultrasound velocity between the atoms, which are situated within the neighbouring parallel planes α and β. That is, this value corresponds to the edge of the primitive cell:

υ1-7 = 2.58×10^{-10} m × 1.287×10^{13} Hz = 3.320×10^{3} m/s, (23)

This value of the ultrasound velocity, which was theoretically calculated on the basis of the space diagonal and on the basis of the primitive cell edge, practically corresponds to the velocity of ultrasound within the copper rod. Difference of the theoretically calculated value of the sound velocity within the rod from the experimentally established value of the ultrasound velocity within the rod (which was determined with the help of the method of molecular acoustics), υ_{ster}= 3,700 m/s, is equal to:

% = [(3700 – 3328) /3700] × 100% = 10%, (24)

This error (10% only) is the allowable deviation due to the error in determination of parameters of the interatomic distance with the help of the method of the X-ray analysis of crystal structure (due to vibration of an atom near the equilibrium position [10]).

We will use the following formula in order to calculate theoretical value of the transversal velocity of ultrasound within the core of the "elementary" cell of the FCC crystal lattice of copper:

υ 2-11 = L 2-11 × ωCu × (1/n), (25)

It is necessary to point out that (in accordance with the accepted restriction in respect of the interatomic interaction through the plane) length L2-11 is the "fictitious" length. Therefore, quantity of the interatomic interactions *n*, which are required for transmission of ultrasound from the monoatomic plane *a* into the monoatomic plane *γ*, will have the value, which will be equal to 2 (two). Then, respectively, the transversal velocity will have the value, which is equal to:

$\upsilon_{transversal}$Cu = 3.62×10^{-10} m × 1.298×10^{13} Hz × (1/2) =2.349×10^{3} m/s, (26)

This theoretically calculated value of the transversal velocity in copper ($\upsilon_{transversal}$Cu = 2.349 m/s) differs from the experimental value of the transversal velocity 2.260 m/s (which was determined with the help of the method of molecular acoustics) by 3.9% only. Therefore, new structure of the primitive cell of the FCC crystal lattice complies with the condition of the ultrasound transmission over edges of the cell, that is, it describes the actually existing interatomic bonds of atoms, which create the entire volume of the constructed cell. Therefore, (in the same manner, as it was in the previous situation with the "elementary" cell of the BCC crystal lattice) it is possible to draw the conclusion that the FCC "elementary cell", which was described in accordance with the Bravais theory as the cube, does not exist.

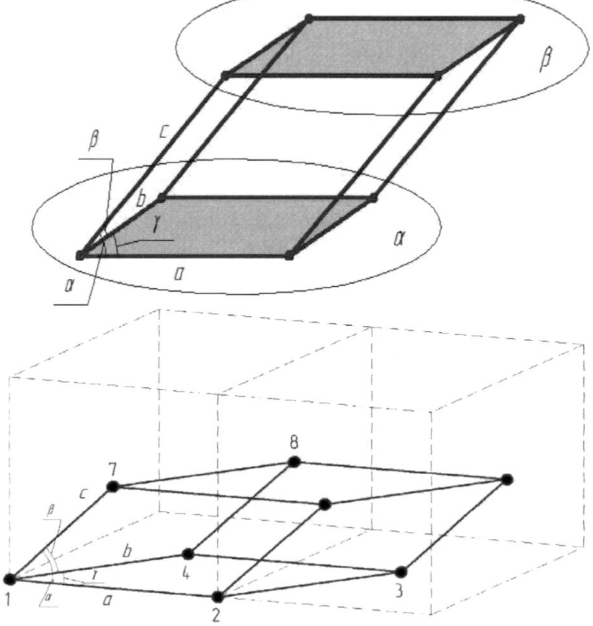

a b

Figure 9. Structures of such "primitive" cells

Both structures of primitive cells of the BCC-lattice (Figure 2a) and the FCC-lattice (Figure 7) comply with the main principle, which was described in accordance with the Bravais theory. This principle was described in the following manner: volume of the elementary cell is created with the help of the interatomic interaction of the atoms, which are situated in two neighbouring parallel monoatomic planes. In the course of thorough analysis of the structure of such "primitive" cells (see Figure 9 a-b), it is possible to determine the main characteristic of such primitive cells through the lengths of edges and relevant angles:

1. For the BCC-cell (Figure 9a) lengths of the edges $a = b \neq c$, angles $\angle\alpha^o = \angle\beta^o \neq \angle\gamma^o$;
2. For the FCC-cell (Figure 9b) lengths of the edges $a \neq b = c$, angles $\angle\alpha^o = \angle\beta^o \neq \angle\gamma^o$.

On the other hand (depending on the accepted names of sides of these primitive cells), it is possible to determine one common characteristic:

$$a = b \neq c, \ \angle\alpha^o = \angle\beta^o \neq \angle\gamma^o, \qquad (27)$$

Development of the new type of the primitive cell confirms the idea of the Academician E. S. Fedorov (which he has formulated in article [14]) in respect of the fact that the process of selection of conditions of the symmetry is the most important factor for subsequent development of the interconnected theory for analysis of all aggregative processes, which occur within the crystal lattice. This theory was clearly demonstrated in this article with the help of selection (that is, construction) of various primitive cells as the core of the "elementary" cell of the BCC-lattice and the FCC crystal lattice.

4. Conclusion

In the course of the theoretical analysis, which was performed in this article, it was established that cores of the "elementary" cells of the body-centred and face-centred crystal lattices (these "elementary" cells were presented in accordance with the Bravais theory, they are used in the models for estimation of various physical and mechanical processes, which occur in the course of deformation, as well as they are the basis of the analytic geometry method) do not describe process of transmission of the ultrasound over the interatomic bonds, that is, over edges of the cell within the selected (that is, within the constructed) volume of the cell core. The discrepancy, which has appeared between structure of the core and actually existing interatomic bonds, has demonstrated that the interatomic bonds through the intermediate plane do not exist. It has also demonstrated that construction of the cubic syngony is implemented by taking into consideration the interatomic interactions of the atoms, which are only situated within two neighbouring parallel planes. This conclusion corresponds to the majority of types of lattices within the same theory of Bravais. Introduction of the additional condition in respect of structure of cells (that is, requirement in respect of transmission of the ultrasound over edges of the selected cell) has made it possible to determine new type of the primitive cell, which has the following characteristics of the crystal lattice: $a = b \neq c$, $\angle\alpha^o = \angle\beta^o \neq \angle\gamma^o$. This type of the primitive cell is used for construction of cores of the BCC and the FCC "elementary" cells. The theoretically calculated values of the ultrasound velocity over faces of the new type of primitive cell (within the alpha iron and cuprum) have coincided with sufficiently high accuracy with the values of the longitudinal sound velocity, transversal sound velocity and with the sound velocity within the rod. These values were experimentally established with the help of the molecular acoustics method.

New approach to construction (selection) of the basics of structure of materials (that is, construction of the primitive cells, edges of which (lengths of the interatomic bonds) are in correspondence with the physical process of the ultrasound transmission) has provided the author of this article with possibility to carry out theoretical calculation of values of the Young's modulus of elasticity and G modulus of transverse elasticity with the help of parameters of primitive cells. Procedure of this theoretical calculation will be presented in the next article of the author, which will be published in future.

References

[1] Pearson W. B. A Handbook of Lattice Spacing's and Structures of Metals and Alloys. Oxford: Pergamon Press, 2013.
[2] Zadiranov A. N. Theoretical foundations of crystallization of metals and alloys. Moscow: MGIU, 2008.
[3] Katsnelson M. I. Graphene: Carbon in Two Dimensions. New York: Cambridge University Press, 2012.
[4] Bravais A. Note sur les polyèdres simétriques de la géometrie. Journal de Mathématiques Pures et Appliquées, 1849, 19, 3.
[5] Vintaykin B. E. Solid State Physics. Moscow: Bauman Moscow State Technical University, 2008. 359 p.
[6] Ashcroft N. and Merlin N. Solid State Physics. Moscow: Mir, 1979.
[7] Mikhailov I. G., Soloviev V. A., Syrnikov Yu. P. The basics of molecular acoustics. Moscow: Nauka, 1964.
[8] Frenkel Ya. I. Introduction to metal theory. Leningrad: Nauka, 1972, 424 p.

[9] Prokhorov, A.M. (ed.). Physics. Great Encyclopedic Dictionary. Moscow: Bolshaya Rossiyskaya entsiklopediya, 1998.
[10] Yurchenko S. A. Refinement of the diatomic model of the interatomic interactions of Frenkel for the improvment of methods of non-destructive quality control. Periodico Tche Quimica, 2018, 15 (Special Issue 1), 14.
[11] Sivukhin D.V. About the international system of physical quantities. *UFN*, 1979, 129(10), 335.
[12] Dirac P.A. The evolution of the physical picture of nature. в Moscow: Nauka, 1965.
[13] Anuryev V.I. Reference designer mechanical engineer. Moscow: Mashinostroyeniye, 2001.
[14] Fedorov E. S. Symmetry and structure of crystals. Leningrad: Nauka, 1949. 630 p.

Security management at facilities vulnerable to attack model development

D A Ukraintceva[1] , **V G Burlov**[1] , **D E Sipovich**[1] , **V V Gryzunov**[1] **and S Y Kapitsyn**[2]

[1] Russian State Hydrometeorological University (RSHU),3 Metallists Av, St. Petersburg, 195027, Russia
[2] Higher School of Technosphere Safety, Peter the Great St. Petersburg Polytechnic University, Polytechnicheskaya, 29, St. Petersburg 195251, Russia

E-mail: ukrainceva2000@yandex.ru

Abstract. The article describes the development of a 'security management at facility vulnerable to attack model'. The result is the creation of a mathematical model of managerial decisions. The concept of synthesis-based management is presented. The approach is based on the resolution of the inverse control problem. It is shown that a human decision lies at the core of security management at vulnerable to attack facilities.

1. Introduction

In modern conditions, ensuring the security of a vulnerable to attack facilities is considered one of the main activities of security officer. There is a large number of facilities falling under the 'vulnerable to attack' category in Russia, including nuclear power plant and radiation hazardous facilities, whose operation is guided by special security regulations.

Neglect of security measures at a vulnerable to attack site can lead to serious irreversible consequences. The facility's level of security is determined based on the probability of prevention of the maximum number of threats in various breach of security scenarios. graphical representation of the threat model is shown in Figure 1.

The subject of protection is the life and health of nuclear power plant employees, enterprise employees, citizens, as well as the integrity of the on-site assets (including information) and preservation of the facility's routine mode of operation, especially in case of an important social institution or a facility with high anthropogenic or environmental risks [1].

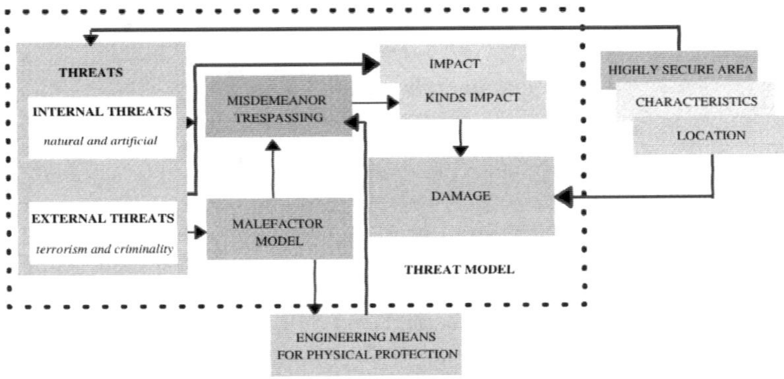

Figure 1. Threat model

Content from this work may be used under the terms of the Creative Commons Attribution 3.0 licence. Any further distribution of this work must maintain attribution to the author(s) and the title of the work, journal citation and DOI.
Published under licence by IOP Publishing Ltd

2. Basic information

Problem statement. When building a security system for a vulnerable to attack facility, the following operation models should be considered:

1. Vulnerable facility operation model
2. Security system operation model

Research Problem. To develop the 'assessment of effectiveness of security measures at a vulnerable facility methodology', an array of decisions made by the Occupational Safety and Health Manager (hereinafter – OSHM) will be considered since decision constitutes the basis of human activities [2, 3].

To solve the above specified problem, a correlation between the 'security management at vulnerable to attack facility models' and the 'management decision-making model' will be established. For the managers of high-security facilities to make viable decisions, a relevant methodology has to be in place.

Enforcement of security at a vulnerable to attack facility is always tied to human decisions made based on the characteristics of the model. An object model is an informative representation of the corresponding object featuring its main properties and characteristics. Therefore, a decision can be presented in a form of a process model where process can be described as an object achieving the intended purpose [4].

For the intended purpose to be fully achieved while complying with the security requirements, it is necessary to develop process models with pre-set properties. In management, decisions are usually generated based on the analysis-based models aimed at solving a direct management problem. However, analysis-based models might not address all the existing conditions. The article presents a synthesis-based management model that factors in all existing conditions. The new approach is based on the inversion of control principle.

A process model can be built only after all its conditions are established, which presents a challenge in case of process models with pre-set conditions [5-6]. Such a limitation, as well as the unavailability of a viable security management methodology, undermines the achievement of the intended purpose [7].

In this article, the management decision is presented in a form of a mathematical model, where:

Δt_{MP} is an average problem acknowledgment time Δt_{IP} – an average problem-solving time Δt_{NP} – an average problem identification time.

The above conditions are an integral part of the security management process at high-security facilities where the key characteristic is the time required for the operation of a vulnerable to attack acility (T_{TP}). A graphical representation of the process is shown in Figure 2 [8].

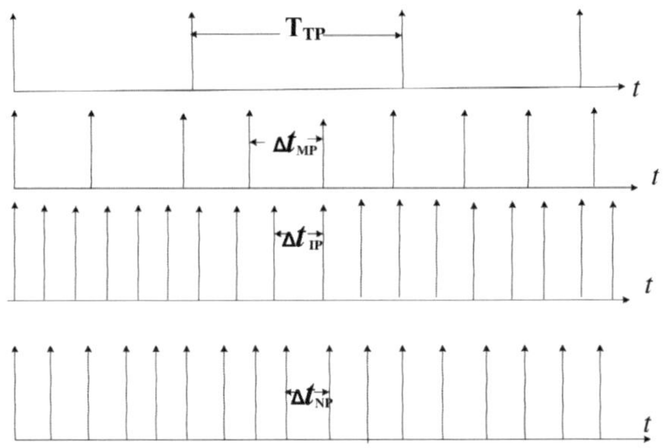

Figure 2. Manifestation of the basic elements of a decision-making model

Let's analyze the main target processes at a vulnerable to attack facility, where a 'target process' is an object achieving the intended purpose. The input parameters characteristic of this target process are: T_{TP} -- an average vulnerable to attack facility officers travel time, Δt_{MP} -- an average time of appearance of a perpetrator, Δt_{IP} -- an average time required to identify the perpetrator using a surveillance system, Δt_{NP} -- an average perpetrator neutralization time.

Synthesis of the 'security management at a vulnerable to attack facility model'. It is advisable to present the 'security management at a vulnerable to attack facility model' in the form of a graph with two states: initial (1) and final (2). The average target task time is set as an average high-security facility officer travel time (T_{TP}), which can be presented as follows: $T_{TP} = f_0(\varkappa_1, \varkappa_{2,\dots,} \varkappa_n)$, where $\varkappa_1, \varkappa_{2,\dots n}$ are the high-security facility operation parameters.

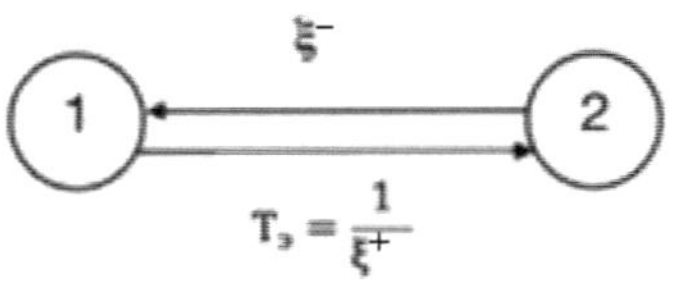

Figure 3. State diagram

Where the average query execution is "$\zeta^+ = 1/\ T_{TP}$". There is a chance, however, that the query will not be satisfied. This can be described as the frequency of query failure, "ζ^+". Operation of a vulnerable to attack facility can be described as the frequency of query execution, or "$\zeta^+ = 1/\ T_{TP}$". Among the repercussion of perpetrator activities are the threat of sabotage, disasters of technological origin (such as chemical disasters or nuclear explosions) and other risks. Hence the problem: "How to link the process of the vulnerable to attack facility operation with the operation of the security system." We will develop our model within the framework of the natural scientific approach (NSA). NSA is determined by the integration of properties of the human mind, the world and cognition.

The decision-making process involves the setting of a goal, data gathering and processing, generation of a range of solutions, coordination, decision-making and monitoring.

The goal is to cut the time a security office needs to make a decision on how to handle the perpetrator while keeping in focus the logic of the security management operation. To reduce the decision-making time, a system of smart video analytics should be introduced. To implement automatic control, it is necessary to identify the conditions of the process and provide feedback. Process diagram (Figure 4):

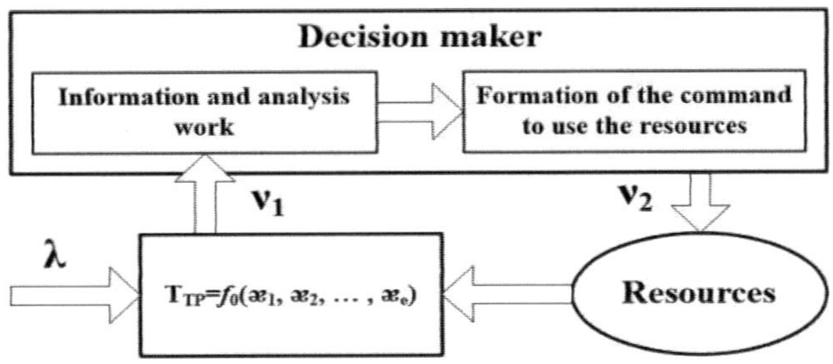

Figure 4. Feedback-based operation of a vulnerable to attack facility

3. Results and Discussion

As a result of the synthesis, a 'security management at facility vulnerable to attack model' was transformed into a mathematical model.

The alleged threat must be identified when the object of control enters state 3 under the influence of λ. It is expected that a security officer will need time Δt to detect a security breach. At this stage, involvement of additional resources required to eliminate the threat is deliberated. Next, the controlled object enters state 4. This is when the security officer acknowledges the threat and decides what resources to involve in order to neutralize it. At the next stage, the threat is neutralized, and the controlled object goes from state 4 to state 2 'problem solved.' At this point, another threat surfaces and has to be analyzed. The process repeats.

To describe the development process in the graph, it is necessary to make assumptions.

1. Human decision-making scheme is considered a form of data management system.
2. Time intervals between problems identification facts are random variables.
3. Facts discovered in time form an arrival close to the Poisson arrival.
4. Time of processing the data on the required characteristic is a random variable.
5. We consider the case, when the time of location of the required characteristics (facts) in the system (person) area is very limited and commensurable with the time required for their identification, as well as for processing the data and taking appropriate actions on these characteristics.

6. The system is prepared for problem identification and elimination.

The transition of the system back to its original state. The frequency of transition from state 1 to state 2 (ζ^+) is equivalent to the inverse of the average target task completion time, which depends on the security officer's response speed. Frequency (ξ^-) is an average frequency of uncompleted target tasks. Usually this value fluctuates within 0.1%.

Following this logic, the following graph can be built (Figure 5):

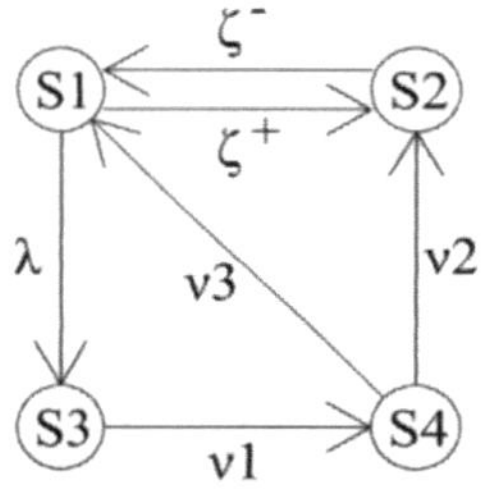

Figure 5. Process of decision-making

A model establishing a correlation between these four processes and the facility performance indicator can be developed now. In this case, the facility performance indicator is equivalent to the probability of identification and neutralization of each security threat.

The Kolmogorov equations (1) can be used:

$$\frac{dP_i(t)}{dt} = \sum_{j=1}^{n} \lambda_{ji}(t) * P_j(t) - P_i(t) * \sum_{j=1}^{n} \lambda_{ji}(t) \quad (1)$$

where i=0,1, 2,…,n.

Nonzero probabilities of occurrence can be calculated by solving a system of linear algebraic equations derived from the Kolmogorov equations, if the derivatives are equal to zero, and the probability functions of states $P_1(t)$, ..., $P_n(t)$in the right part of the equations pass into the unknown finite probabilities P_1 , ..., P_n.. To find the value of P_1 , ..., P_n, a normalizing condition is added to the equations$P_0 + P_1 + \cdots + P_n = 1$.

For the graph shown in Figure 4, the Kolmogorov equations will look like this (2):

$$\begin{cases} \frac{dP_1(t)}{dt} = -(\xi^+ + \lambda) * P_1(t) + \xi^- * P_2(t) + v_3 * P_4(t) \\ \frac{dP_2(t)}{dt} = \xi^+ * P_1(t) - \xi^- * P_2(t) + v_2 * P_4(t) \\ \frac{dP_3(t)}{dt} = \lambda * P_1(t) - v_1 * P_3(t) \\ \frac{dP_4(t)}{dt} = v_1 * P_3(t) - (v_3 + v_2) * P_4(t) \end{cases} \quad (2)$$

Then the nonzero probabilities of occurrence can be calculated by solving a system of linear algebraic equations (3).

$$\begin{cases} 0 = -(\xi^+ + \lambda) * P_1 + \xi^- * P_2 + v_3 * P_4 \\ 0 = \xi^+ * P_1 - \xi^- * P_2 + v_2 * P_4 \\ 0 = \lambda * P_1 - v_1 * P_3 \\ 1 = P_1 + P_2 + P_3 + P_4 \end{cases} \quad (3)$$

The system solution will look like this (4):

$$P_1 = \frac{v_1 * v_2 * \xi^- + v_1 * v_3 * \xi^-}{\lambda * v_1 * v_2 + \lambda * v_1 * \xi^- + \lambda * v_2 * \xi^- + v_1 * v_2 * \xi^+ + v_1 * v_2 * \xi^-}$$

$$P_2 = \frac{\lambda * v_1 * v_2 + v_1 * v_2 * \xi^- + v_1 * \xi^- * v_2}{\lambda * v_1 * v_2 + \lambda * v_1 * \xi^- + \lambda * v_2 * \xi^- + v_1 * v_2 * \xi^+ + v_1 * v_2 * \xi^-}$$

$$P_3 = \frac{\lambda * v_1 * \xi^- + \lambda * v_3 * \xi^-}{\lambda * v_1 * v_2 + \lambda * v_1 * \xi^- + \lambda * v_2 * \xi^- + v_1 * v_2 * \xi^+ + v_1 * v_2 * \xi^-}$$

$$P_4 = \frac{\lambda * v_1 * \xi^-}{\lambda * v_1 * v_2 + \lambda * v_1 * \xi^- + \lambda * v_2 * \xi^- + v_1 * v_2 * \xi^+ + v_1 * v_2 * \xi^-} \qquad (4)$$

The probability of detecting and neutralizing a capacity problem is determined by the following correlation (5):

$$P_2 = \frac{\lambda * v_1 * v_2 + v_1 * v_2 * \xi^+ * v_3}{\lambda * v_1 * v_2 + \lambda * v_1 * \xi^- + \lambda * v_2 * \xi^- + v_1 * v_2 * \xi^+ + v_1 * v_2 * \xi^-} \qquad (5)$$

Three parameters are present in this relation. Thus, we have established an analytical dependence of the generalized characteristics of the security breach occurrence (Δt_{MP}), acknowledgment (Δt_{IP}) and neutralization (Δt_{NP}) [9].

4. Conclusion.

Application of this solution to synthesis-based modeling opens a way to the development of a system of security management at a high-secure facility with the required performance efficiency. The key advantage of such a system is that it is void of the major pitfall -- discrepancy between the OSHM's expectations and results [10]. This solution allows to evaluate any decision from the perspective of time and resource costs, as well as to establish a clear scientifically based relationship between the decision and the result.

This model can be built for a range of different states, including a threat to an economic entity, law enforcement agencies, as well as to the life and health of facility employees. In addition, the model allows to consider different threat origins. For example, errors of the security system development, automated image evaluation failures, human factor or deliberate violation of the site access policy, etc.

References

[1] Lazorkin V I. , Mikhailin A I and Silnikov M V 2015 *Optimization of measures to ensure the survivability of objects* // Questions of defense equipment. Ser. 16. Technical means of countering terrorism. **3-4**. pp. 3-10

[2] Anokhin P K . *System mechanisms of higher nervous activity*. The science. Moscow. p 453

[3] Burlov V G 2018 *The methodological basis for solving the problems of the information warfare and security protection.* // 13th International Conference on Cyber Warfare & Security ICCWS 2018 8-9 March 2018 National Defense University, Washington DC, USA pp 64-74

[4] Burlov V G 2007 *Fundamentals of modeling socio-economic and political processes (methodology, methods).* Saint Petersburg: SPbGPU publishing house. p. 265

[5] Tsvetkov V Y and Matchin V T. *Aggregation of information models* Slavic forum. **2(6)**. pp 77-81

[6] Goode H H Machol R E 1957 *System Engineering: An Introduction to the Design of Large-Scale Systems.* McGraw-Hill Book Co. New York, pp. 551

[7] Burlov V G, Popov N N and Garcia Escalona H A 2018 *Managing the process of using the space geoinformation system in the interests of ensuring the environmental safety of the region* //Scientific notes of RSHU **50**. Scientific and theoretical journal. - SPb: RSHU. p 226

[8] Burlov V G , Grobitski A M and Grobitskaya A M 2016 *Construction management in terms of indicator of the successfully fulfilled production task. Magazine of Civil Engineering*, **63(3)**, pp. 68-71

[9] Burlov V, Andreev A , Gomazov F and Somga-Bichoga N *2018 System integration of*
[10] security maintenance processes in knowledge management. Proceedings of the European
[11] *Conference on Knowledge Management*, ECKM pp. 112-122
[12] Burlov V, Andreev A and Gomazov F 2018 *Development of a model for the management of*
[13] *environmental safety of the region, taking into account of the GIS capacity.* MATEC Web of
[14] Conferences **193**, 02038

A multistage heat pump unit model for reducing energy consumption of space heating at low ambient temperatures

A S Kosoi[1,2], Y A Antipov[1], K V Shkarin[1], I K Shatalov[1] and D D Sokolov[1]

[1] Peoples' Friendship University of Russia, Miklukho-Maklaya street bld. 6, Moscow, 117198, Russia;
[2] Joint Institute for High Temperatures, Russian Academy of Sciences, Izhorskaya st. 13 bld.2, Moscow, 125412, Russia

E-mail: rudn-tit@yandex.ru

Abstract. Air heat pump units (HPUs) for space heating are used more widely than geothermal ones, mainly due to lower installation costs. The main disadvantages of geothermal installations are their low productivity and coefficient of performance (COP). In this work, a multistage heat pump unit with an air source was modeled, which lead to reduce in the energy consumption of the compressors in all of the HPU stages at an ambient temperature of -30 ° C and a heating temperature from 20 ° C to 65 ° C. Wherein, the COP of the modeled multistage HPU in all modes is 20% higher in comparison with a two-stage HPU considered in the literature review. In addition, the influence of the decrease in the ambient temperature on the increase in the energy consumption of the first stage of a multistage HPU was presented in this work.

1. Introduction

Currently, due to the simplicity and low cost of installation, air HPUs are mainly used for space heating. However, at negative ambient temperatures, the use of HPU is limited by a decrease in the COP. Three options to find a solution to this problem are reflected in the work [1]. The configurations for improving the HPU design were analyzed and the results showed an insignificant decrease in the COP gradient with decreasing ambient temperature. The configurations are as follows: a two-stage HPU with intermediate cooling, a HPU with an economizer and a multistage HPU of a cascade cycle. The two-stage HPI with intercooling proved a decrease in the temperature of the refrigerant at the compressor inlet of the high-pressure stage, which consequently lead to an increase in the COP, when compared to a single-stage HPU. Instead of an HPU with intermediate cooling, a two-stage HPU with a closed economizer can be implemented. The economizer feeds a predetermined amount of two-phase refrigerant to the mixing chamber in the suction line of the high-pressure compressor stage, where it mixes with the hot discharge gas of the low-pressure compressor stage. This leads to a decrease in the temperature of the refrigerant at the compressor inlet of the high-pressure stage, which slightly increases the COP. The difference between the multistage cascade HPU in comparison with the other two-stage HPUs, is that two different refrigerants can be used in these units, while each refrigerant operates in its optimal range of operating conditions. The main disadvantage of the multistage cascade HPU with an air evaporator is the sharp decrease in COP at low ambient temperatures, where at an ambient temperature of -30 ℃ and a heated water temperature of 50 ° C the COP = 2.3.

In the work [2], research was carried out on the use of refrigerant injection at several points of the HPU spiral compressor channel, which made it possible to reduce the energy consumption of the compressor. This configuration of the cycle ensures minimal overheating of the refrigerant in the HPU compressor. However, with an increase in the number of refrigerant injection points, only the first three points give an increase in COP to 1.33. While, at other points, an insignificant increase in COP does not pay off by complicating the configuration of the HPU.

In the work [3], in order to increase the COP, an option of replacing the throttle valve with a refrigerant ejector was considered. The potential energy of the refrigerant compression is converted into kinetic energy of

Content from this work may be used under the terms of the Creative Commons Attribution 3.0 licence. Any further distribution of this work must maintain attribution to the author(s) and the title of the work, journal citation and DOI.
Published under licence by IOP Publishing Ltd

IOP Conf. Series: Materials Science and Engineering 1100 (2021) 012045 doi:10.1088/1757-899X/1100/1/012045

the flow in the throat of the nozzle, where the refrigerant is ejected at the outlet of the evaporator and further in the mixing chamber which leads to an increase in the pressure of the refrigerant at the compressor inlet. This configuration of the HPU allows to reduce the operation of the compressor, especially at negative ambient temperatures, which will ultimately lead to an increase in COP by 14.5%.

In this work, a multistage HPU model was developed to reduce the total energy consumption of the compressors due to the redistribution of the refrigerant flow between the sequentially installed compressors [4]. The aim of this work is to develop an efficient multistage heat pump unit model to reduce energy consumption of space heating at low ambient temperatures.

2. Methods and calculations

As already mentioned, the multistage HPU developed in this work was able to reduce the total energy consumption of the compressors due to the redistribution of the refrigerant flow between the sequentially installed compressors. The scheme of the multistage HPU developed is shown in Figure 1.

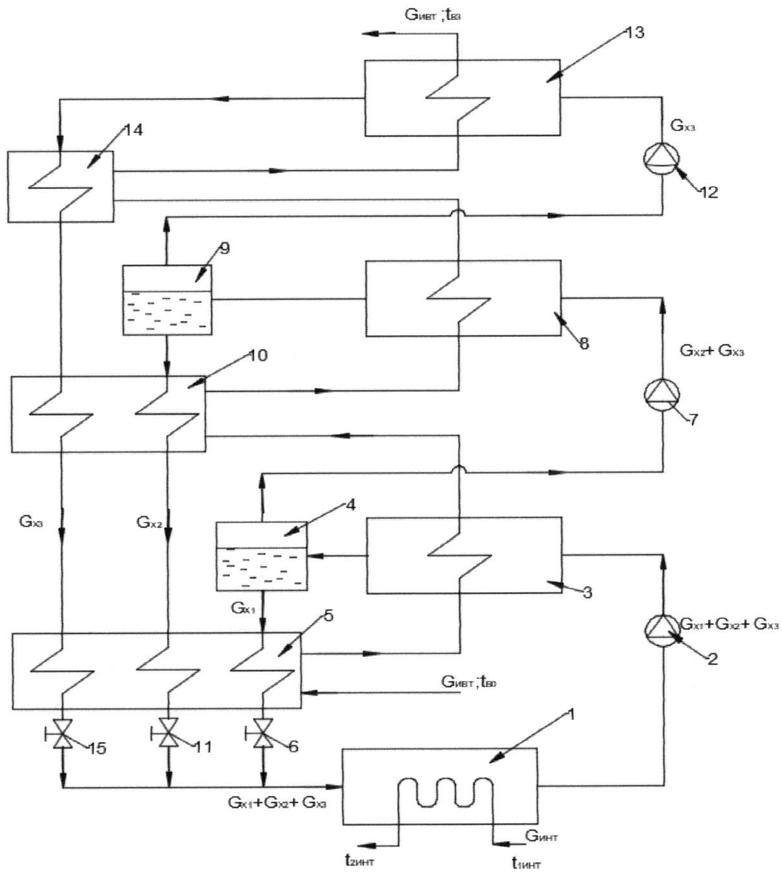

Figure 1. The scheme of the multi-stage HPU model.

The refrigerant from the evaporator 1, after being compressed in the first-stage compressor 2, sequentially enters the condenser 3 and the separator 4 of the first stage, where the refrigerant is separated into liquid and vapor phases. The liquid phase of the refrigerant through the subcooler 5 and the throttle 6, enters the evaporator 1. The vapor phase of the refrigerant is compressed by the second-stage compressor 7. Then the refrigerant sequentially enters the condenser 8 and the separator 9. The liquid phase of the refrigerant from the separator 9 through the subcoolers 10, 5 and the throttle 11 enters the evaporator 1. The vapor phase of the refrigerant is compressed in the third stage compressor 12 and then, sequentially passing the condenser 13, the subcoolers 14, 10, 5 and throttle 15, enters the evaporator 1.

The thermodynamic cycle of the multistage HPU in the P-H diagram is shown in Figure 2. G1, G2 and G3 are the refrigerant flow, respectively, through throttles 6, 11, and 15. From the P-H diagram it can be deduced that each of the compressors 2, 7 and 12, pumps the refrigerant with a flow rate equal to (G1 + G2 + G3), (G2 + G3) and G3. Thus, the heat mainly obtained as a result of the phase transition from vapor to liquid of the refrigerant with the flow rate G1, G2 and G3, respectively, in condensers 3, 8 and 13 of each stage goes to the stepwise heating of the high potential heat source (HHS). Compared with the known schemes of multistage HPI, where the entire refrigerant flow is sequentially pumped to the condensation pressure in the last stage of the HPU developed, with the same heating power, the total energy consumption of the compressors decreased approximately by 30%.

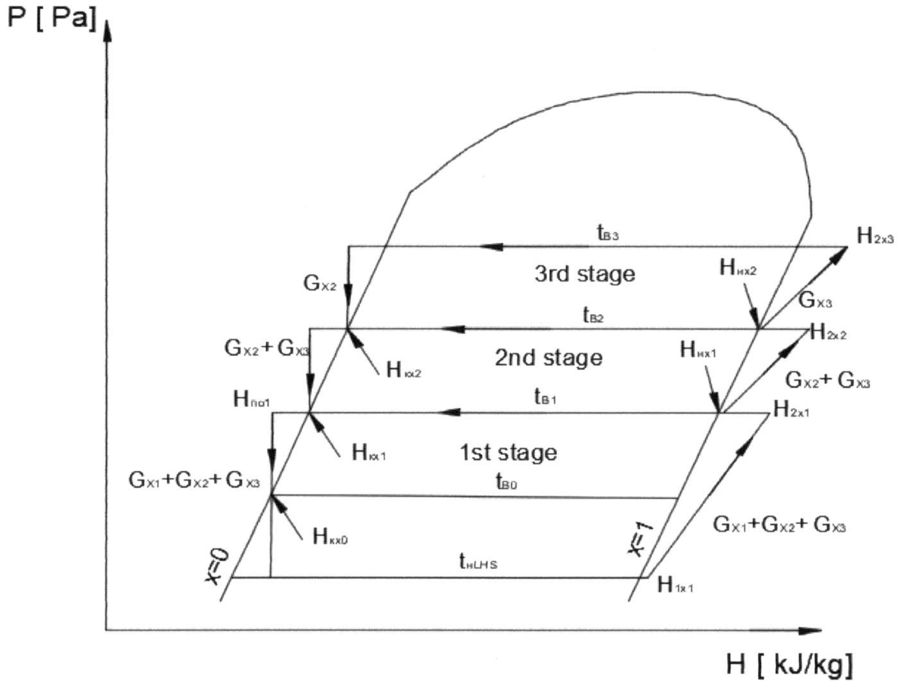

Figure 2. The P-H diagram of the multistage HPU

Thermal calculation of single- two- and three-stage HPUs was carried out according to the method described in the patent [4]. A calculation example of a three-stage HPI is presented in this patent, where the calculation is based on the method of determining the proportion of the refrigerant consumption from the flow rate of one kg of the heated HHS, determined from the heat balance of each HPU stage.

$$\alpha_z = \frac{G_{xz}}{G_{HHS}} = \frac{c_{P_{HHS}}\left(t_{Hz} - t_{H(z-1)}\right)}{\left(H_{2z} - H_{H(z-1)}\right)\eta_{eff}} \tag{1}$$

$$\alpha_{x2} = \frac{G_{x2}}{G_{HHS}} = \frac{c_{p_{HHS}}\left(t_{B_2} - t_{B_1}\right) - \alpha_{x2}\left(H_{2x_2} - H_{HX_2}\right) - \alpha_{x2}\left(H_{kx_2} - H_{kx_1}\right)}{\left(H_{2x_2} - H_{kx_1}\right)\eta_{eff}} \tag{2}$$

$$\alpha_{x1} = \frac{G_{x1}}{G_{HHS}} = \frac{c_{p_{HHS}}\left(t_{B_1} - t_{B_0}\right) - (\alpha_{x3} + \alpha_{x2})\left(H_{2x_1} - H_{HX_1}\right)}{\left(H_{2x_1} - H_{kx_0}\right)\eta_{eff}} - \frac{(\alpha_{x3} + \alpha_{x2})\left(H_{kx_1} - H_{kx_0}\right)}{\left(H_{2x_1} - H_{kx_0}\right)\eta_{eff}} \tag{3}$$

where G_{x1}, G_{x2}, and G_{x3} are the refrigerant flow rate, respectively, of the first, second and third stages; $c_{p_{HHS}}$ is the isobaric heat capacity of the HHS working fluid; t_{B_0} is the temperature of the working fluid of the HHS before entering the three-stage HPU; $t_{\text{в}_1}$, $t_{\text{в}_2}$ and $t_{\text{в}_3}$ are the temperatures of the HHS working fluid at the outlet of condensers 3, 8 and 13, respectively, of the first, second and third stage of the HPU; H_{2x_1}, H_{2x_2} и H_{2x_3} are the enthalpies of the refrigerant at the compressor outlets 2, 7 and 12, respectively, of the first, second and third stages of the HPU; H_{kx_0}, H_{kx_1} and H_{kx_2} are the enthalpies of the refrigerant at the subcooler outlets 5, 10 and 14, respectively, of the first, second and third stage; $H_{\text{н}x_1}$ and $H_{\text{н}x_2}$ are the enthalpies of the refrigerant at the saturation line (x = 0 Steam Dryness Fraction) with a pressure equal to the pressure of the refrigerant after the compressors 2 and 7, respectively, of the first and second stages of the HPU; η_{eff} is the effective efficiency of the condensators 3, 8, and 13.

The energy consumption by the compressor of each stage is calculated by the following formulas:

$$N_1 = G_{HHS}L_1(\alpha_{x1} + \alpha_{x2} + \alpha_{x3}) \quad (4)$$

$$N_2 = G_{HHS}L_2(\alpha_{x2} + \alpha_{x3}) \quad (5)$$

$$N_3 = G_{HHS}L_3(\alpha_{x3}) \quad (6)$$

where L_1, L_2 and L_3 is the work spent on compressing one kg of refrigerant by compressors 2, 7 and 13, respectively, of the first, second and third stages of the HPU. The COP value is calculated by the following formula:

$$\text{COP} = \frac{G_{HHS}c_{pHHS}(t_{\text{в}_3} - t_{\text{в}_0})}{N_1 + N_2 + N_3} \quad (7)$$

R401a was chosen as the refrigerant. The multistage HPU heats the HHS (water) from $t_{\text{во}}$ = 20℃ to $t_{\text{в}3}$ = 65℃ in the range of the low-potential heat source (LHS) (ambient air) $t_{\text{н}}$ from -30℃ to +15℃. The results of the calculations of COP_1, COP_2 and COP_3, respectively, of single- two- and three-stage HPUs are shown in Table 1. As an example, the distribution by stages and the total energy consumption of all stages of the compressors of a three-stage HPU is shown in Table 2.

Table 1. The COP of single-, two- and three-stage HPU

	$t_{\text{н}}$, ℃									
	-30	-25	-20	-16	-10	-5	0	5	10	15
COP_1	1.77	1.98	2.17	2.36	2.68	2.99	3.32	3.77	4.27	4.87
COP_2	2.90	3.11	3.36	3.53	3.95	4.35	4.78	5.33	6.00	6.83
COP_3	2.97	3.201	3.46	3.66	4.11	4.54	5.01	5.62	6.36	7.31
ΔCOP_2, %	38.9	36.4	35.5	33.2	32.5	31.2	30.0	29.1	28.8	28.7
ΔCOP_3, %	40.4	38.1	37.3	35.5	34.8	34.2	33.8	32.9	32.8	33.3

Table 2. The total energy consumption of all stages of the compressors of a three-stage HPU

	$t_{\text{н}}$, ℃									
	-30	-25	-20	-16	-10	-5	0	5	10	15
N_1	52.48	48.00	43.54	40.6	34.9	30.75	26.66	22.69	18.70	14.86
N_2	7.20	7.28	7.28	7.28	7.28	7.28	7.28	7.28	7.28	7.28
N_3	3.64	3.64	3.64	3.64	3.64	3.64	3.64	3.64	3.64	3.64
ΣN	63.4	58.9	54.46	51.5	45.86	41.67	37.58	33.55	29.62	25.78

3. Results and discussion

The results in Table 1 shows that with a decrease in the ambient temperature +15℃ to - 30℃, ΔCOP_2 increased from 28.7 % to 38.9 %, while ΔCOP_3 increased fromт 33.3 % to 40.4 %, where $\Delta COP_2 = (COP_2 - COP_1)100\%/COP_2$ and $\Delta COP_3 = (COP_3 - COP_1)100\%/COP_3$. Therefore, when water is heated from 20℃ to 65℃, comparing the energy consumption of a two-stage HPU and a three-stage HPU to a single-stage HPU, with similar initial parameters, energy saving of a two-stage HPU is from 11.1 kW to 41.5 kW, while for a three-stage HPU energy saving is from 12.9 kW to 43 kW.

The results in Table 2 showed the energy consumption of the second and third stages compressors does not depend on the ambient air temperature, while the energy consumption of the first stage compressor id directly related to the ambient air temperature, for an ambient temperature of +15℃ N_1=14.86 kW, while for –30℃ N_1=52.48 kW.

4. Conclusions

In this work, a multistage air HPU model was presented. The main highlights and conclusions of this work are presented below:

1. The reduction of energy consumption for space heating at low ambient temperatures is achieved by the modeled multistage HPU in this work, in which the total energy consumption of the compressors is reduced by redistributing the refrigerant flow between the sequentially installed compressor, which lead to. an increase in the COP by 20%;
2. Heating the HHS from t_{B0}=const$_1$ to t_{B3}=const$_2$, lead to an increase in the energy consumption of first-stage compressor only with decreasing ambient temperature.

Acknowledgments

This work was prepared with the support of the RUDN University Program "5-100"

References

[1] Bertsch S S and Groll E A 2008 *International Journal of Refrigeration* **31** 1282–9
[2] Mathison M M, Braun J E and Groll E A 2011 *International Journal of Refrigeration* **34** 234–42
[3] Elbel S and Hrnjak P 2008 *International Journal of Refrigeration-revue Internationale Du Froid - INT J REFRIG* **31** 411–22
[4] Antipov Y, Shatalov I and Shkarin K 2018 Patent 2705696 RU

Features of high-temperature synthesis in mechanoactivated gamma-irradiated mixtures of triple systems Ti-Al-C and Ti-Al-Nb

A V Sobachkin[1], A Yu Myasnikov[1], A A Sitnikov[1], M V Loginova[1], V I Yakovlev[1] and A V Gradoboev[2]

[1] Polzunov Altai State Technical University, 46, Lenin ave., Barnaul, 656038, Russia
[2] National Research Tomsk Polytechnic University, 30, Lenin ave., Tomsk, 634050, Russia

E-mail: anicpt@rambler.ru

Abstract. Effects of gamma-irradiation on the process of SHS of mechanoactivated mixtures Ti-Al- C and Ti-Al-Nb were studied. Mechanical activation was performed for 7 minutes in planetary ball mill at energy intensity 40 g. SHS was initiated by induction heating with varying time of annealing. Gamma-irradiation was performed on ^{60}Co isotope to absorbed dose $3 \cdot 10^4$ Gy. Structural parameters of cells of components were calculated. Gamma-irradiation affects components of triple systems in different ways, leading to unsystematic changes in elementary cells. After SHS in powder mixture 80% Ti + 12% Al + 8% C without gamma-irradiation, it was found that product contains Ti_2AlC, Ti_3AlC_2, TiC and unreacted C. Irradiated mixture react fully – diffractogram identifies Ti_2AlC, Ti_3AlC_2 and TiC. Annealing for 2 minutes produces a product with dominant Ti_2AlC. SHS in mixture 45% Ti + 12% Al + 43% Nb revealed that product consists of Ti_2AlNb, B2-phase, α_2-phase and residual Nb and Ti. During synthesis in irradiated mixture without annealing, diffractograms show decrease in level of diffuse background and increase in values of reflection intensities, structural state of products is stabilized. Effect of gamma-irradiation contributes to growth of phases and stabilization of structure synthesized product.

1. Introduction

The development of titanium-based powder materials is an urgent task for space, aviation, automotive and energy industries. Titanium aluminides are heat-resistant, have a high elastic modulus and have a low density [1, 2]. To improve the properties of titanium aluminides, they are doped with Hf, Mo, Nb, Ta, V, W, C and some other elements. For example, addition of niobium to Ti-Al system makes it possible to obtain hydride-forming material Ti_2AlNb, which can be used in alternative energy as a hydrogen storage material [3, 4]. If carbon is used as an additive, MAX-phases Ti_2AlC and Ti_3AlC_2 can be synthesized, which combine advantages of intermetallic and ceramic [5, 6].

One of the ways to obtain new materials using combustion processes is the method of self-propagating high-temperature synthesis (SHS) in thermal explosion mode [7-10]. The main advantages of SHS technologies include: simplicity of equipment, low energy consumption and short duration of synthesis process. SHS is based on exothermic chemical reaction of initial reagents in the form of combustion, where the target product of combustion is solid chemical compounds (carbides, nitrides, borides, oxides, etc.) and materials based on them.

There are known publications that studied effect of gamma-irradiation on mechanocomposites Ti-Al binary system [11-13], but the question of the effect of gamma-quanta on powder composite mixtures of triple systems is practically unexplored.

2. Materials and methods

The object of study was titanium, aluminum, niobium and carbon powders. Mixtures of the following compositions were prepared from them in the ratio by weight %:

Content from this work may be used under the terms of the Creative Commons Attribution 3.0 licence. Any further distribution of this work must maintain attribution to the author(s) and the title of the work, journal citation and DOI.
Published under licence by IOP Publishing Ltd

1) 80% Ti + 12% Al + 8% C;
2) 45% Ti + 12% Al + 43% Nb.

For mechanical activation processing a planetary ball mill AGO-2 was used. The ratio of mass of initial powder mixture to the mass of grinding media was 1: 20. The grinding modes are selected as follows: centrifugal acceleration – 40 g, mechanical activation time – 7 minutes.

To protect against oxidation, air was pumped out of the cylinders, then they were filled with argon at a pressure of 0.3 MPa. After mechanical activation, powders were extracted from cylinders in a special box in argon atmosphere.

To implement high-temperature synthesis, the installation shown in figure 1 was used.

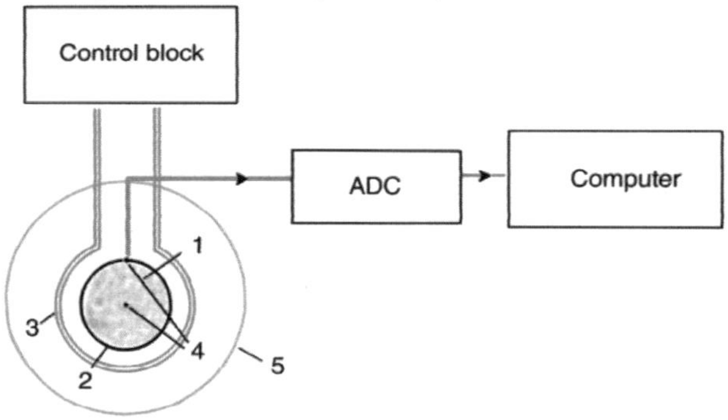

Figure 1. Experimental complex of high speed induction heating: 1 – powder composition; 2 – graphite melting pot; 3 – induction filament; 4 – thermocouple units; 5 – vacuum lid.

Powder mixture 1 was poured into cylindrical graphite crucible 2. Temperature control was carried out using chromel-aluminum thermocouples 4. Microwave heating of crucible was carried out by induction coil 3. Signal from thermocouples was fed to computer via analog-to-digital converter. The system was located under a vacuum cap 5. To prevent interaction with oxygen, the volume under the hood was continuously purged with argon. The advantages of this method include: first, possibility of rapid heating of mixture to high temperatures; second, possibility of prolonged exposure of system at high temperatures immediately after synthesis (high-temperature annealing); and third, possibility of rapid cooling of mixture to room temperatures. Thus, there is a possibility of a continuous transition from rapid reaction process to high-temperature annealing process, followed by rapid cooling, which provides a controlled thermal effect on powder mixture.

Structural-phase analysis of the samples was performed on diffractometer DRON-6 with CuKα-radiation (λ = 15.418 nm). Diffractograms of all samples were recorded under identical conditions, which made it possible to compare obtained values more correctly. Scan step h = 0.05°, exposure time t = 3 seconds, shooting angle range 2Theta = 20°...80°. The analysis of diffraction peak shapes, as well as calculation of second moments and intervals for their determination, were performed using preprocessing program PDWin. The intensities were adjusted for Lorentz factor and polarization factor. The unit cell parameters were refined using standard least squares method using PDWin software package. The broadening of diffraction lines of all samples was calculated with an adjustment for instrumental broadening, using experimental data from reference material, as well as the broadening calculated for main instrumental aberrations (goniometer parameters and linear absorption coefficient of sample were set).

Experimental studies on irradiation of mechanoactivated mixtures were carried out on a certified stationary installation «Explorer» (isotope ^{60}Co) (Tomsk) under normal climatic conditions. The dose rate of gamma radiation in this case was 1 Gy/s. The level of exposure to gamma-quanta was characterized by absorbed doses, which were $3\cdot10^4$ Gy.

3. Results and discussion

Figure 2 shows radiographs of the composite mixture after mechanical activation (bottom) and after mechanical activation with additional exposure to gamma-quanta with accumulated dose of $3\cdot10^4$ Gy (top).

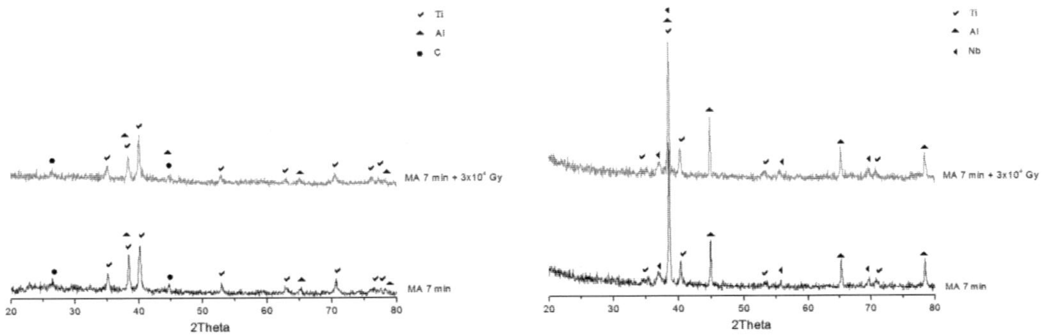

(a) (b)

Figure 2. Diffractograms of powder mixture after mechanical activation (bottom) and exposured to gamma quanta (top): (a) Ti-Al-C system; (b) Ti-Al-Nb system.

Analyzing the appearance of diffractograms, we can say that after gamma irradiation, intensity values of diffraction maxima for both systems remain at the same level. Broadened diffraction reflections indirectly indicate the preservation of nanostructural state of crystallites and presence of residual micro-deformations that occurred during mechanical activation.

The structural parameters of cells of components of both systems were calculated from obtained radiographs. The calculated values are summarized in tables 1-4.

Table 1. Structural parameters of titanium

cell parameters (nm)	after MA	after MA and gamma irradiation	Ti Reference
		system Ti-Al-C	
a	29.56	29.55	29.5
c	46.84	46.91	46.83
		system Ti-Al-Nb	
a	29.3	29.4	29.5
c	47.2	46.8	46.83

Table 2. Structural parameters of aluminum

cell parameters (nm)	after MA	after MA and gamma irradiation	Al Reference
		system Ti-Al-C	
a	40.47	40.49	40.41
		system Ti-Al-Nb	
a	40.3	40.4	40.41

Table 3. Structural parameters of niobium

cell parameters (nm)	after MA	after MA and gamma irradiation	Nb Reference
a	33.1	33.0	33.0

Table 4. Structural parameters of carbon

cell parameters (nm)	after MA	after MA and gamma irradiation	C Reference
a	24.5	24.54	24.56
c	67.55	67.18	66.96

At the next stage, it was carried out high-temperature synthesis in mechanoactivated and mechanoactivated irradiated mixtures of specified compositions of Ti-Al-C and Ti-Al-Nb systems. During experiment, synthesis modes were varied. Synthesis was performed with the source switched off when the system reached maximum temperature and with isothermal annealing ($\tau = 2$ min.)

Figure 3 shows the results of determining phase composition of SHS materials in Ti-Al-C system.

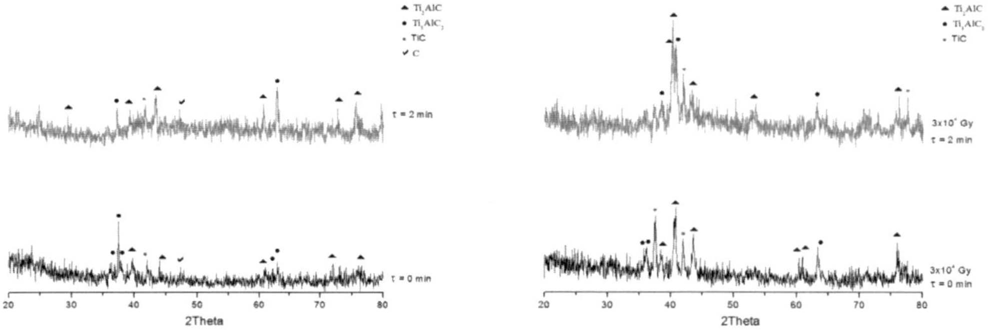

(a) (b)

Figure 3. Diffractograms of synthesis products after thermal explosion in mechanoactivated mixture of 80% Ti + 12% Al + 8% C: (a) mixture without irradiation, (b) mixture with accumulated dose of $3 \cdot 10^4$ Gy.

X-ray phase analysis of sample, synthesized from mechanically activated non-irradiated mixture of Ti-Al-C system without additional annealing (τ = 0 min), showed that reaction products contain several phases: MAX-phases Ti_2AlC and Ti_3AlC_2, titanium carbide TiC and unreacted carbon C. If the system is held for τ = 2 min annealing immediately after reaction is completed, phase composition of SHS sample is preserved. However, the system is in non-equilibrium state – amplitude and intensity of diffraction reflections change, and diffusion background increases.

When SHS is implemented in mechanoactivated and irradiated Ti-Al-C mixture (figure 3, b), MAX-phases Ti_2AlC and Ti_3AlC_2, and titanium carbide TiC are observed on diffractograms. There are no unreacted components. However, the system is still in non-equilibrium state – diffraction maxima are broadened, and their intensity is at the same level as in experiment on synthesis of a mixture without irradiation. However, when the system is kept for additional annealing for τ = 2 min, intensity values increase, and phase composition of the product is characterized by dominant Ti_2AlC phase.

Figure 4 shows the results of study of phase compositions of products obtained by high-temperature synthesis of Ti-Al-Nb powder mixtures.

According to results of X-ray phase analysis after synthesis of mechanically activated non-irradiated mixtures Ti-Al-Nb without additional annealing (τ = 0 min) the reaction products contain several phases: O-phase (Ti_2AlNb), β0(B2)-phase and small content of α_2-phase (Ti_3Al). Reflections of residual niobium and titanium are also recorded. When exothermic annealing is performed for τ = 2 min. after chemical reaction is completed, phase composition is preserved, but new additional reflections corresponding to the O-phase appear. However, there is a decrease in intensity of diffraction maxima, their broadening and an increase in level of diffuse background. This may indirectly indicate that holding system for 2 minutes leads to thermal vibrations of atoms, which reduces amplitude and intensity of peaks with transition of the system to non-equilibrium state.

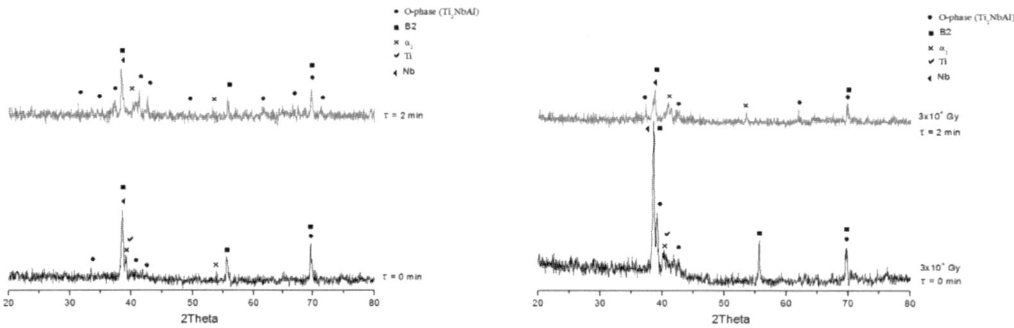

(a) (b)

Figure 4. Diffractograms of synthesis products after thermal explosion in mechanoactivated mixture of 45% Ti + 12% Al + 43% Nb: (a) mixture without irradiation, (b) mixture with accumulated dose of $3 \cdot 10^4$ Gy.

After SHS without subsequent annealing in mechanoactivated and gamma-irradiated powder mixture Ti-Al-Nb system, final product is multiphase (figure 4, b): O-phase (Ti_2AlNb), β0(B2)-phase and small content of α_2-phase (Ti_3Al). Reflections of residual niobium and titanium are also recorded, which is identical to experiment without gamma-irradiation. However, diffraction patterns show decrease in level of diffuse background and increase in values of reflection intensities (compared to synthesis without gamma-irradiation) by almost two times. Thus, preliminary effect of gamma-irradiation on powder mixture contributes to growth of phases and stabilization of structure of synthesized product [14, 15]. Further exposure of system for additional annealing (τ = 2 min.) leads to transition of system to non-equilibrium state.

4. Conclusion

1. After SHS in powder mixture 80% Ti + 12% Al + 8% C without gamma irradiation, it was found that reaction product contains MAX-phases Ti_2AlC and Ti_3AlC_2, titanium carbide TiC and unreacted C. When system is annealed for 2 minutes, system transitions to non-equilibrium state. In case of synthesis of irradiated mixture, components react fully – diffractogram identifies reflections of Ti_2AlC, Ti_3AlC_2 and TiC. Annealing for 2 minutes allows obtaining a product with dominant Ti_2AlC phase.

2. After high-temperature synthesis in powder mixture 45% Ti + 12% Al + 43% Nb without gamma-irradiation, it was found that O-phase (Ti_2AlNb), β0(B2)-phase, α_2-phase (Ti_3Al) and reflections of residual niobium and titanium are fixed in final product. When annealing time is increased to τ = 2 min. the number of O-phase reflections increases, and system goes into non-equilibrium state. When synthesized in an irradiated mixture without annealing, structural state of products is stabilized. Diffraction patterns of final product show low level of diffuse background and increase in intensity of main diffraction reflections, compared to synthesis product of non-irradiated mixture. Exposure for additional annealing (τ=2 min.) leads to transition of system to non-equilibrium state.

3. Gamma-irradiation affects the components of powder mixture of triple systems in different ways, leading to unsystematic changes in elementary cells. This is due to both different quantitative content of components in initial mixture and structure of activated mixture, which depends on degree of plasticity of components. Also, preliminary effect of gamma irradiation on powder mixture contributes to growth of phases and stabilization of structure of synthesized product.

Acknowledgments

This work was supported by Ministry of Education and Science of Russian Federation (Zadanie no. FZMM-2020-0002) and grant of the President of the Russian Federation (agreement 075-15-2020-234).

References

[1] Lasalmonie A 2006 *Intermetallics* **14** 1123–29.
[2] Planck S K and Rosenberger A H 2002 *Materials Science and Engineering* **325, 1-2** 270–280.
[3] Chen Yu-yong, Yu-feng Si, Fan-tao Kong, Zhi-guang Liu and Jun-wen Li 2006 *Transaction Nonferrous Metals Society of China* **16** 316–320.
[4] Qian M and Froes F H *Titanium Powder Metallurgy: Science, Technology, and Applications* (Oxford: Butterworth)
[5] Potanin A Yu, Loginov P A, Levashov E A, Pogozhev Yu S, Patsera E I and Kochetov N A 2015 *EurasianChemico-Technological Journal* **17** 233–242
[6] Aiguo Zhou, Chang-an Wang, Zhenbin Ge and Lifeng Wu 2001 *Journal of materials science letters* **20** 1971–73
[7] Merzhanov A G and Abramov V G 1981 *Propellants and explosives* **6** 130–148.
[8] Zhu X, Zhang T, Morris V and Marchant D 2010 *Intermetallics* **18** 1197–1204.
[9] Naplocha K and Granat K 2009 *Journal of Alloys and Compounds* **480** 369–375.
[10] Lagosa M A, Agotea I, Gutiérreza M, Sargsyana A and Pambaguian L 2010 *International Journal of Self-Propagating High-Temperature Synthesis* **19, 1** 23–27.
[11] Loginova M V, Sobachkin A V, Sitnikov A A, Yakovlev V I, Filimonov V Yu, Myasnikov A Yu, Sharafutdinov M R, Tolochko B P and Gradoboev A V 2019 *Journal of Synchrotron Radiation* **26** 1671–78

[12] Sobachkin A V, Loginova M V, Sitnikov A A, Yakovlev V I, Filimonov V Yu and Gradoboev A V 2018 IOP Conference Series: Materials Science and Engineering 327 032051
[13] Loginova M V, Yakovlev V I, Filimonov V Yu, Sitnikov A A, Sobachkin A V, Ivanov S G and Gradoboev A V 2018 Letters on Materials 8(2) 129–134
[14] Loginova M V, Yakovlev V I, Sitnikov A A, Sobachkin A V, Ivanov S G, Negodyaev A Z and Gradoboev A V 2017 Physics of Metals and Metallography 118, 2 170–175.
[15] Loginova M V, Yakovlev V I, Sitnikov A A, Filimonov V Yu, Sobachkin A V and Gradoboev A V 2018 Journal of Surface Investigation: X-ray, Synchrotron and Neutron Techniques 12, 3 480–484

IOP Conf. Series: Materials Science and Engineering 1100 (2021) 012047 doi:10.1088/1757-899X/1100/1/012047

Triangular finite element in a mixed formulation for a flate problem of elasticity theory

N A Gureeva[1] and V V Ryabukha[2]

[1]Financial University under the Government of the Russian Federation, 49 Leningradsky Prospekt, Moscow, 125993, Russia.
[2]Volgograd state agrarian University, 26 Universitetskiy Prospekt, Volgograd region, 400002, Russia.

E-mail: Vitalik30090@mail.ru

Annotation. The algorithm for obtaining the stress-strain state matrix of a triangular finite element in a mixed formulation for calculating elastic structures under flat stress and flat deformation is analyzed. The nodal unknowns are assumed to be displacements and stresses. Approximation of desired quantities through the nodal values was carried out pelenanie functions. The stress-strain state matrix is obtained on the basis of a functional that reflects the equality of the actual work of external and internal forces when replacing the actual work of internal forces with the difference between the total and additional work of these forces. The calculation example shows the specificity of the algorithm.

Introduction

The theory of analysis of the stress-strain state(SSS) of solids by the finite element method (FEM) is currently developed quite fully[1]. FEM is used to solve a very wide range of problems in construction mechanics. When constructing a discrete model of a continuous quantity, the domain of its definition is divided into a finite number of elements of various forms. Since it is impossible to discretize the object under study with elements of only a quadrangular shape, the development of a triangular finite element is an urgent task. The predominant number of finite elements is developed in the formulation of the displacement method[4-8] and their use allows us to obtain a satisfactory result in most cases. However, the use of finite elements in the mixed formulation [9-11] with nodal unknown displacements and stresses has significant advantages, one of which is the possibility of using the simplest polynomials for approximating the desired values via

1. Basic relations of the plane problem of the theory of elasticity

1.1. Flat deformation. The solid has the shape of a prism of an transverse section in the form of triangle. Axis Oy directed along the length of the solid, axis Ox and Oz - in the cross-sectional flat. The acting loads are functions of coordinates x, z. Displacements, stresses and strains are also functions of these coordinates and have the form

$$u = u(x,z);\; w = w(x,z);$$
$$\sigma_{xx} = \sigma_{xx}(x,z);\; \sigma_{zz} = \sigma_{zz}(x,z);\; \sigma_{xz} = \sigma_{xz}(x,z); \qquad (1.1)$$
$$\sigma_{yy} = \sigma_{yy}(x,z);$$
$$\varepsilon_{xx} = \varepsilon_{xx}(x,z);\; \varepsilon_{zz} = 0;\; \varepsilon_{xz} = \varepsilon_{xz}(x,z).$$

In case of flate deformation, the following relations are fulfilled:

$$\varepsilon_{\mathrm{xx}} = \frac{\partial u}{\partial x};\; \varepsilon_{\mathrm{zz}} = \frac{\partial w}{\partial z};\; 2\varepsilon_{\mathrm{xz}} = \frac{\partial u}{\partial z} + \frac{\partial w}{\partial x};$$
$$\varepsilon_{\mathrm{xx}} = \frac{1}{E}\left(\sigma_{\mathrm{xx}} - \nu\sigma_{\mathrm{zz}} - \nu\sigma_{\mathrm{yy}}\right); \qquad (1.2)$$
$$\varepsilon_{\mathrm{yy}} = \frac{1}{E}\left(\sigma_{\mathrm{yy}} - \nu\sigma_{\mathrm{xx}} - \nu\sigma_{\mathrm{zz}}\right);$$
$$\varepsilon_{\mathrm{zz}} = \frac{1}{E}\left(\sigma_{\mathrm{zz}} - \nu\sigma_{\mathrm{xx}} - \nu\sigma_{\mathrm{yy}}\right);\; 2\varepsilon_{\mathrm{xz}} = \frac{\varepsilon_{\mathrm{xz}}}{G},$$

where E, G - elastic tensile and shear moduli; ν - lateral deformation ratio.

Based (1.2) matrix relations are compiled

Content from this work may be used under the terms of the Creative Commons Attribution 3.0 licence. Any further distribution of this work must maintain attribution to the author(s) and the title of the work, journal citation and DOI.
Published under licence by IOP Publishing Ltd

$$\underset{3x1}{\{\varepsilon\}} = \underset{3x2}{[L]}\underset{2x1}{\{v\}};\ \underset{3x1}{\{\varepsilon\}} = \underset{3x3}{[C_1]}\underset{3x1}{\{\sigma\}} \tag{1.3}$$

where

$\underset{1x3}{\{\varepsilon\}^T} = \{\varepsilon_{xx}\varepsilon_{zz}2\varepsilon_{xz}\}$- string of deformations;

$\underset{1x2}{\{v\}^T} = \{u\ w\}$- string of displacements;

$\underset{1x3}{\{\sigma\}^T} = \{\sigma_{xx}\sigma_{zz}\sigma_{xz}\}$- string of stresses;

$\underset{3x2}{\{L\}}$- matrix of differential operators;

$\underset{3x3}{\{G\}}$- Hooke's law matrix for flat deformation.

1.2. Flat stress state.

In this case, the differential relations between deformations and displacements remain unchanged, and deformations through stresses are determined by the expressions

$$\varepsilon_{xx} = \frac{1}{E}(\sigma_{xx} - \nu\sigma_{zz});\ \varepsilon_{zz} = \frac{1}{E}(\sigma_{zz} - \nu\sigma_{xx}); \tag{1.4}$$
$$2\varepsilon_{xz} = \frac{\sigma_{xz}}{G},$$

or in matrix form

$$\underset{3x1}{\{\varepsilon\}} = \underset{3x3}{[C_2]}\{\sigma\}. \tag{1.5}$$

2. Algorithm for obtaining the matrix of the stress-strain state of a triangular finite element. The finite element is a prism with a cross-section in the form of an arbitrary triangle with nodes $i,\ j,\ k$, located in the loading plane of the body under consideration. To perform numerical integration over the area of an arbitrary triangle, it is mapped onto an right-angled triangle in the coordinate system $\xi,\ \eta$, within changing limits $0 \leq \xi,\ \eta \leq 1$. Cartesian coordinates $x,\ z$ of an arbitrary triangle are determined in terms of the coordinates of the nodes of the triangle by the linear relations

$$\lambda = \underset{1x3}{\{\varphi(\xi,\ \eta)\}^T}\underset{3x1}{\{\lambda_y\}}, \tag{2.1}$$

where λ - coordinate $x$or z;

$\underset{1x3}{\{\varphi(\xi,\ \eta)\}^T} = \{1 - \xi - \eta\quad \xi\quad \eta\};\ \underset{1x3}{\{\lambda_y\}^T} = \{\lambda^i\quad \lambda^j\quad \lambda^k\}.$

Differentiation (2.1) derivatives $x_\xi, x_\eta, z_\xi, z_\eta$, Cartesian coordinates in the local system $\xi,\ \eta$ and derivatives $\xi_x, \xi_z, \eta_x, \eta_z$, local coordinates in Cartesian coordinates x, z.

The displacements of the interior point of the finite element are approximated through the nodal values also by linear relations (2.1), where under the symbol λ understood displacement u, w.

The derivatives of displacements are determined by the relations

$$u_{,x} = \underset{1x3}{\{\varphi_{,x}\}^T}\underset{3x1}{\{u_y\}};\ u_{,z} = \underset{1x3}{\{\varphi_{,z}\}^T}\underset{3x1}{\{u_y\}};$$
$$w_{,x} = \underset{1x3}{\{\varphi_{,x}\}^T}\underset{3x1}{\{w_y\}};\ w_{,z} = \underset{1x3}{\{\varphi_{,z}\}^T}\underset{3x1}{\{w_y\}}; \tag{2.2}$$

where

$$\underset{1x3}{\{\varphi_{,x}\}^T} = \underset{1x3}{\{\varphi_{,\xi}\}^T}\xi_{,x} + \underset{1x3}{\{\varphi_{,\eta}\}^T}\eta_{,x};$$
$$\underset{1x3}{\{\varphi_{,z}\}^T} = \underset{1x3}{\{\varphi_{,\xi}\}^T}\xi_{,z} + \underset{1x3}{\{\varphi_{,\eta}\}^T}\eta_{,z};$$

The stresses at the inner point of the finite element are also determined by the relations (2.1), where the symbol λ implied tension σ_{xx} , σ_{zz}, σ_{xz}.

For the inner point of the finite element, matrix expressions are formed

$$\underset{2x1}{\{v\}} = \underset{2x6}{[A]}\underset{6x1}{\{v_y\}};\ \underset{3x1}{\{\sigma\}} = \underset{3x9}{[S]}\underset{9x1}{\{\sigma_y\}}; \tag{2.3}$$

where

$$\underset{1x6}{\{v_y\}^T} = \{u^i\ u^j\ u^k\ w^i\ w^j\ w^k\};$$

$\underset{1x9}{\{\sigma_y\}^T} = \{\sigma_{xx}^i\ \sigma_{xx}^j\ \sigma_{xx}^k\ \sigma_{zz}^i\ \sigma_{zz}^j\ \sigma_{zz}^k\ \sigma_{xz}^i\ \sigma_{xz}^j\ \sigma_{xz}^k\ \}.$

Deformations (1.3) taking into account the (2.2) can be represented by a matrix expression

$$\underset{3x1}{\{\varepsilon\}} = \underset{3x2}{[L]} \underset{2x1}{\{v\}} = \underset{3x2}{[L]} \underset{2x6}{[A]} \underset{6x1}{\{v_y\}} = \underset{3x6}{[B]} \underset{6x1}{\{v_y\}}. \quad (2.4)$$

To form the matrix of the stress-strain state of a finite element using the mixed functional

$$\Phi_S = \int_F \{\sigma\}^T [L] \{\sigma\} \, dF - \frac{1}{2}\int_F \{\sigma\}^T \{\varepsilon\} \, dF - \frac{1}{2}\int_S \{v\}^T \{q\} \, dS \quad (2.5)$$

where F - area of a triangle; S - load contour length;

$\underset{1x2}{\{q\}^T} = \{q_1 \; q_2\}$- components of the external load vector.

Given the expressions (1.3), (1.5.) and approximating relations (2.3), (2.4) function (2.5) will written in the form

$$\Phi_S = \underset{1x9}{\{\sigma_y\}^T} \int_F \underset{9x3}{[S]^T} \underset{3x6}{[B]} \, dF \, \{v_y\} -$$

$$- \frac{1}{2} \underset{1x9}{\{\sigma_y\}^T} \int_F \underset{9x3}{[S]^T} \underset{3x3}{[C_n]} \underset{3x9}{[S]} \, dF \, \underset{9x1}{\{\sigma_y\}} -$$

$$- \frac{1}{2} \underset{1x6}{\{\sigma_y\}^T} \int_S \underset{6x2}{[A]^T} \underset{2x1}{\{q\}} dS, \; n = 1,2 \quad (2.6)$$

After minimization with respect to the nodal unknowns from the functional (2.6) systems of equations are obtained

$$\frac{\partial \Phi_S}{\partial \{\sigma_y\}^T} \equiv \underset{9x6}{[Q]} \underset{6x1}{\{v_y\}} - \underset{9x9}{[H]} \underset{9x1}{\{\sigma_y\}} = 0; \quad (2.7)$$

$$\frac{\partial \Phi_S}{\partial \{v_y\}^T} \equiv \underset{6x9}{[Q]^T} \underset{9x1}{\{\sigma_y\}} - \underset{6x1}{\{f_q\}} = 0; \quad (2.8)$$

where

$$\underset{9x9}{[H]} = \int_F \underset{9x3}{[S]^T} \underset{3x3}{[C_n]} \underset{3x9}{[S]} \, dF; \; \underset{9x6}{[Q]} = \int_F \underset{9x3}{[S]^T} \underset{3x6}{[B]} \, dF;$$

$\underset{6x1}{\{f_q\}} = \int_S \underset{6x2}{[A]^T} \underset{2x1}{\{q\}} \, dS$; $n = 1$ in flat deformation $n = 2$- in a flat stressed state.

Systems (2.7) and (2.8) are presented in traditional finite element formulation

$$\underset{15x15}{[k]} \underset{15x1}{\{z_y\}} = \underset{15x1}{\{F_y\}}, \quad (2.9)$$

where

$$\underset{15x15}{[k]} = \begin{bmatrix} -\underset{9x9}{[H]} & \underset{9x6}{[Q]} \\ \underset{9x6}{[Q]^T} & \underset{6x6}{[O]} \end{bmatrix}; \; \underset{1x15}{\{z_y\}^T} = \left\{ \underset{1x9}{\{\sigma_y\}^T} \underset{1x6}{\{v_y\}^T} \right\};$$

$$\underset{1x15}{\{F_y\}^T} = \left\{ \underset{1x9}{\{O\}^T} \underset{1x6}{\{f_q\}^T} \right\}.$$

Example. The stress-strain state of a cantilever beam loaded at the end of the cantilever by a concentrated force P is considered (Drawing.1)

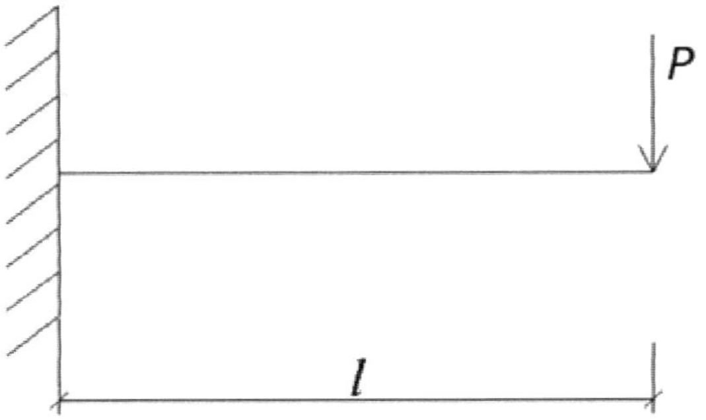

Figure 1. Calculation diagram of the beam

The following initial data are accepted: P = 0,1 kN; l = 0,2 m; h =0,01m (the height of the beam cross section); b = 0,01 m (the width of the cross section of the beam); E = $2*10^5$ MPa (elastic modulus of the beam material); ν =0,3 (lateral deformation ratio).
The design scheme of the beam is shown on drawing 2.

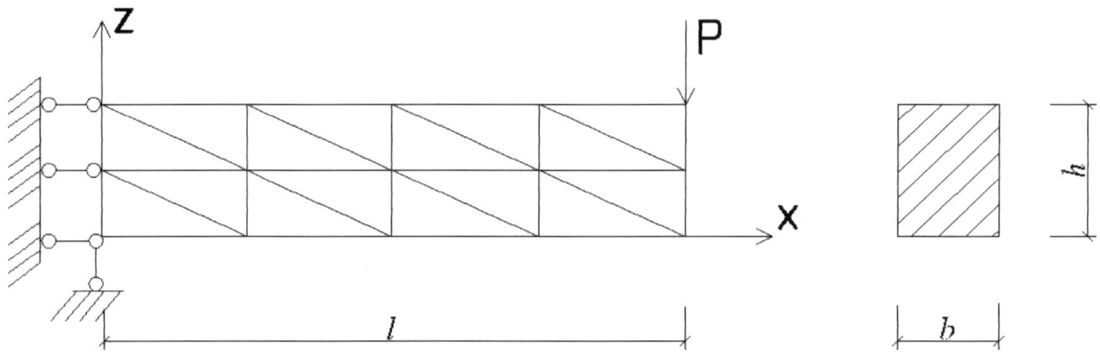

Figure 2. The design scheme of the beam

When discrediting the beam on the final elements, the notation is used: N,M - number of nodes along the span of the cantilever beam; NN - number of nodes in the beam cross-section height; NE=(NM-1)×(NN-1)×2 - the number of finite elements for discretization of the beam.
The results of calculations for various options for discrediting the beam to the finite elements are presented in table 1.

Table 1

Coordinates of the point, cm	Stress σ_{xx}, MPa			
	NM=21 NN=3 NE=80	NM=41 NN=3 NE=160	NM=81 NN=5 NE=640	NM=161 NN=9 NE=2560
x = 0.0; z =0.0	-98.34	-111.86	-120.51	-127.51
x = 0.0; z =1.0	126.15	131.54	125.91	122.91
x = 0.5; z =0.0	-101.22	-109.66	-116.95	-116.83
x = 0.5; z =1.0	104.31	113.75	121.96	117.88
x = 1.0; z =0.0	-106.89	-112.,73	-113.44	-118.81
x = 1.0; z =1.0	102.98	123.64	115.05	114.40
x = 1.5; z =0.0	-108.14	-112.08	-110.36	-110.91
x = 1.5; z =1.0	113.26	108.81	110.71	110.90
x = 2.0; z =0.0	-106.56	-106.49	-107.95	-108.00
x = 2.0; z =1.0	120.93	106.32	107.79	108.04
	w=-0.158cm= =-0.00158m	w=-0.158cm= =-0.00158m	w=-0.160cm= =-0.0016m	w=-0.160cm= =-0,0016m

As can be seen from the table, the convergence of the computational process is already performed at a point located at a distance of 0.5h from the other end of the beam.
Calculations based on the technical theory of beams (taking into account the hypothesis of plane sections) lead to the following results.
The deflection under the concentrated force is determined by the formula

$$w = \frac{Pl^3}{3EI_2} = \frac{0{,}1*0{,}2^3}{3*2*10^5*\frac{1*0{,}01^3}{12}} = 0{,}0016м = 16*10^{-4}m.$$

As you can see this result coincides with the results of table 1.

Moment at a point with coordinates x=0,5 cm defined by a numeric value
$M = P \ * \ (0{,}2 - 0{,}005) = 0{,}1 \ * \ 0{,}195 = 0{,}0195kN * m.$
The normal stresses at this point are

$$\sigma_{xx} = \pm\frac{M}{w} = \pm\frac{0{,}0195}{0{,}1 \ * \ 0{,}01^2 \ * \frac{1}{6}} = \pm 117MPa,$$

which is almost the same as the table values.
The results show the correctness of the developed prismatic finite element in a mixed formulation with a cross section in the form of a triangle.

References

[1] Petrov V. V. Nonlinear incremental construction mechanics. - M.:
[2] Infra-Engineering, 2014. - 480c.
[3] Galimov K. 3, Paimushin V. N. Theory of shells of complex geometry.
- Kazan: Publishing house of Kazan state University, 1985. - 164c.
[4] Yakupov N. M., Serazutdinov M. N., Nuh N. M. Calculation of elastic
[5] thin-walled structures of complex geometry. - Kazan: IMIRAN, 1993. - 206s.
[6] Zenkevich O. method of finite elements in engineering. - M.: Mir, 1975, 541 p.
[7] Auden J. Finite elements in nonlinear continuum mechanics. Per. s Engl. - M.: Mir, 1976, 464 p.
[8] Postnov, V. A. Method of superelements in calculations of engineering
[9] structures / V. A. Postnov, S. A. Dmitriev, B. K. Eltyshev, A. A. Rodionov. Under the General editorship of V. A. Postnov. - L.: Sudostroenie, 1979. - 288 p.
[10] Golovanov, A. I., Konoplev, Yu. G. Numerical study of finite deformations of hyperelastic bodies IV finite element implementation. Examples solving problems / A. I. Golovanov, U. G. Konoplyov, L. Y, Sultanov //Kazan. Zap. Cauldron.UN-TA. Ser: Phys. - Matem. nauki, -152:4(2010), 115-126 p.
[11] Nikolaev, A. P. Vector interpolation of displacement fields in finite
[12] element calculations of shells monograph (A. P. Nikolaev, Yu. V. Klochkov, A. P.Kiselev, N. A. Gureeva,). - Volgograd: fgbou VO Volgogradsky GAU, 2012- 264 p.
[13] Bandurin N. f. determination of the plane stress state of shells
[14] based on a mixed formulation of the finite element method taking into account geometric nonlinearity, N. A. Gureeva // Cosmonautics and rocket science. 2013. - Vol. 1. - No. 70. - P. 69-75.
[15] Gureeva, N. A. Calculation of an axisymmetrically loaded rotation shell taking into account geometric nonlinearity based on a mixed FEM/ N. A.
[16] Gurseva, Yu. V. Klochkov, A. P. Nikolaev / / Izv.vuzov. Aviation technology
[17] 2014. - no. 3. - P. 14-19.

IOP Conf. Series: Materials Science and Engineering 1100 (2021) 012048 doi:10.1088/1757-899X/1100/1/012048

Complex diffusion saturation of carbon steel 1045 with boron, chromium, titanium and silicon

A M Guryev[1,2], S G Ivanov[2], M A Guryev[2], S Mei [1] and Z Quan [1,2]

[1]Wuhan Textile University, No.1 Sunshine Avenue, Wuhan, Hubei, 430200, China
[2]Polzunov Altai State Technical University, 46 Lenin Ave., Barnaul, 656038, Russia

E-mail: gurievam@mail.ru

Abstract. The article presents the results of studies on the formation of a diffusion layer upon saturation of steel 1045 with boron together with chromium, titanium and silicon. The results of tests of diffusion coatings on steel 1045 obtained by simultaneous diffusion saturation with boron, chromium, titanium and silicon under the action of a sulfuric acid solution under conditions of abrasive wear are presented. A comparison is made of the service life of knives for shredding lead-acid batteries made of steel HARDOX 600 and steel 1045 after diffusion saturation with boron, chromium, titanium and silicon. It was found that the simultaneous strengthening with boron, chromium, titanium and silicon made it possible to increase the operational durability of knives for shredding lead-acid batteries by 2.4 times. The possibility of highly efficient use of products made of simple carbon steel with a multicomponent (B + Cr + Ti + Si) diffusion coating instead of products made of hardened alloy steel is shown.

1. Introduction

Wear of the working surfaces of machine parts and tools is the cause of breakage in 75% of cases. The problem of increasing wear resistance is becoming increasingly important. The trend in the development of chemical-thermal treatment is due to the problem of saving alloying materials (manganese, chromium, cobalt, tungsten, vanadium, molybdenum, niobium, etc.). The use of diffusion hardening and protective coatings can significantly reduce manufacturing costs and improve the performance properties of the created products. In some cases, chemical heat treatment is the only possible means of obtaining the required performance properties not only of the surface, but also of the product as a whole.

Boriding occupies a special place among the processes of chemical thermal treatment. The increased interest in the boriding process is caused by the possibility of obtaining layers with a unique set of properties in the surface zone of the processed products. In addition, diffusion boriding can be applied to a wide range of structural and tool steels and alloys. Boron-based coatings lead the way in improving properties such as wear resistance, heat resistance and hardness. As shown [1-20], the simultaneous complex diffusion saturation with boron together with other elements makes it possible to further enhance the performance properties of boride coatings.

Our studies [15–22] on the complex saturation of steels with boron together with chromium and titanium showed a significant increase in such characteristics of the coating as wear resistance, corrosion resistance when heated and in aggressive media (for example, solutions of acids and alkalis). Leadership in durability belongs to complex diffusion coatings based on boron, chromium and titanium at the same time. However, the resistance of such coatings in sulfuric acid solutions is much lower than their resistance in solutions of other acids (for example, nitric or hydrochloric). The coatings obtained by siliconizing [15–16] have the maximum resistance in acidic solutions. But acid-resistant silicide coatings have high fragility, and the simultaneous diffusion saturation with silicon and boron did not solve the problem of resistance in sulfuric acid solutions. In this regard, we see it as logical to obtain diffusion coatings based on boron, chromium, titanium and silicon.

2. Methods

Content from this work may be used under the terms of the Creative Commons Attribution 3.0 licence. Any further distribution of this work must maintain attribution to the author(s) and the title of the work, journal citation and DOI.
Published under licence by IOP Publishing Ltd

Experimental studies on diffusion saturation were carried out on samples of steel 1045 with dimensions of 15x25x40 mm. The saturation was carried out in a SNOL 30/13-C1 chamber furnace equipped with a PID controller. The saturation was carried out at a temperature of 950 ° C, the holding time at the saturation temperature was 1.5 h. Saturating pastes of the following composition were used as saturating media:

1) paste for simultaneous hardening with boron and chromium:
25% CrB2 + 50% B4C + 5% NaF + 3% Fe2O3 + 7% Na2B4O7 + 10% graphite.

2) paste for simultaneous hardening with boron and titanium:
25% TiB2 + 50% B4C + 5% NaF + 3% Fe2O3 + 7% Na2B4O7 + 10% graphite.

3) paste for simultaneous hardening with boron, chromium and titanium:
15% CrB2 + 15% TiB2 + 50% B4C + 5% NaF + 3% NH4Cl + 7% Na2B4O7 + 5% graphite.

4) paste for simultaneous hardening with boron, chromium, titanium and silicon:
10% CrB2 + 10% TiB2 + 10% Si + 50% B4C + 5% NaF + 3% NH4Cl + 7% Na2B4O7 + 5% graphite.

Powdered saturating media of the above compositions were diluted in water to a paste state and spread on the surface of the samples with a layer of 3-4 mm. After drying, the samples were placed in a preheated oven. At the end of the high-temperature holding, the samples were removed from the oven and cooled in air. After complete cooling of the samples, they were washed in water to remove the remains of the saturating medium.

Metallographic studies were carried out on transverse samples cut with a MICRACUT-201 precision metallographic cutting machine and pressed into a bakelite compound using a METAPRESS metallographic press. Grinding was carried out on an automatic grinding and polishing machine "DIGIPREP" using diamond wheels with a grain size of 54, 15 and 6 microns, the force on each sample was 25, 20 and 18 N, respectively, with individual clamping of the samples. Polishing was carried out on polishing cloths "ALUPOL" using monocrystalline diamond suspensions on a water basis with a grain size of 3 and 1 μm at a pressure of 15 N on the sample in the mode of individual clamping. Superfinish polishing was carried out on AQUA BLACK cloth using an acidic suspension of silica with a grain size of 0.05 μm. Metallographic analysis was performed using the "Thixomet PRO" software package and a Carl Zeiss Axio Observer Z1m optical microscope according to the methods [23–30]. To reveal the microstructure, an etchant of the following composition was used: ethyl alcohol - 95 ml, hydrochloric acid - 2 ml, selenic acid - 3 ml.

3. Results and discussion

In all cases, diffusion coatings were obtained with a morphology characteristic of boride layers. Structures and their description for complex boride coatings doped with chromium and titanium are presented in [15–22, 31–34]. The structure of the diffusion coating obtained by simultaneous saturation with boron, chromium, titanium and silicon is shown in figure 1.

The diffusion coating shown in figure 1 has an average thickness of 125 ± 2 μm and a morphology characteristic of boride layers. The microhardness measured on boride needles at a load of 100 g (0.98 N) was 2542–2776 HV_1. The microhardness of the white phase adjacent to the boride needles was 623–575 HV_1. The microhardness of the sample core is 375–385 HV_1. The light phase, is adjacent to the boride needles is morphologically similar to ferrite, however, compared to ferrite, it has a significant hardness that exceeds that of pearlite.

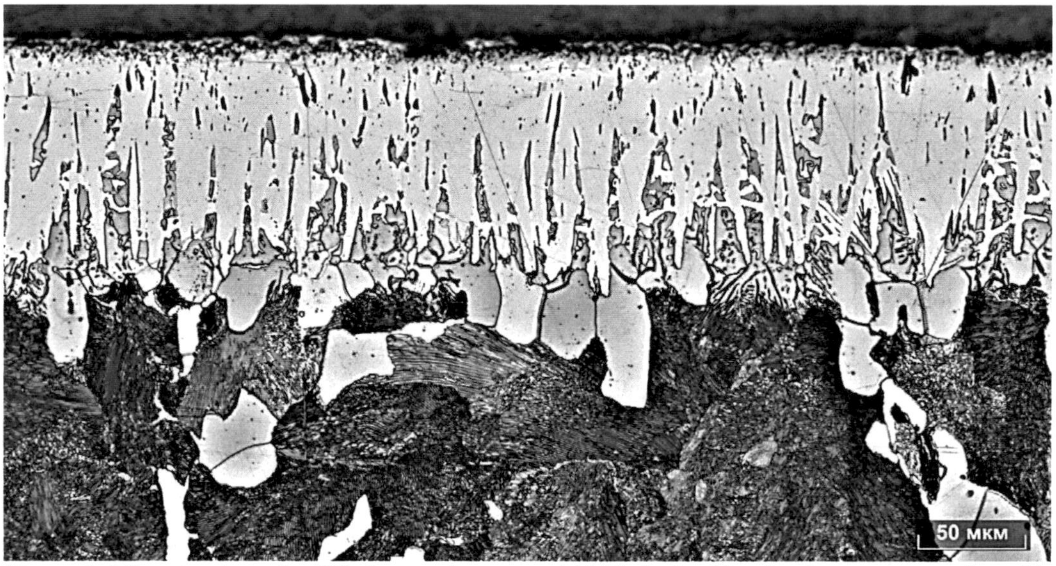

Figure 1. Microstructure of a diffusion coating on steel 1045 obtained by simultaneous saturation with boron, chromium, titanium and silicon.

On closer inspection, can see dark and light concretion in this phase (figure 2).

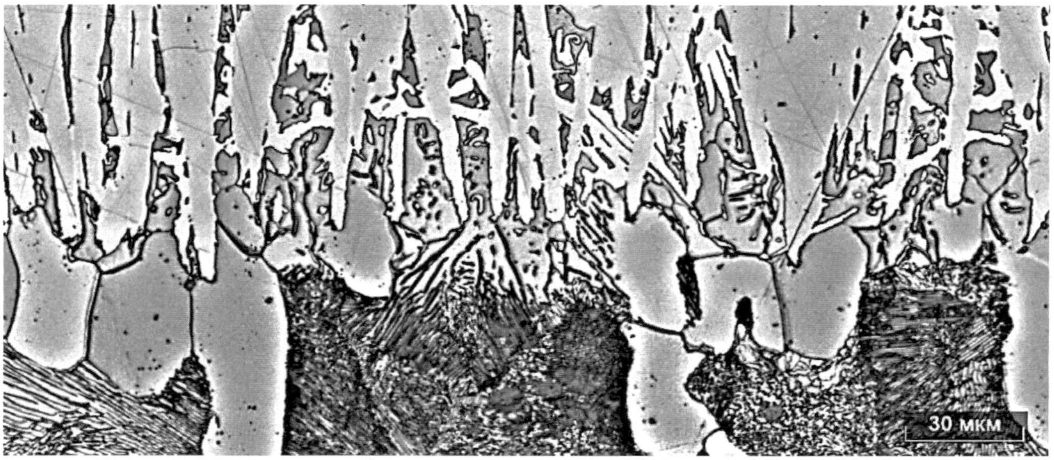

Figure 2. Microstructure of the lower part of the boride layer and the transition zone.

Previous studies [17–22, 33–34] of similar concretion have shown that these are carbides, carboborides and borides with a complex composition. Also, these phases have high hardness values, reaching values of 3200–3700 HV. The average size of these secretions is 1.2–1.5 μm, the weighted average form factor according to Ferret (F2) is 0.87, which indicates that these secretions have a shape close to globular. It should be emphasized that precipitates of borides and carboborides with a high boron content are localized mainly along grain and interphase boundaries. The concretion of borides and carboborides with a high boron content inside ferrite grains is also possible and is located between the boride needles. The concretion of carbides and carboborides are located in the lower part of ferrite precipitates. Elemental analysis of the boride layer showed the following composition (wt.%): Fe - 56; B - 26; Cr - 7; Ti - 8; Si - 1.7; the rest is impurities. Elemental analysis of the ferrite phase, including concretions in it: Fe - 77; B - 8; Cr - 9; Ti - 2; Si - 2.7. Elemental analysis of the pearlite phase did not reveal any particular deviations from the natural elemental composition characteristic of this steel. The exception is in this case chromium - its content varies from 1.5-1.7 mass. % at the border with ferrite and gradually decreases to values of approximately 0.17 mass. % at a depth of 450–500 μm from the saturation surface.

Thus, the diffusion coating on steel 1045 obtained by simultaneous saturation with boron, chromium, titanium, and silicon contains all the introduced elements. Corrosion tests of this coating in laboratory conditions have shown a high resistance of this coating, which exceeds the resistance of coatings obtained by saturation from pastes 1–3 by 3–5 times. In addition to tests for corrosion resistance, full-scale tests were carried out in real conditions, combining the effect of a sulfuric acid solution and abrasive wear. Knives for crushing used lead–acid battery for further recycling were used as test samples. These knives were made of steel 1045 and their surface was diffusion-hardened from saturating pastes 1–4. The service life of the hardened knives was measured by the real weight of the shredded batteries. Additionally, knives made of HARDOX 600 steel, made using water-jet cutting, were used as reference knives.

The production tests carried out have confirmed the correctness of our assumptions: diffusion coatings were obtained that have high indicators of resistance to abrasion in a sulfuric acid solution. Tests of knives, hardened in various ways, made of steel 1045 for shredding lead–acid battery showed the following results:

1) simultaneous strengthening with boron and chromium - 18 tons of shredded batteries
2) simultaneous strengthening with boron and titanium - 16 tons of shredded batteries
3) simultaneous hardening with boron, chromium and titanium - 54 tons of shredded batteries
4) simultaneous strengthening with boron, chromium, titanium and silicon - 183 tons of shredded batteries.

The durability of the reference knives made of HARDOX 600 steel was 77 tons of crushed batteries. It should be especially noted that the wear of the knives was predominantly corrosive. That is, the greatest damage to all surfaces of the crushing knives occurred as a result of exposure to the sulfuric acid solution contained in the battery.

4. Conclusion

Thus, the simultaneous diffusion saturation with boron, chromium, titanium and silicon of carbon steel made it possible to increase the operational durability of knives for shreding lead–acid battery by 2.4 times. The possibility of highly efficient use of products made of simple carbon steel with a multicomponent (B + Cr + Ti + Si) diffusion coating instead of products from hardened alloy steel is shown.

References

[1] Matuschka, A.G., Boronizing, Philadelphia: Heyden and Son Inc., 1980.
[2] Sudarshan T S, Khor K A, Jeandin M 1999 Review of Surface Modification Technologies XII (ASM International, Am. Techn. Publ. Ltd)
[3] L. G. Voroshnin and L. S. Lyakhovich, Boriding Steels [in Russian], Metallurgiya, Moscow (1978)
[4] Bindal C, Ucisik A H 1999 *Surf. Coat. Technol.* **122** 208
[5] Voroshnin L G, Lyakhovich L S 1978 Boriding Steels (Metallurgiya) (in Russian)
[6] Kuz'ma Yu B 1983 Crystal Chemistry of Borides (Vishcha Shk) (in Russian)
[7] Campos I, Oseguera J, Figueroa U, Garcı J A, Bautista O, Kelemenis G 2003 *Mater. Sci. Eng. A* **352** 261
[8] Campos I, Ramirez G, Figueroa U, Martinez J, Morales O 2007 *Appl. Surf. Sci.* **253** 3469
[9] Kayali Y, Gunes I, Ulu S 2012 *Vacuum* **86** 1428
[10] Sinha A K 1991 *ASM handbook. J. Heat Treat.* **4** 437
[11] Mao M X, Ge S H, Chen Z H, Chen G L, Zhang C L, Zhang Y D, Hines WA, Budnick J I 1994 *Magn. Mater.* **138** 301
[12] Li F, Li Z 2014 *J. Alloy. Compd.* **587** 267
[13] Van Rompacy T, Hari Kumar K C, Wollants P 2002 *J. Alloy. Compd.* **334** 173
[14] Filonenko N Yu, Beryoza O Yu, Bezrukava O G 2013 *PAST* **87** 168
[15] *ASM Handbook Vol 4: Heat Treating* p 2173
[16] *ASM Handbook Vol 5: Surface Engineering* p 2535
[17] Ivanov S G, Guriev A M, Starostenkov M D, Ivanova T G, Levchenko A A 2014 *Russian Physics Journal* **57** 266
[18] Ivanov S G, Garmaeva I A, Guriev M A, Guriev A M, Starostenkov M D 2015 *Lecture Notes in Mechanical Engineering* **22** 49
[19] Garmaeva I A, Guriev A M, Ivanova T G, Guriev M A, Ivanov S G 2016 *Letters on Materials* **6**(4) 262 (in Russian)
[20] Gur'ev A M, Ivanov S G, Gur'ev M A, Chernikh E V, Ivanova T G 2015 *Steel in Translation* **45**(8) 555
[21] Mei S, Dong Y, Guriev A M, Zhang W, Liu C, Ivanov S G, Chernykh E V 2016 *4th Annual International*

Conference on Material Science and Engineering (ICMSE) AER-Advances in Engineering Research **101** 308
[22] Guriev A M, Ivanov S G, Guriev M A, Chernykh E V, Mei S 2019 *IOP Conference Series-Materials Science and Engineering* **479** 012077
[23] Kazakov A A, Ryaboshuk S V, Lyubochko D A, Chigintsev L S 2015 *Microscopy and Microanalysis* **21**(3) 1755
[24] Kazakov A, Kiselev D 2015 *Microscopy and Microanalysis* **21**(3) 457
[25] Kazakov A, Kovalev S, Ryaboshuk S 2012 *CIS Iron and Steel Review* **4**
[26] Kazakov A, Lyubochko D 2015 *Microscopy and Microanalysis* **21**(3) 1749.
[27] Kazakov A, Kiselev D 2016 *Metallography, Microstructure, and Analysis* 294
[28] *ASM Handbook Vol 9: Metallography and Microstructures* p 2733
[29] Kazakov A A, Kiselev D 2016 *Metallography, Microstructure, and Analysis*
[30] Vander Voort G F 1991 *Materials characterization* **27** 241
[31] Gurev A M, Greshilov A D, Kosheleva E A, Ivanov S G, Gurev M A, Ivanov A G, Dolgorov A A 2010 O*brabotka metallov-metal working and material science* **3** 35
[32] Gurev A M, Ivanov S G, Greshilov A D, Zemlyakov S A 2011 O*brabotka metallov-metal working and material science* **2** 19
[33] Guriev A M, Guriev M A, Ivanov S G, Zheng Q, Mei S Q 2019 *SSP* **298** 32
[34] Guriev A M, Guriev M A, Ivanov S G, Zhang Y W, Mei S Q 2019 *SSP* **298** 75

Structure, phase composition and properties of vacuum pressureless sintered ZrO_2-based composites reinforced with carbon nanotubes

A A Leonov[1,2], Jing Li[1], M P Kalashnikov[1,3], M A Rudmin[1] and O L Khasanov[1]

[1] National Research Tomsk Polytechnic University, Tomsk, Russia
[2] Institute of High Current Electronics SB RAS, Tomsk, 634055, Russia
[3] Institute of Strength Physics and Materials Science SB RAS, Tomsk, 634055, Russia

Email: laa-91@yandex.ru

Abstract. In this work, composites based on 3 mol. % yttria-stabilized zirconia (3YSZ) with additives of 1, 5 and 10 wt. % multi-walled carbon nanotubes (MWCNT) were investigated. Samples were obtained by cold pressing followed by vacuum pressureless sintering at a temperature of 1500 °C. It was found that MWCNTs retain their structure after high-temperature sintering, they are located along the zirconia grain boundaries and they lead a grain refining effect for zirconia matrix. Nanotubes can significantly affect the phase composition of composites. Additives 5 wt. % and 10 wt. % MWCNT partially limits the monoclinic-tetragonal phase transition of ZrO_2 and leads to the formation of a new phase of zirconium carbide. An increase of MWCNTs content in composites leads to a smooth decrease of the relative density from 99.2 % to 81.2 %. To obtain fully dense 3YSZ/MWCNT composites, longer sintering at higher temperatures is required compared to the conditions for sintering 3YSZ ceramics without additives.

1. Introduction

In recent years, carbon nanotubes (CNTs), due to their small size, large aspect ratio (100-1000), low weight, high mechanical strength, and high electrical and thermal conductivity, have been considered as attractive filler for composites materials [1-3]. They occur as single-walled carbon nanotubes (SWCNTs) consisting of a single graphene sheet rolled-up into a cylinder with diameters as small as one nanometer, and multi-walled carbon nanotubes (MWCNTs) composed of concentric graphene layers. However, MWCNTs can be prepared in large scale and thus are much cheaper than SWCNTs at present time [4-6]. Several applications were proposed for CNTs many of which are concerned with conductive or high strength composites. The inclusion of CNTs in a brittle ceramic matrix is expected to create composites exhibiting improved mechanical properties compared to the ceramic material without additives [7-9]. The CNTs dispersed in a ceramic matrix should serve as sites for fracture energy dissipation through mechanisms such as crack deflection and nanotube debonding/breakages. 3 mol. % yttria stabilized zirconia ceramics (3YSZ) is a technologically important material due to its high stability at high temperatures, chemical inertness, abrasion resistance, large energy bandgap, and it may be used in applications such as solid-oxide fuel cells, membrane oxygen separation technology, components of furnaces and engines, blade cutting tools, to name a few [10-12]. Therefore, in this work, we prepared 3YSZ/MWCNT composites with different MWCNT concentrations and analyzed changes in microstructure, phase composition, density and mechanical properties obtained composites, to improve their properties and exploit new functional applications.

2. Materials and methods

The initial materials commercially available were nanopowder of 3 mol. % yttria stabilized zirconia (TZ-3YS, Tosoh Corporation, Japan) and multi-walled carbon nanotubes (Taunit, NanoTechCenter, Tambov, Russia). 3YSZ nanopowder was mixed with MWCNTs in ethanol using an ultrasonic bath and a magnetic

Content from this work may be used under the terms of the Creative Commons Attribution 3.0 licence. Any further distribution of this work must maintain attribution to the author(s) and the title of the work, journal citation and DOI.
Published under licence by IOP Publishing Ltd

stirrer [13, 14]. The concentration of MWCNTs in composite powders were 1, 5 and 10 wt. %. The morphologies of initial MWCNTs and composite powder were observed by transmission electron microscopy (TEM) JEM-2100F. The prepared composite powders and the initial ZrO_2 powder were pressed into compacts by uniaxial pressing in a steel mold 10 mm in diameter at a pressure of 100 MPa. Green compacts were sintered in a high-temperature vacuum furnace for 2 h at a temperature of 1500 °C with a heating rate of 300 °C/h. Theoretical density values for composites were calculated by the rule of mixtures assuming density values of 6.1 g/cm^3 for 3YSZ [15] and 2.1 g/cm^3 for MWCNTs [16]. Bulk densities were measured using Archimedes' method, with distilled water as immersion medium. The microstructure of the sintered samples was analysed by scanning electron microscope (SEM) TESCAN VEGA 3 SBU. The phase composition of samples were determined by X-ray diffraction (XRD) by using X-ray diffractometer XRD-7000S with Cu-Kα radiation (λ = 1.54056 Å). The tube voltage and current were 40 kV and 30 mA, respectively. Scan range and sampling pitch were 10 ° – 120 ° and 0.03 °, respectively. The microhardness H_V of sintered samples was measured with using of microhardness tester PMT-3. The samples were indented using a Vickers diamond pyramid with a load of 4.9 N applied on the surface for 15 s and dimension of the formed indent was observed by using installed optical microscope. To obtain reliable data, indents were made along the line with a step of 0.5 mm. For each sample, 15 measurements were taken. A Vickers hardness tester TP-7R-1 was used for fracture toughness measurements using a load of 49 N. The fracture toughness values K_{IC} were calculated using the Anstis equation [17].

3. Results and discussion

Figure 1a shows a TEM image of the initial MWCNTs. The diameter of an individual nanotube is about 20 nm. The investigated MWCNTs have a bamboo structure of several types. There are multilayer structures in which graphene cones are nested within each other.

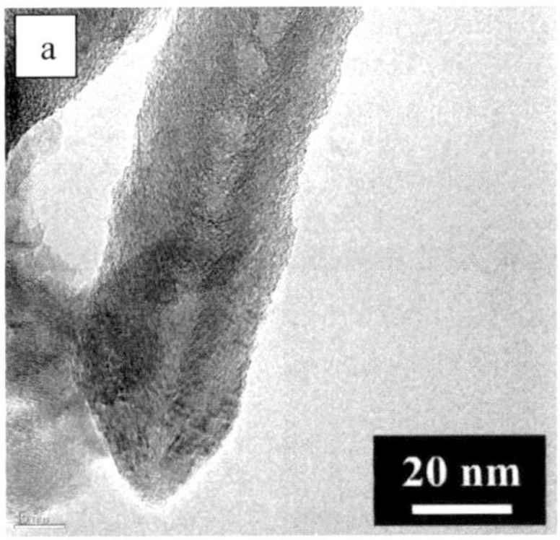

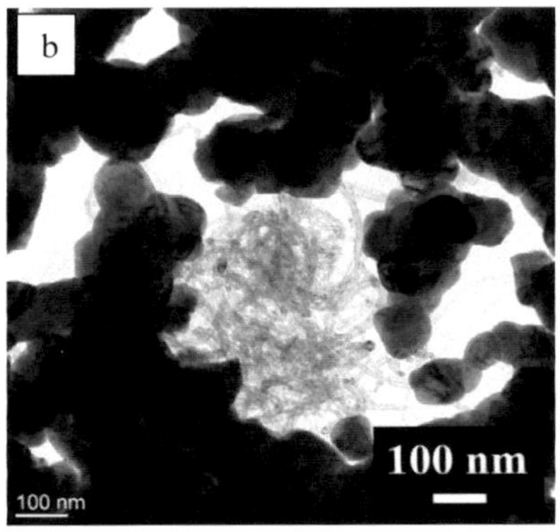

Figure 1. (a) TEM image showing morphology of initial MWCNTs, (b) TEM image showing morphology of composite powder based on 3YSZ with 1 wt. % MWCNTs.

Nanotubes with internal isolated compartments formed by bent graphene layers are also found. Such bamboo-structured MWCNTs believed to offer better mechanical properties than hollow MWCNTs due to better connectivity with matrix, higher pull-out resistance and lower inner-wall slippage of the former in composites [18]. During TEM studies of composite powder mixture it was found that some nanotubes were individually dispersed in 3YSZ matrix (Figure 1b), the other part was presented clustering of MWCNTs due to enhanced effect of Van dar Waal's attraction force among tubes and eventually, a lots of CNTs remained agglomerated (Figure 1b). As it can be observed in TEM image (Figure 1b), the flexibility of MWCNTs allows them to bend and pass through space between nanoparticles and wrap around.

Sintered 3YSZ ceramics showed dense microstructure, comprising well developed and nearly equiaxed ZrO_2 grains with d = 329 nm (Figure 2a, Table 1). After high-temperature sintering of composites in vacuum, MWCNTs retained their structure (Figure 2a and 2c), which was in its initial state. The distribution of both individual MWCNTs and their aggregates along the zirconia grain boundaries was observed [19, 20]. Presence of MWCNTs showed formation of smaller ZrO_2 grains (320 nm and 287 nm) than pure 3YSZ sintered under identical condition (Figure 2a and 2c, Table 1). Therefore, MWCNTs definitely possessed a grain refining effect for zirconia matrix that could help in enhancing mechanical properties of resulting composites [21, 22]. The fracture mode of composites is mainly intergranular and debonded MWCNTs from the matrix can be observed on the fracture surfaces (Figure 2a and 2c).

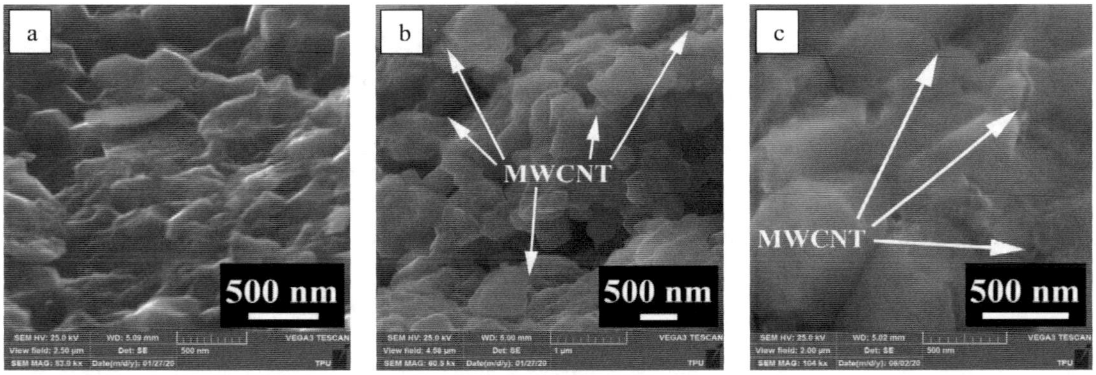

Figure 2. SEM images of the cross-sections of the studied (a) 3YSZ, (b) 3YSZ/1 wt. % MWCNT, and (c) 3YSZ/5 wt. % MWCNT samples.

The X-ray diffraction patterns of the initial 3YSZ powder, sintered 3YSZ ceramics, and sintered composites with different contents of MWCNTs are given in Figure 3. After sintering of the 3YSZ nanopowder, the resulting ceramic occur a complete transformation of the m-ZrO_2 phase into t-ZrO_2, and an additional peak appears at $2\theta = 30$ ° with the plane index (111), which is characteristic of the cubic zirconia phase (c-ZrO_2). The c-ZrO_2 phase is present in all studied composites. When MWCNTs are introduced into the ZrO_2 matrix, various phase transformations occur. So, in composites with 5 wt. % and 10 wt. % MWCNTs, a partial limitation of the monoclinic-tetragonal transition of zirconia is observed, and weak peaks of the m-ZrO_2 phase appear, which was present in the initial 3YSZ nanopowder. Similar phenomena of limitation of the monoclinic – tetragonal transition in ZrO_2/CNT composites were observed in previously published papers [23, 24]. In addition, in composites with a high MWCNT content, additional peaks appear, belonging to the cubic phase of zirconium carbide (c-ZrC). Formation ZrC phase in composites ZrO_2/CNT noted in several studies [25-27].

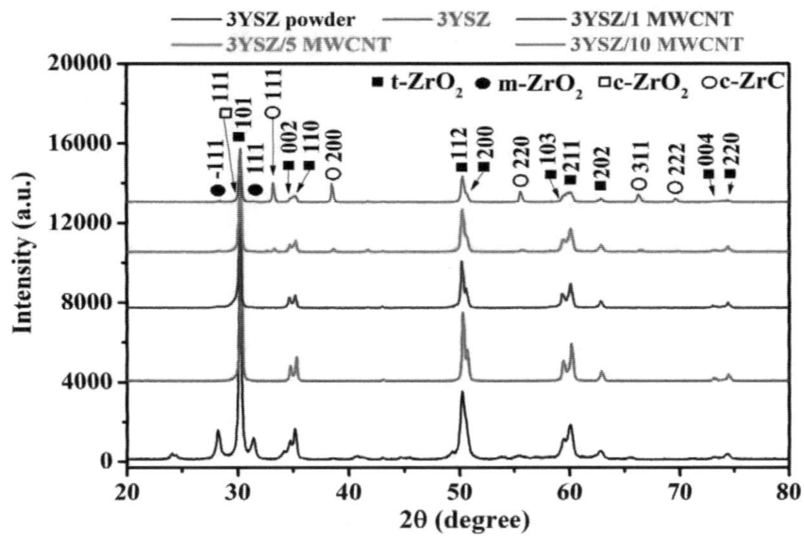

Figure 3. X-ray diffraction patterns of the initial 3YSZ powder, sintered 3YSZ ceramics, and composites with various MWCNT contents.

It can be noted that an increase in the content of MCNTs in composites leads to a smooth decrease in the relative density from 99.2 % to 81.2 % (Table 1). Apparently, the lower compaction of zirconia powders in presence of MWCNTs due to the fact that the ceramic powder sliding and rearrangement have been suppressed. This is in line with TEM observation of composite powder (Figure 1b), which shows entangled and kinked nanotubes around the 3YSZ nanopowder hindering the powders from rearranging freely during compaction. The lower densification rate of MWCNT reinforced composites with respect to the pure 3YSZ could be attributed to a slower grain boundary sliding during sintering. In other words, MWCNTs located in the grain-boundaries (Figure 2b and 2c) could pin them and hinder their movement (sliding or migration) and consequently hinder the densification rate. For example [28], SPS-sintering at 1300 °C resulted in a fully dense (99.2 % relative density) monolithic 3YSZ, while 2 wt. % and 6 wt. % MWCNTs reinforced composites have not shown densities higher than 96.4 % and 89.7 %, respectively. Therefore, in order to obtain a fully dense ZrO_2/MWCNT composite, longer sintering at higher temperatures is required compared to the conditions for sintering ZrO_2 ceramics without additives.

Table 1. Weight content of MWCNTs in samples; theoretical and relative density; average grain size; microhardness; fracture toughness.

C (wt. %)	ρ_{theor} (g/cm^3)	ρ_{rel} (%)	D (nm)	H_V (GPa)	K_{IC} (MPa*m$^{1/2}$)
0	6.1	99.2	329 ± 79	12.8 ± 0.3	4.1 ± 0.1
1	6.0	98.7	320 ± 64	11.7 ± 1.3	4.3 ± 0.3
5	5.6	87.5	287 ± 69	10.1 ± 1.5	2.7 ± 0.1
10	5.1	81.2	-	1.7 ± 0.8	-

Data in Table 1 shows that the microhardness values H_V decreased from 12.8 ± 0.3 GPa to 1.7 ± 0.8 GPa with increasing the MWCNTs content from 0 wt. % to 10 wt. %, respectively. Such decrement in mechanical properties can be attributed to the decreased relative density when incorporating MWCNTs into 3YSZ matrix. In addition, the existing agglomerates of MWCNTs in the composite structures are a “soft” phase, which also contributes to a decrease in microhardness. Fracture toughness increased by only 5 % for the composite with 1 wt. % MWCNT. However, composites sintered by the spark plasma sintering (SPS) method from these powders had an increased 46 % fracture toughness [27]. SPS-composites had a higher relative density (99.0-96.1 %) and hardening mechanisms were implemented in them, such as crack deflection and crack branching.

4. Conclusions

In this work, the influence of MWCNTs on the microstructure, phase composition, compaction and mechanical properties of zirconia-based composites obtained by cold pressing followed by vacuum pressureless sintering was analyzed. After high-temperature sintering (1500 °C) of composites in vacuum, MWCNTs retained their structure, which was in its initial state. MWCNTs possessed a grain refining effect for zirconia matrix, grain size decreased from 329 nm to 287 nm for a composite with 5 wt. % MWCNT. Additives 5 wt. % and 10 wt. % MWCNT partially limits the monoclinic-tetragonal phase transition of ZrO_2 and leads to the formation of a cubic phase of zirconium carbide. The lower densification rate of composites could be attributed to a slower grain boundary sliding during sintering because, MWCNTs located in the grain boundaries could pin them and hinder their movement (sliding or migration). Due to the low relative density obtained in composites, it is impossible to improve the mechanical properties. To obtain by vacuum pressureless sintering fully dense 3YSZ/MWCNT composites with improved mechanical properties, longer sintering at higher temperatures were required.

Acknowledgments
The authors are grateful to Galunin E V, Melezhik A V and Tkachev A G for providing «Taunit» multi-walled carbon nanotubes. The study was performed in part on the basis of Nano Center of National Research Tomsk Polytechnic University.

References

[1] Zhang S, Qian L, Zhao Q, et al. 2020 *Sci. China Mater.* **63** 16–34
[2] Cho J, Boccaccini A R, Shaffer M S P 2009 *J. Mater. Sci.* **44** 1934–1951

[3] Guo C, Luo X, Shah W A, et al. 2020 *Ceram. Int.* **46** 17449–17460
[4] Monthioux M, Serp P, Flahaut E, et al. 2010 *Introduction to carbon nanotubes* Bhushan B (Ed.) Springer Handbook of Nanotechnology (3rd ed.) (Berlin: Springer) pp. 47–118
[5] Zhang J, Terrones M, Park C R, et al. 2016 *Carbon* **98** 708–732
[6] Bocanegra-Bernal M H, Echeberria J, Ollo J, et al. 2011 *Carbon* **49** 1599–1607
[7] Jia J, Liu D, Zhang S, et al. 2019 *Ceram. Int.* **45** 19956–19961
[8] Leonov A, Kalashnikov M, Li J, et al. 2020 *2020 7th International Congress on Energy Fluxes and Radiation Effects (EFRE)* 1169–1173. doi: 10.1109/EFRE47760.2020.9242000
[9] Leonov A A, Ivanov Yu F, Kalashnikov M P, et al. 2020 *IOP Conf. Ser.: Mater. Sci. Eng.* **731** 012015
[10] Poyato R, Macías-Delgado J, García-Valenzuela A, et al. 2016 *Ceram. Int.* **42** 6713–6719
[11] Yin L, Nakanishi Y, Alao A R, et al. 2017 *Procedia CIRP* **65** 284–290
[12] Glukharev A G and Konakov V G 2018 *Rev,Adv. Mater. Sci.* **56** 124–138
[13] Leonov A 2019 *Materials Today: Proceedings* **11** 66–71
[14] Leonov A A and Abdulmenova E V 2019 *IOP Conf. Ser.: Mater. Sci. Eng.* **511** 012001
[15] Cano-Crespo R, Moshtaghioun B M, Gómez-García D, et al. 2018 *J. Eur. Ceram. Soc.* **38** 3994–4002
[16] Poyato R, Morales-Rodríguez A, Gutiérrez-Mora F, et al. 2017 *Ceram. Int.* **43** 16560–16568
[17] Anstis G R, Chantikul P, Lawn B N and Marshall D B 1981 *J. Am. Ceram. Soc.* **64** 533–538
[18] Olek M, Ostrander J, Jurga S, et al. 2004 *Nano Lett.* **4** 1889–1895
[19] Leonov A A, et al. 2017 *IOP Conf. Ser.: Mater. Sci. Eng.* **286** 012034
[20] Yi J, Wang T, Xie Z, et al. 2013 *J. Alloys Compd.* **581** 452–458
[21] Ivanov Y, Shugurov V, Kalashnikov M, et al. 2018 *AIP Conference Proceedings* **2051** 020110
[22] Leonov A A, Kuzichkin E E, Shugurov V V, et al. 2018 *J. Phys.: Conf. Ser* **1115** 032040
[23] Bocanegra-Bernal M H, et al. 2012 *Int. J. Refract. Met. Hard Mater.* **35** 315–318
[24] Leonov A A, Dvilis E S, Khasanov O L, et al. 2019 *Nanotechnologies in Russia* **14** 118-124
[25] Hassan R, Nisar A, Ariharan S, et al. 2017 *Mater. Sci. Eng. A* **704** 329–343
[26] Leonov A A, Ivanov Yu F, Kalashnikov M P, et al. 2019 *J. Phys.: Conf. Ser.* **1393** 012106
[27] Leonov A A, Abdulmenova E V and Kalashnikov M P 2020 *Perspektivnyye materialy* **10** 56–68 (in Russian). doi: 10.30791/1028-978X-2020-10-56-68
[28] Yi J, Xue W, Wang T and Xie Z 2015 *Ceram. Int.* **41** 9157–9162

Multi-phase ion-plasma Cu-Ti coatings deposited on copper and copper beryllium alloy

A V Kolubaev[1], O V Sizova[1], Yu A Denisova[1,2], A A Leonov[2], N V Teryukalova[1] and O S Novitskaya[1]

[1] Institute of Strength Physics and Materials Science, Siberian Branch of the Russian Academy of Sciences, 2/4, pr. Akademicheskii, Tomsk, 634055, Russia
[2] Institute of High Current Electronics Siberian Branch of the Russian Academy of Sciences, 2/3, pr. Akademicheskii, Tomsk, 634055, Russia

E-mail: kav@ispms.ru

Abstract. The paper presents outcomes of vacuum-arc plasma-assisted deposition of titanium, titanium nitride and copper on copper (C11000) and the hardened copper beryllium alloy (C17200) at a temperature of 320÷330^{0}C. The study has revealed that multi-phase coatings form when varying discharge currents generated by arc evaporators and supply sequences of producer gas (argon) and nitrogen. An X-ray phase analysis has demonstrated these coatings to consist of copper, titanium and CuTi, $CuTi_2$ and TiN compounds. Another important outcome to emerge from the study is that ion-plasma processing results in ageing of copper beryllium alloy and furthers the formation of CuBe particles. The microhardness of Cu-Ti coatings is as high as 400 ÷ 540 $HV_{0.02}$ provided that the elements are vacuum-arc deposited at a temperature of 320÷330^{0}C. The microhardness of a nitrogenized Cu-TiN surface layer is measured to be 760 $HV_{0.05}$. The thickness of deposited coatings is within several micrometers, so the measured hardness hardly complies with the real characteristic of coatings because the indenter breaks a thin hard layer. In addition, vacuum-arc sputtering of titanium and copper cause metal drops on the surface, as a consequence, the hardness of coatings is below referential values.

1. Introduction

To date, ion-plasma coatings based on titanium, chromium, other materials and their compounds have been widely applied [1]. State-of-the-art deposition technologies of ion-plasma coatings are intended to provide high operation properties of tools and components of modern equipment. The mainstream processes in ion-plasma processing include plasma-magnetron sputtering [2] and vacuum-plasma sputtering by electric arc evaporators of metals [3]. The combination of functional properties in hard coatings makes machine parts more resistant to corrosion, erosion, abrasive and contact wear.

In general, coatings are produced on ferrous metals and alloys in industry. Non-ferrous metal are less sensitive to above mentioned factors. Nevertheless, a great number of items produced of copper and its alloys and used in electrical and chemical engineering for manufacturing radiators, pipelines, heat exchangers, etc. with high electrical and thermal conductivity also require corrosion, erosion and wear protection. In addition, copper alloys are characterized with relatively low hardness and fail to resist to adhesion and abrasive wear.

An application field of copper alloys in extreme operation conditions can be broadened by means of surface modification methods. An example is diffusion saturation of copper surface with titanium [4]. A binary copper-titanium phase diagram demonstrates that several intermetallic Ti–Cu phases can be produced in this system and significantly improve surface properties of copper alloys. This approach seems to be rather promising, especially for the enhancement of copper alloys surfaces, e.g. to make sliding electric contacts more wear resistant. The main point of methods to deposit coatings for protection of electric contacts from wear and corrosion is not to raise the electric resistance of a contact. For this purpose a component with high conductivity and filler with high electrical erosion resistance are introduced into the coating composition. For instance, electrical explosion sputtering of the CdO-Ag system on a copper surface [5] results in a multi-phase coating, components of which include submicro-crystalline copper and nano-dimensional inclusions of copper oxides, cadmium and Cd-Cu compounds.

Content from this work may be used under the terms of the Creative Commons Attribution 3.0 licence. Any further distribution of this work must maintain attribution to the author(s) and the title of the work, journal citation and DOI.
Published under licence by IOP Publishing Ltd

The surface hardening process of future is suggested to be ion-plasma deposition of hard coatings, e.g. those based on nitrides of different elements [6, 7]. As a result of this processing hard layers form on the alloy surface and enhance significantly operation properties of products. However, there are some limitations in the application of hard coatings deposited on relatively soft substrates and intended to be used in extreme mechanical loading. A copper substrate tends to deformations and causes cracking and lamination of hard coatings. To minimize such negative effects hardness of a plastic copper substrate should be increased, what is a greater challenge in comparison to iron-based or hard alloys. Pre-hardening of a substrate surface is a method to make hard coatings more compatible with a plastic metal. It is a diffusion product of alloying elements into the substrate metal surface [8]. A number of hardening elements for copper and its alloys is limited [9], therefore composite layers [10] and multi-layered coatings where layers of copper alternate with those of its compounds [11] are produced to enhance hardness and tribotechnical properties of coatings via reducing residual stresses and increasing coating-substrate adhesion.

To sum up, ion-plasma alloying makes it possible to vary broadly surface properties of different alloys, e.g. copper-based ones and allows their application in the deposition of wear and corrosion resistant coatings on machine elements manufactured of copper alloys and operated in extreme loading conditions. The most promising products in this respect are multi-phase and multi-component functional coatings on the base of metal nitrides, oxides and carbides which form a complex "substrate-modified surface layer" with smoothly varying physical and mechanical properties responsible for high operation characteristics.

2. Materials and Methods of Research

In the research 8.0 mm high cylindrical samples of copper and beryllium alloy (C17200) were used which were fabricated from a hot-rolled rod with a diameter of 5 mm, water-quenched at a temperature above 800^0C and quenched with the subsequent ageing at a temperature of 315^0C within 2 hours. Samples of copper (11000) represented 15×15×5 mm plates.

The vacuum-arc plasma-assisted deposition of coatings was carried out using a unit "KVADRO", a part of unique electro-physical facilities at the Institute of High Current Electronics, Siberian Branch of the Russian Academy of Sciences. To generate a metal plasma flow we used two electric arc evaporators with cylindrical titanium (Grade 2) and copper (C1100) cathodes 80 mm in diameter, as well as a gaseous plasma source with a heated and hollow cathode "PINK". The gaseous plasma source was used for pre-cleaning of product surfaces from leftover dielectric, e.g. oxide films, with the help of ion etching, heating the substrate, as well as additional ionization of gas and assisting in the process of coating deposition. To obtain a necessary element composition of the Ti-Cu coating we selected a balance between discharge currents of electric arc evaporators. Samples were placed in the center of the chamber on the holder at a level of output apertures of plasma sources.

Prior experimenting, the pressure of 10^{-2} Pa was generated in the 650×650×650 mm vacuum chamber then argon was supplied through the plasma generator "PINK", and the pressure of 0.3 Pa was generated. When starting a gaseous discharge with a current of about 40A and setting the shift voltage of (600) V substrates on the holder were heated up to a temperature of 320-330°C. As the surface of samples was cleaned via ion bombardment and chemically activated, charges were started in the electric arc evaporators, and a coating was sputtered. When depositing nitrides we used nitrogen, whereas we applied argon for sputtering compounds of pure metals. Coatings were produced according to three modes presented in Table 1. The samples were rotated on the holder in the center of the chamber at a speed which furthered smooth deposition of several nanometer thin titanium and copper layers, their remixing and mutual diffusion.

Table 1. Deposition parameters of coatings

Mode	Current of evaporator 1 (Ti)	Current of evaporator 2 (Cu)	Current of gaseous plasma source	Shift voltage when sputtering	Process temperature	Operation gas	Period of deposition
	I_1, A	I_2, A	I_{GAS}, A	U_{sh}, V	T, °C		t, *min*
№1	90	90	≈ 40	100	320-330	Ar	120
№2	(30÷90)*	90	≈ 30	100	320-330	Ar	240
№3	(30-90)*	90	≈ 50	100	320-330	Ar, (10%Ar+90%N2)**	240

* a start current value 30 A increased by 5 A every 20 min to get a value of 90 A.
** 40 min sputtering of TiCu first in the Ar atmosphere and then in the 10 % Ar + 90% N_2 gas mixture at the initial pressure of 0.1 Pa, increased by 0.02 Pa every 20 min to get a value of 0.3 Pa.

The phase composition, morphology and microstructure of coatings were explored using the methods of optical and scanning electron microscopy with a laser confocal microscope LEXT OLS4000 in optical and laser modes, a scanning electron microscope Philips SEM 515 tooled with an energy-dispersive X-ray micro-analyzer "Genesis" in the Collective Resource Center of National research Tomsk State University and with a scanning electron microscope "LEO EVO 50" in the Core Facilities Center of Institute of Strength Physics and Materials Science, Siberian Branch Russian Academy of Sciences "NANOTECH". The structure of coatings was analyzed with an X-ray diffractometer DRON–3.0 in the CuKα radiation. To determine the phase composition software PowderCell was used.

3. Results and Discussion

The metallographic research of coated copper samples revealed an insignificant thickness difference of samples fabricated in different modes. The coatings were deposited as a solid uniform in thickness layer with a distinct coating-substrate boundary line. Figure 1 (a) demonstrates as an example a coating produced in the first mode. A microanalysis (Fig. 1, b) detected copper and titanium in the surface layer and suggested it was a result of the subsequent deposition of several nanometer thick titanium and copper layers when rotating samples in the chamber, their remixing and mutual diffusion. Owing to this method of Ti and Cu sputtering on a copper sample the surface hardness was higher (400 $HV_{0.02}$) than that of pure copper (75 $HV_{0.02}$). A coating produced according to the second mode is similar in appearance to that produced in the first mode. There is also copper and titanium in the coating but its hardness is significantly higher – 540 $HV_{0.02}$. The reason might be a thinner copper layer on the coating surface (Fig. 2) in comparison to that sputtered in the first mode. These differences depend on uncontrolled deviations when coating.

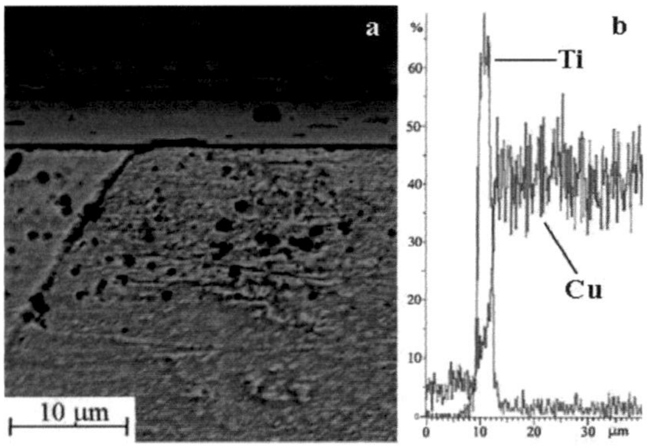

Figure 1. The Cu-Ti coating on copper – product of the first mode (a), the pattern of elements in the coating (b).

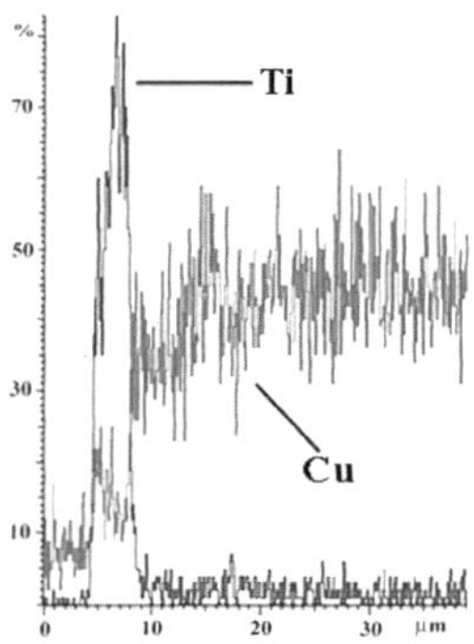

Figure 2. The pattern of elements in the coating – product of the second mode.

The effect of nitrogen plasma on the structure, phase composition and hardness of Cu-Ti coatings was revealed via sputtering thin nitride layers on copper and copper beryllium substrate in conditions of plasma-assisted vacuum-arc deposition (mode 3). When plasma nitriding with switched on Cu and Ti electric arc evaporators we used a gaseous plasma source "PINK". An X-ray phase analysis disclosed a multi-phase layer (Figure 3) consisting of Cu, $CuTi_3$ and TiN (Figure 4) – a product of the nitride deposition on a copper substrate. According to the X-ray spectral analysis the coating thickness was below 6 µm. The micro-hardness of the azotized coating was as high as 760 $HV_{0.05}$. Importantly, it was concluded [12] a soft copper substrate might influence measured micro-hardness.

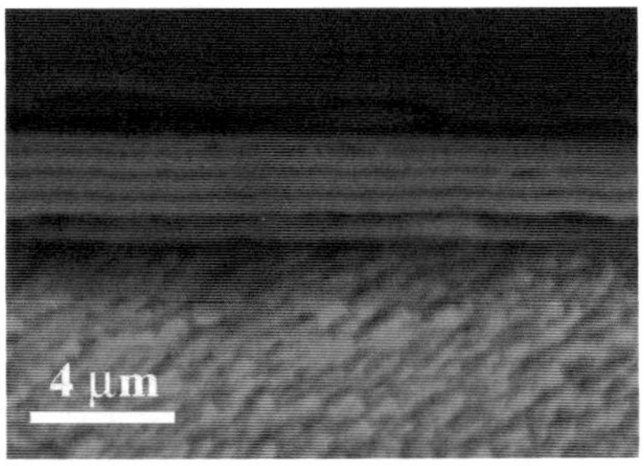

Figure 3. The Cu-Ti-N coating on copper.

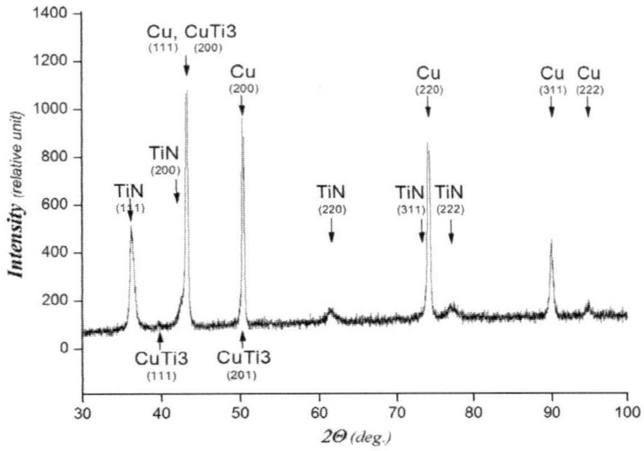

Figure 4. The X-ray diffraction pattern of a Cu-Ti-N coating on copper.

Unlike copper, strength and antifriction characteristics of copper beryllium alloys, as well as their corrosion resistance are quite good. The specified operation qualities of the alloy are possible due to thermal treatment – quenching and ageing – resulting in the two-phase alloy comprising α-solid solution with a low beryllium percentage and sporadic decomposition zones – ultra-dispersive γ'-phase particles (CuBe). Ageing in the range of 300-400°C makes the alloy harder and stronger. Ion-plasma coatings deposited at a temperature of 315°C on the quenched alloy (C17200) brings about both its ageing and a hard layer on its surface.

Figure 5 presents a cross section of a beryllium bronze sample coated with copper and titanium according to mode 1 and its element composition. The thickness of a composite coating was measured to be 5÷7 μm; a layerwise microanalysis detected almost the same patterns of Ti and Cu in the coating section. It is a result of the sequential deposition of copper and titanium in the process of rotating samples in the chamber, their remixing and mutual diffusion. An element microanalysis of the coated alloy under consideration showed there is titanium in the substrate material since titanium diffuses into it when depositing a coating layer. CuBe particles forming in the bronze ageing were also found in the substrate material. Their maximal size was 300 nm.

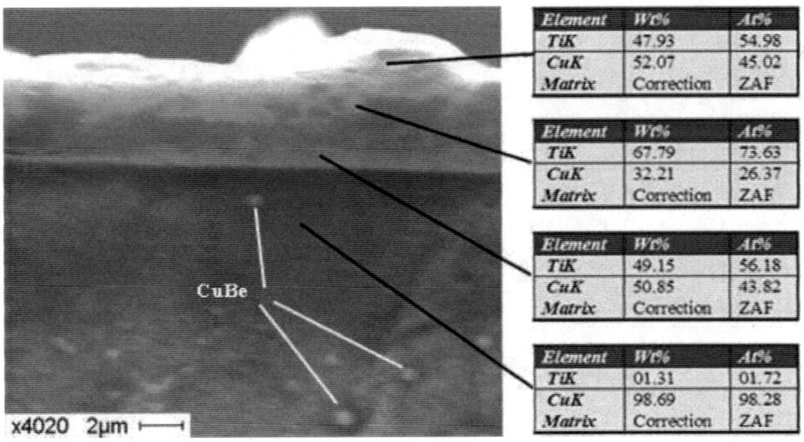

Figure 5. The cross-section of the beryllium bronze with a Cu–Ti coating deposited according to the first mode.

Unlike the coating produced in the first mode the second one displays a gradient pattern of elements in the section; this is caused by the conditions of sputtering titanium. In the substrate material, similarly to the first case, there is titanium which might form a solid titanium solution in copper. The thickness of this surface layer is up to 8 μm (Figure 6). The coatings produced according to the first and second modes had the hardness of 530-540 $HV_{0.02}$ like that of a coating deposited on copper. These coatings are assumed to have the same phase composition because of relatively equal hardness values. It is worth saying the hardness of the aged substrate when sputtering the coatings was 350 $HV_{0.02}$.

An X-ray phase analysis of the coatings demonstrated elements to be deposited form a multi-phase structure on the copper beryllium alloy surface; its components are copper, titanium and CuTi and $CuTi_2$ compounds (Figure 7). Since it is not thick, X-rays penetrate the substrate material; therefore a copper reflex can be formed either by substrate reflexes or pure copper of the coating.

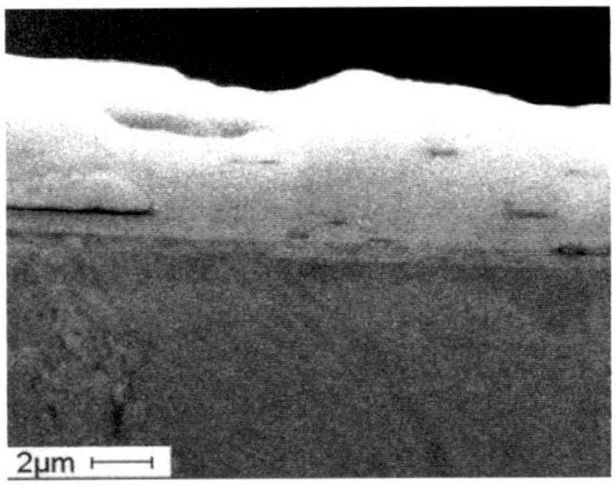

Figure 6. The Cu-Ti coating on C17200 deposited in the second mode.

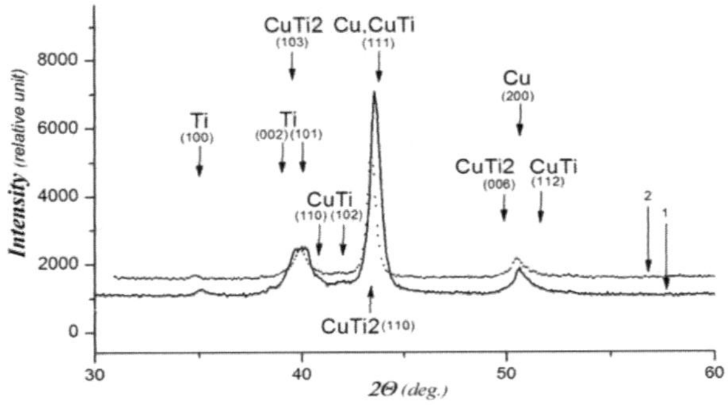

Figure 7. The X-ray diffraction pattern of the Cu-Ti coating on C17200. 1 – a coating in the first mode, 2 – a coating in the second mode.

4. Conclusion

The results presented demonstrate copper and titanium coatings deposited on copper and copper beryllium alloy consist mostly of copper, titanium and compounds on the Cu-Ti base. Intermetallic Ti – Cu compounds on the surface of copper and beryllium bronze formed using the vacuum-arc plasma-assisted method have a significant increasing effect on hardness up to 530-540 $HV_{0.02}$, irrespectively to a deposition mode. Owing to the Cu-TiN coating formed on copper and the C17200 alloy the hardness is 760 $HV_{0.05}$, therefore it can be used in sliding friction units.

Acknowledgment

The deposition of coatings on copper (C11000) was studied within the Program of State Order. The research on the structure and properties of C17200 alloy was granted by Russian Foundation for Basic Research (project # 20-58-00048 Bel_a).

References

[1] Zhang X, Tian X-B, Zhao Z-W, Gao J-B, Zhou Y-W, Gao P. and Lv Z 2019 Surface & Coatings Technology **364** 135
[2] Wei R, Vajo JJ, Matossian JN and Gardos MN 2002 Surface & Coatings Technology **158–159** 465
[3] Akhmadeev YH, Denisov VV, Koval NN, Kovalsky SS, Lopatin IV, Schanin PM and Yakovlev VV 2017 Plasma Phys. Rep. **43**(1) 67
[4] Cherenda NN, Basalai AV, Uglov VV, Laskovnev AP, Astashynski VM and Kuzmitski AM 2019 Vacuum **167** 452
[5] Romanov DA, Moskovskii SV, Gromov VE, Ivanov YF, Semin AP and Filyakov AD 2019 Fundamental'nye problemy sovremennogo materialovedenia **16**(1) 62
[6] Audisio SC and Leidheiser H, Jr. 1972 J. Electrochem. Soc. **119** 408
[7] Carrasco C, Segers L, Benavente B and Vergara V 2004 J. of Materials Processing Technology **145**(3) 371
[8] Li Y, Ye Q, Zhu Y, Zhang L, He Y, Zhang S and Xiu J 2019 Surface & Coatings Technology. **362** 27
[9] Britchi M, Branzei M and Gheorghe D 2002 International Journal of Materials and Product Technology **17**(8) 659
[10] Yan MF, Zhu YD, Zhang CS, Zhang YX, Wang YX and Yang L 2015 Materials and Design **84** 10
[11] Kim GS, Lee SY, Hahn JH, Lee BY, Han JG, Lee JH and Lee SY 2003 Surface & Coatings Technology **171**(1–3) 83
[12] Iost A, Guillemot G, Rudermann Y and Bigerelle M 2012 Thin Solid Films **524** 229

Obtaining and painting properties of red iron oxide pigment from spent pickling solutions

T A Xolmurodov [1], A P Svetlakov[1], O O Mirzaev [1], V E Katnov [1,2], M A Mirzayeva[3]

[1]Department of technology of paints and coatings, Kazan National Research Technological University, 68 Karl Marks Street, Kazan, 420015, Russia
[2]Kazan Federal University, 18 Kremlin Street, Kazan, 420008, Russia
[3]Ferghana Polytechnic Institute, Ferghana, Uzbekistan.

E-mail: tima.xolmuradov@mail.ru

Abstract. A technique has been developed for producing iron oxide pigment based on galvanic waste. The effect of the norm, concentration and nature of the precipitant, temperature and time on the deposition of $Fe(OH)_3$ was studied. To determine the effect of the nature of the neutralizer on the phase composition of the pigments, an X-ray phase and thermogravimetric analysis was carried out. It was established that the main phase is $Ca_2Fe_2O_5$, and β-Ca_2SiO_4 is present as an impurity. The basic physicochemical properties of the obtained pigment were studied.

1. Introduction

Today, iron oxides are one of the most widespread powder materials that are used in various industries in the production of pigments and fillers, information carriers, catalysts, sorbents, etc. [1-7]. There are various methods of obtaining iron oxides [8-11], including from the waste of various industries [12,13], which is currently, when the whole world pays special attention to the problem of processing waste and secondary raw materials, is very important.

In the last ten years, environmentally friendly coating technologies have appeared. However, galvanic production is still relevant, in which it is impossible to refuse electrolysis and pickling. Each of these processes requires not only modern equipment for production and energy consumption, but also careful disposal of galvanic production wastes, since often these wastes are chemically quite aggressive and toxic.

The possibility of obtaining pigments from hydrochloric acid etching solutions containing iron chloride by the thermal method was previously shown [14]. However, the pigment obtained in this way, as a rule, has increased hardness and, in addition, very high particle aggregates are formed during high-temperature processing, which complicates the grinding process.

2. Results and discussion

We have investigated the possibility of obtaining a red iron oxide pigment from the spent solution formed during the etching of printed circuit boards by the sediment-calcining method. Marble dust, ammonia, soda ash were used as a neutralizer. For comparison, we studied the optimal deposition parameters of Fe (III) in model systems $FeCl_3$-$CaCO_3$-H_2O и FeCl-NH_4OH-H_2O.

An analysis of the data obtained showed that in model and real systems the best neutralizer is ammonia and that the norm of the neutralizer is the main factor affecting the completeness of Fe (III) deposition.

A study of the filterability of suspensions formed during the neutralization of spent pickling solution (SPS) by marble dust, soda, and ammonia showed that the filtration performance of iron (III) hydroxide by dry sediment is 4,83 times higher than in the second and 6.08 times higher than in the third. The filtration performance of the filtrate in the first case is 1.63 tons 2.34 times greater than in the second and third, respectively. The specific resistance of the sediment in the first case is $3{,}1 \cdot 10^{15}$ $м^{-2}$, and in the second and third case, respectively, $8{,}6 \cdot 10^{15}$ and $12{,}9 \cdot 10^{15}$ $см^{-2}$, i.e. more than 2.8 and 4.1 times, and the filtration rate,

Content from this work may be used under the terms of the Creative Commons Attribution 3.0 licence. Any further distribution of this work must maintain attribution to the author(s) and the title of the work, journal citation and DOI.
Published under licence by IOP Publishing Ltd

respectively, less than in the first case, 1,8 and 2,5 times. Such a big difference in the filtration rate between the suspensions obtained by neutralizing SPS with marble dust, soda and ammonia is explained by the presence of an undecomposed part of the marble dust in the first suspension and, accordingly, an increase in the concentration of the suspension, which leads to the formation of a precipitate with larger pores and lower resistivity.

The study of the physicochemical and painting properties of the pigments obtained from SPS showed that the synthesized pigments are not inferior to the K-2 brand pigment by the main characteristics (Table 1). The exception is the pigment of sample 1, which contains a certain amount of marble dust, however, there is evidence of using calcium carbonate as a filler to increase the protective properties of paint coatings and to obtain lighter weather-resistant coatings.

Table 1. Physico-chemical characteristics of iron oxide pigments obtained from etching solutions.

The name of indicators	Pigment sample			Red iron oxide pigment
	1	2	3	
FeO_3 content % of mass	51,5	88,3	91,0	89,0
The content of water-soluble salts,% mass	1,15	0,66	0,35	1,00
The moisture content,% of the mass	0,70	0,40	0,20	1,00
pH of the aqueous extract	6,80	6,90	6,90	6,00
suspension pH	7,30	7,00	6,90	7,00
Spreading in terms of dry pigment, g / m^2	13,50	8,00	6,80	8,00
Oil absorption in grams of oil per 100 g of pigment	30,2	36,9	29,0	40,0
Density, kg / m^3	3160	4800	4950	4900
Specific surface, m^2 / g	0,4	12,6	27,5	-
Bulk weight, kg / m^3	1147	914	1217	1500
Bulk volume, l / kg	0,87	1,09	0,82	-
Volume after manual shaking, l / kg	0,78	0,87	0,73	-
Mass loss on ignition,% mass	26,16	3,16	2,58	-

The study of the physicochemical and painting properties of the pigments obtained from SPS showed that the synthesized pigments are not inferior to the K-2 brand pigment by the main characteristics. The exception is the pigment of sample 1, which contains a certain amount of marble dust, however, there is evidence of using calcium carbonate as a filler to increase the protective properties of paint coatings and to obtain lighter weather-resistant coatings.

X-ray phase (Figure 1) and thermogravimetric (TG) analysis was performed. From the analysis of the diffraction patterns it follows that samples 2 (Figure 1b) and 3 (Figure 1c) are almost identical in composition and content: the main phase is- α Fe_2O_3 hematite. Sample 1 is more complex (Figure 1a). The main phases of these samples are α-Fe_2O_3 hematite and $CaCO_3$ - calcite, in addition to the above phases, the maxima related to α-SiO_2 – quartz and $CaCO_3 \cdot Cu(OH)_2$ are noted in the diffractogram of sample 1.

To establish the effect of the nature of the neutralizer on the phase composition of the pigments, an Differential thermal analysis (DTA) shows (Figure 2) that the general appearance of the DTA curve and the TG gravimetric curve for samples 2 (Figure 2b) and 3 (Figure 2c) are quite close, and the derivatogram of pigment 1 differs from samples 2 and 3. As can be seen from the TG curve, mass loss is observed throughout the studied temperature range (40-450°C). For samples 1,2 and 3, it is 26%, respectively; 5% and 4.4%. Mass loss is associated with the removal of hygroscopic moisture, as well as with the decomposition of compounds present as impurities. Significant weight loss in the range of 530-800°C (sample 1, Figure 2a) is mainly due to the decomposition of $CaCO_3$.

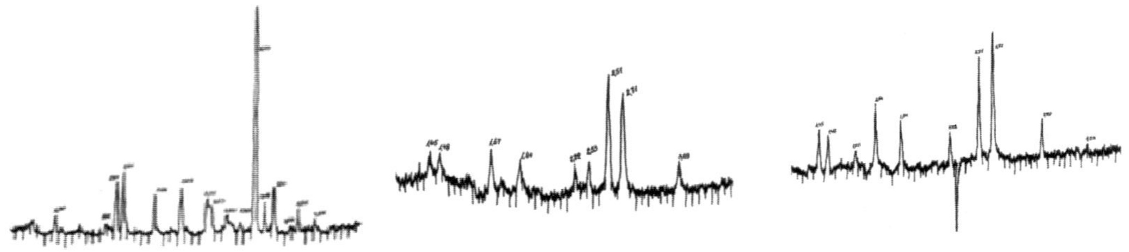

(a) (b) (c)

Figure 1. The x-ray red iron oxide pigment obtained from SPS

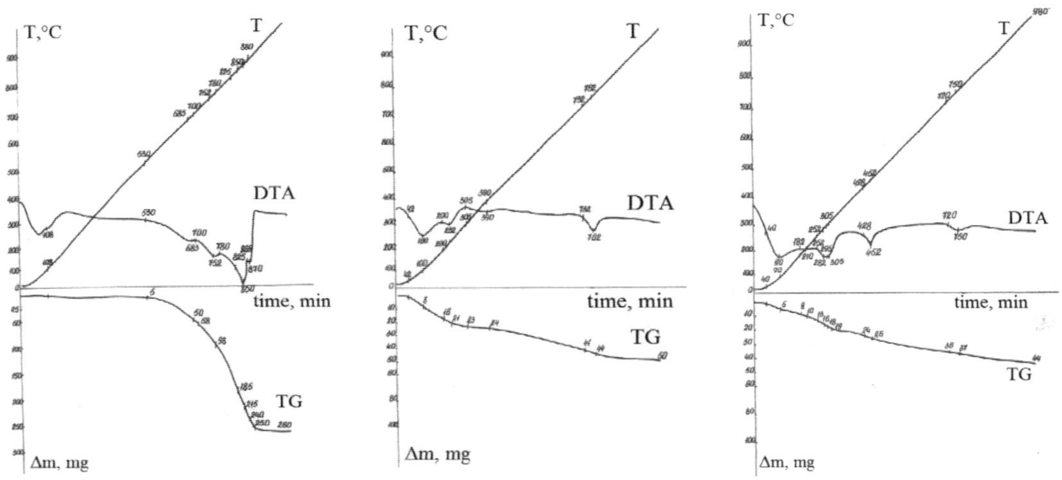

(a) (b) (c)

Figure 2. Derivatogram of red iron oxide pigment obtained from SPS

In addition, it is known that structural water entering $Fe(OH)_3$ is completely removed only at 940°C, i.e. when calcined at 550-600°, part may remain as an impurity in the pigment. Dehydration is a multi-stage process that occurs in several stages with the formation of iron compounds of different hydration. The products of iron dehydration are X-ray amorphous, which explains the absence of lines in X-ray diffraction patterns. A number of thermal effects in the temperature range 200–400° C, apparently, can be caused by the melting of ammonium chloride impurities and the decomposition of copper and iron carbonates. Sample 1, in addition to α-Fe_2O_3, contains a rather significant amount of $CaCO_3$. When considering the thermogram of this sample, one should take into account the possibility of a solid-phase reaction between Fe_2O_3 and $CaCO_3$.

The authors found [15] that in the Fe_2O_3 – CaO system, calcium ferrites $CaFe_2O_4$ and $Ca_2Fe_2O_5$ exist in the temperature range 1000°C and that the formation of ferrite occurs at a temperature above 600°C by the reaction:

$CaCO_3 + Fe_2O_3 \rightarrow CaFe_2O_4 + CO_2$

Taking into account these data, one can explain the complex form of endothermic effects at 612°C. One should also take into account the possibility of polymorphic transformation of α-quartz α- tridylite at 870°C and the possibility of a reaction between SiO_2 and CaO with the formation of silicates. Indeed, an X-ray phase analysis of sample 1, carried out after heating to 1000° C, established that the main phase is $Ca_2Fe_2O_5$, a β-Ca_2SiO_4 is present as an impurity.

3. Conclusion

Thus, the results of laboratory studies and pilot tests have shown the possibility of obtaining red iron oxide pigment from spent pickling solutions, which in its performance is not inferior to pigment produced by industry.

IOP Conf. Series: Materials Science and Engineering 1100 (2021) 012051 doi:10.1088/1757-899X/1100/1/012051

References

[1] Bouarab R, Bennici S, Mirodatos C and Auroux A 2014 *Journal of Catalysts* **2014** 1–6
[2] Stepin S N et al 2018 Int. Conf. SGEM2018, 417
[3] Voss L, Hsiao I-L, Ebisch M, Vidmar J, Dreiack N, Böhmert L, Stock V, Braeuning A, Loeschner K, Laux P, Thünemann A F, Lampen A and Sieg H 2020 *Food Chem.* **327** 127000
[4] Zheng Y and Liu Z 2011 *Powder technol.* **207(1-3)** 335-42.
[5] Voss L, Yilmaz K, Burkard L, Vidmar J, Stock V, Hoffmann U, Pötz O, Hammer H S, Peiser M, Braeuning A, Löschner K, Böhmert L and Sieg H 2020 *Arch. Toxicol.* 1-13
[6] Demirci H, Slimani N, Pawar M, Kumon R E, Vaishnava P and Besirli C G 2019 *Transl. Vis. Sci. Technol.* **8** 18
[7] Zeng D, Liu S, Wang G, Qiu J and Chen H 2014 *Asia-Pacific Journal of Chemical Engineering* **9** 737–42
[8] Schwertmann U and Cornell R M 2008 *Iron Oxides in the Laboratory: Preparation and Characterization* (John Wiley & Sons)
[9] Faivre D 2016. *Iron oxides: from nature to applications* (John Wiley & Sons)
[10] Shabanova N A, Popov V V and Sarkisov P D 2006 *Chemistry and technology of nanodispersed oxides: Textbook for universities* (Moscow: Akademkniga)
[11] Turevskaya E P, Yanovskaya M I and Ya. Turova N 2000 *Inorg. Mat.* **36** 260–70
[12] Kargin D B, Konyukhov Y V, Biseken A B, Lileev A S and Yu. Karpenkov D 2020 *Izvestiya. Ferrous Metallurgy* **63** 146–54
[13] Motovilov I Y, Yu. Motovilov I, Luganov V A, Chepushtanova T A, Guseynova G D and Itkulova S S 2017 *Solutions, Treatments and Opportunities II* 191–6
[14] Zyulkov B N and Khoroshkin V M Patent RU2211191C2, 2001.
[15] Lietard J-M and Guido M. U.S. Patent 4,066,748, 1978.

Energy recovery strategies as a sustainable solutions for municipal solid waste in Egypt

M Koraiem Handawy[1,2], A Yu Snegirev[1], V V Stepanov[1], V A Talalov[1]

[1] Saint-Petersburg State Polytechnic University, Department of Thermal Physics, Polytechnicheskaya 29, Saint-Petersburg 195251, Russia
[2] Minia University, Faculty of Engineering, Department of Mechanical Power Engineering and Energy, 61111, Egypt.

E-mail: mohkoraiem@mu.edu.eg

Abstract. Plastic is highly durable, strong, elastic, and less expensive to produce, and widely used globally. The result of excessive use causes a high accumulation of plastic waste for communities in developing countries Like Egypt. Egypt is one of the largest countries in population in Africa and the middle east. According to Environments Affairs Agency (EEAA), Egypt produces over 22 million tons of municipal waste per year, and waste plastic represents 13% of the total waste. Therefore, plastic recycling is critical in our life. But mechanical recycling is unable to recycle all plastic, as it is an expansive and challenging process. It exhibits a series of degenerated properties like lowering the impact resistance characterizers of polymers. The second recycling process is chemical methods and thermochemical recycling. Thermochemical degradation has a promising substitutional for recycling polymers. For example, pyrolysis can deal with high molar mass organic molecules since the plastic owns a calorific value comparable to fuel. Energy Recovery from waste Plastic using the thermal treatment for plastics as incineration, gasification, and pyrolysis represents a suitable process as discussed. So, the fuel production from pyrolysis polymer would be a better and more promising alternative for a country like Egypt.

1. Introduction

The significant increase in the world's population and changing lifestyle and consumption patterns, such as rapid civilization and manufacturing, are causing an increase in the generation and accumulation of municipal solid waste at a rate that exceeds the natural environment's capacity to absorb and manage the wastes—the risk represented in developing countries such as Egypt. Egypt is one of the most populous countries with a population of more than 100 million, which means increased demand for traditional energy sources, resulting in accumulating municipal solid waste, affecting the global environment like the greenhouse effect and global warming. The economic benefits of modernization, industrialization, population growth, and technology transfer have always been attractive to increase municipal solid waste [1]—Egypt one of the most significant waste plastic sources in the Mediterranean and the Middle East. Indeed, According to the Egyptian Environments Affairs Agency (EEAA), Egypt produces over 22 million tons of waste per year. The composition of municipal waste depends mainly on the population's affluence, contributing to the MSW amount. For example, Cairo-capital city-alone leads the highest percentage, with a waste volume of about 6 million tons annually, representing 45% of the total municipal solid waste, according to the Ministry of the Environment[2–5]. The Municipal solid waste composition in Egypt shown in Fig. 1, according to the Egyptian Environmental Affairs Agency 2015 [3]

The problem of municipal solid waste in Egypt is represented in the lack of possibilities. The increase in its quantity annually is expected to increase at a rate of 3.4% per year. The municipal solid waste grows at a high rate because less than 20 percent of this is appropriately disposed of or recycled. This waste is exposed to uncontrolled incinerators, which negatively affects Egypt and can hazard soil, air, and water source, the same for waste. Therefore, it is necessary to find appropriate and economical ways to solve it and be based on

Content from this work may be used under the terms of the Creative Commons Attribution 3.0 licence. Any further distribution of this work must maintain attribution to the author(s) and the title of the work, journal citation and DOI.
Published under licence by IOP Publishing Ltd

sustainable development. An effort must be made to conserve non-renewable resources to the maximum extent possible.

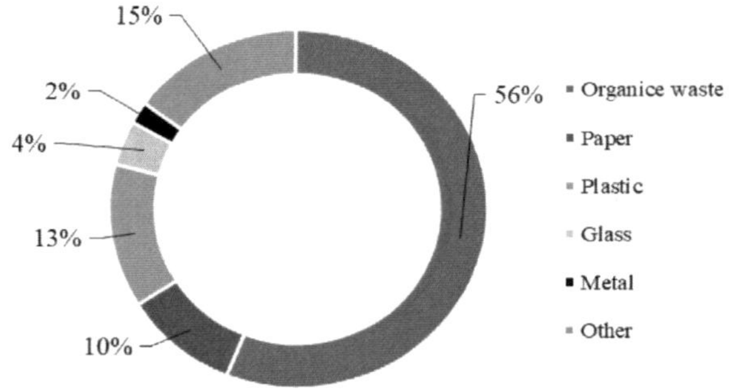

Figure 1. Typical composition of solid waste in Egypt according to EEAA 2015 [3]

The increase of solid waste accumulations everywhere in Egypt can hazard soil, air, and water source. This increase in the solid wasted leads to a negative impression of public health and the environment [5]. Uncontrolled plastic incineration cause emissions, toxic gases, and ash containing lead and cadmium [6]. The treatment and recycling of solid municipal waste in Egypt are estimated at 12%. About 81% of the waste is disposed of in public and random landfills, while 7% is disposed of in safe landfills. This causes more difficulties on the road to reducing GHG emissions and combating climate change regarding the management of the waste MSW resources in Egypt [4]. As shown in Fig. 1.1, plastic represents 13% of solid waste in Egypt, this means there are approximately million of a ton of waste plastic is generated in Egypt every year [3].

As known, plastic has become a big problem due to the difficulty of dealing with its wastes, and what exacerbates the issue even more, is the significant increase in the daily use of plastic. Also, there is an increase in disposable materials (packaging and packing of food), difficult to recycle. So, getting rid of plastic is difficult because of the variety of uses and materials. And this matter becomes more difficult because of mixing many types of plastic, which may cause great difficulty in sorting and recycling the types. Now, about 9% of plastic can be recycled, and the remaining causes many problems for the environment. The rapid growth of Egypt's population led to an increase in the demand for plastic until it became the first in Africa and the Middle East to consume plastic. This increase has turned into a significant challenge for the local authorities responsible for solid waste management. Egypt as a country, produce and use different types of plastic. The percentage of Egypt's consumption of plastics is shown in Fig. 2 [8].

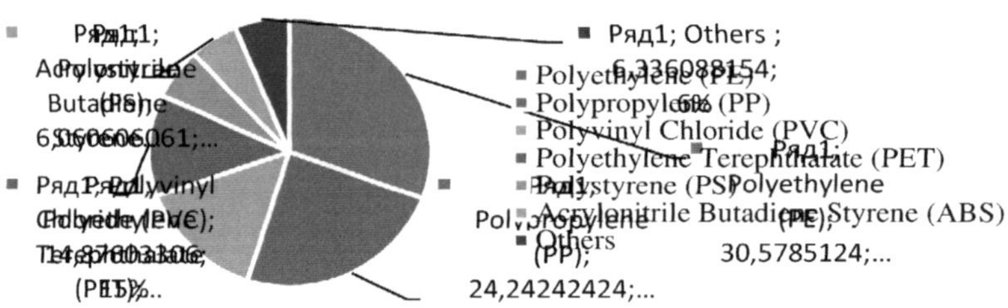

Figure 2. Percentage of Egypt consumption of plastics by polymer type [8]

Plastic recycling includes four categories. The first is mechanical reprocessing, which is the return to a product that has comparable properties, which is called closed-loop recycling. The secondary mechanical recycling is treated mechanically to obtain products with fewer characteristics, and the third is the return to the essential chemical components, Which is known as chemical recycling. The fourth case is energy recovery from plastic waste, which is done through several effective value processes to obtain fuel or energy from plastic waste [9]. In general, it found that mechanical and chemical recycling is uneconomic without significant subsidy due to the lower price of petrochemical feedstocks than recycled materials due to the process costs of producing monomers from waste plastics.

Plastic waste is characterized as one of the useful materials for the production of recycling fuel, energy from it, due to the high combustion temperature and heating value as shown in table 1, and due to the increasing availability in local communities. Methods for converting waste plastic into fuel depend on the types of plastic to be targeted and the other characteristics of the waste used in the process. Besides, efficient conversion requires selecting appropriate technologies according to local economic, environmental, social, and technical aspects.

2. Material and method of recovery energy

Plastic

Plastic, or polymer, consists of a high molecular mass chain and often contains many materials such as carbon, hydrogen, oxygen, nitrogen, chlorine, and Sulphur. Today's plastics have become an essential ingredient in our life since Alexander Parkes unveiled Parkesine in 1862. Plastics are utilized in our day by day lives in various applications. This is because plastics are versatile and affordable and useful for manufacturers to use from nurseries, mulches, covering and wiring, bundling, films, spreads, packs, and holders. Food waste and paper squander. These materials' peculiarities, which have lightness, manufacturability, low thermal conductivity, and high chemical resistance [6,10].

3. Energy Recovery as recycling methods

The polymer recycling process is divided into three main classes: First, the mechanical recycling (direct) process, but it exhibits a series of degenerated properties like lowering the impact resistance. The second recycling process is chemical methods and thermochemical recycling.

Chemical recycle such as hydrolysis, ammonolysis, and alcoholics can yield a high percentage of the monomer; nevertheless, these processes required a particular separation material, expansive and complex, had harmful series effects of environmental and safety problems, and toxic organic solvents[11,12] For example, Hydrolysis required critical condition water [13–15].

Fig 3 shows the alternative thermal treatment of waste plastics (incineration, gasification, pyrolysis) technologies and their products. Plastic has a rapid development production and can be considered an alternative energy resource for substituting fossil fuels. So, using these methods able to produce a significant amount of energy. [16].

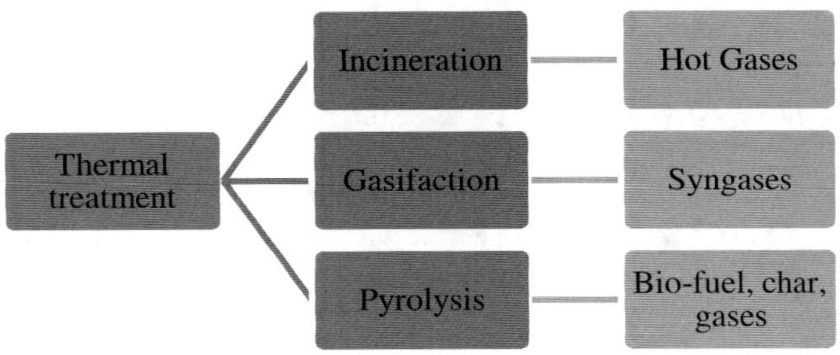

Figure 3. Thermal treatments of waste plastics

IOP Conf. Series: Materials Science and Engineering 1100 (2021) 012052 doi:10.1088/1757-899X/1100/1/012052

4. Incineration

Incineration as a method to get energy from waste plastic can reduce the CO_2. Plastic is a material derived from petroleum and contains a hydrocarbon material and minimal moisture content, so plastic calorific values are similar to those of LPG, petrol, and diesel as given in Table 1. On the other hand. Direct combustion/ incineration of polymers may be detrimental to the environment. The incinerate polymers/plastics results in residues of noxious substances such as light hydrocarbons. NOx, sulfur oxides, and dioxins. [17–21].

5. Gasification or Partial oxidation

As known as partial oxidation, using oxygen or steam could generate a mixture of hydrocarbons and synthesis gas (CO and H2) and produce bulk chemicals, such as acetic acid, from polyolefins via oxidation NO, which is also possible. This method had developed to produce high calorific value purified gas. With efficiency, reach 70% in hydrogen production from polymer waste. Co-gasification of biomass with waste plastic showed increased hydrogen produced while the CO content reduced [22].

Table 1. Comparison of calorific values of plastics and different types of fuels

Material	Calorific value, MJ/kg
Polyethylene	46.3
Polypropylene	46.4
Polystyrene	41.4
Polyvinyl chloride	18.0
Coal	24.3
Liquefied petroleum gas	46.1
Petrol	44.0
Kerosene	43.4
Diesel	43.0
Heavy fuel oil	41.1

6. Pyrolysis

Pyrolysis is known as thermal cracking, which is defined as controlled thermal degradation of materials in the absence of an oxygen environment. The process is to feed material decomposed into char and volatiles, and volatiles gas is condensed to oil. The non-condensable fraction can collect as a high calorific value gas at the reactor's exit [22–25].

In plastics pyrolysis, polymer macromolecular structures are cleaved into smaller molecules, oligomers, or sometimes monomeric units. Subsequent decomposition of these to low molar mass molecules is dependent on several process conditions and the chemistry of depolymerization. These include but are not limited to temperature, heating rate, the sample's residence time, and released volatiles in the reactor, particle size, inherent or added catalysts, reactor type, and feed composition. Generally, plastic polymers reactivity is influenced by the substituent's size in the side chain [9].

In degradation processes, part of the species generated directly from the initial degradation reaction is transformed into secondary products due to inter and intramolecular reactions. The extent and nature of these reactions depend on the reaction temperature and the residence of the products in the reaction zone, which is primarily affected by the reactor design [16,17,22–24,26].

Pyrolysis has many advantages as an emerging technology that can help reduce fossil fuel resource consumption by recycling plastics for fuel and valuable chemical recovery. It is a suitable method to recycle mixed plastic waste, which cannot be effectively recycled by other means, such as mechanical recycling. Also, pyrolysis can process plastic films and multi-layer packaging films that are not recycled using conventional reprocessing techniques. It is characterized by reducing waste and reducing pollutant emissions since the process is carried out without air [9,24,25,27].

7. Conclusion

Plastic is highly durable, strong, elastic, and less expensive to produce, and widely used globally. The result of excessive use causes a high accumulation of plastic waste for communities in developing countries Like Egypt. According to Environments Affairs Agency (EEAA), Egypt produces over 22 million tons of municipal waste per year, and waste plastic represents 13% of the total waste. Therefore, plastic recycling is essential in our life. But mechanical recycling is unable to recycle all plastic, as it is an expansive and challenging process. It exhibits a series of degenerated properties like lowering the impact resistance characterizers of polymers. The second recycling process is chemical methods and thermochemical recycling. Thermochemical degradation (pyrolysis) has a promising substitutional for recycling polymers, as pyrolysis can deal with high molar mass organic molecules since the plastic owns a calorific value comparable to fuel. Therefore, Energy Recovery from waste Plastic using the thermal treatment for plastics as incineration, gasification, and pyrolysis represents a suitable process as discussed. So, the fuel production from pyrolysis polymer would be a better and more promising alternative for countries like Egypt.

Acknowledgment
This work was performed for grant 16-49-02017 of the Russian Science Foundation, Russia. Collaboration with the Boeing company and support of the Supercomputer Centre "Polytechnic" (SPbPU), authors are gratefully acknowledged

References
[1] Anwar W a 2003 Environmental health in Egypt *Int. J. Hyg. Environ. Health* **206** 339–50
[2] Ibrahim M I M and Mohamed N A E M 2016 Towards Sustainable Management of Solid Waste in Egypt *Procedia Environ. Sci.* **34** 336–47
[3] EEAA 1983 State of the environment report *Environ. Policy Law* **11** 15
[4] Nakhla D A, Hassan M G and Haggar S El 2013 Impact of biomass in Egypt on climate change *Nat. Sci.* **05** 678–84
[5] Elfeki M and Tkadlec E 2015 Treatment of municipal organic solid waste in Egypt *J. Mater. Environ. Sci.* **6** 756–64
[6] Shent H, Pugh R J and Forssberg E 1999 A review of plastics waste recycling and the flotation of plastics *Resour. Conserv. Recycl.* **25** 85–109
[7] Das P and Tiwari P 2018 Valorization of packaging plastic waste by slow pyrolysis *Resour. Conserv. Recycl.* **128** 69–77
[8] Kerir S and Co P 2013 Overview on Egyptian Petrochemicals Market (A) Egypt Basic Information . (B) Petrochemicals in Egypt : Egyptian opportunities . Case studies (C) Conclusion .
[9] Patni N, Shah P, Agarwal S and Singhal P 2013 Alternate Strategies for Conversion of Waste Plastic to Fuels *ISRN Renew. Energy* **2013** 1–7
[10] Panda A K, Singh R K and Mishra D K 2010 Thermolysis of waste plastics to liquid fuel. A suitable method for plastic waste management and manufacture of value added products-A world prospective *Renew. Sustain. Energy Rev.* **14** 233–48
[11] Antonakou E V. and Achilias D S 2013 Recent advances in polycarbonate recycling: A review of degradation methods and their mechanisms *Waste and Biomass Valorization* **4** 9–21
[12] Siddiqui M N, Redhwi H H, Antonakou E V. and Achilias D S 2018 Pyrolysis mechanism and thermal degradation kinetics of poly(bisphenol A carbonate)-based polymers originating in waste electric and electronic equipment *J. Anal. Appl. Pyrolysis* **132** 123–33
[13] Pant D 2016 Polycarbonate Waste Management using Glycerol *Process Saf. Environ. Prot.* **100** 281–7
[14] Pant D and Kumar S 2014 GREEN CHEMICAL MODIFICATION : AN ECOFRIENDLY WAY TO MATERIAL *J. Sci. Res. Tech* **2** 58–61
[15] Grause G, Sugawara K, Mizoguchi T and Yoshioka T 2009 Pyrolytic hydrolysis of polycarbonate in the presence of earth-alkali oxides and hydroxides *Polym. Degrad. Stab.* **94** 1119–24
[16] Ouda O K M, Raza S A, Nizami A S, Rehan M, Al-Waked R and Korres N E 2016 Waste to energy potential: A case study of Saudi Arabia *Renew. Sustain. Energy Rev.* **61** 328–40
[17] Williams E A and Williams P T 1997 The pyrolysis of individual plastics and a plastic mixture in a fixed bed reactor *J. Chem. Technol. Biotechnol.* **70** 9–20
[18] Hazrat M A, Rasul M G, Khan M M K, Azad A K and Bhuiya M M K 2014 Utilization of polymer wastes as transport fuel resourcesa recent development *Energy Procedia* vol 61 (Elsevier B.V.) pp 1681–5
[19] Begum S, Rasul M G and Akbar D 2012 An Investigation on Thermo Chemical Conversions of

SolidWaste for Energy Recovery *World Acad. Sci. Eng. Technol. Int. J. Environ. Ecol. Eng.* **62** 624–30
[20] Ramos A, Monteiro E, Silva V and Rouboa A 2018 Co-gasification and recent developments on waste-to-energy conversion: A review *Renew. Sustain. Energy Rev.* **81** 380–98
[21] Handawy M K, Hegazy A A, Koura N S and Dahab O M 2020 Experimental study of Partially Premixed-Diffusion Turbulent Flame with Circular Plate Bluff Body *J. Phys. Conf. Ser.* **1565** 012103
[22] Garforth A A, Ali S, Hernández-Martínez J and Akah A 2004 Feedstock recycling of polymer wastes *Curr. Opin. Solid State Mater. Sci.* **8** 419–25
[23] Li H, Zhang S, Zhao X and Eiji S 2005 Pyrolysis characteristics and kinetics of municipal solid waste *Trans. Tianjin Univ.* **11** 353–9
[24] Jung C G and Fontana A 2006 Production of Gaseous and Liquid Fuels by Pyrolysis and Gasification of Plastics: Technological Approach *Feedstock Recycling and Pyrolysis of Waste Plastics: Converting Waste Plastics into Diesel and Other Fuels* pp 249–83
[25] Snegirev A Y, Handawy M K, Stepanov V V. and Talalov V A 2019 Pyrolysis and combustion of polymer mixtures: Exploring additivity of the heat release rate *Polym. Degrad. Stab.* **161** 245–59
[26] Gilbert M 2017 Plastics Materials *Brydson's Plastics Materials* pp 1–18
[27] Handawy M K, Snegirev A Y, Stepanov V V and Talalov V A 2020 Kinetic Modeling of Pyrolysis and Combustion of High-Density Polyethylene-Based Polymer Mixtures *J. Phys. Conf. Ser.* **1565** 012097

Kinetic modeling and analysis of pyrolysis of polymethyl methacrylate using isoconversional methods

M Koraiem Handawy[1,2], A Yu Snegirev[1], V V Stepanov[1] and V A Talalov[1]

[1] Saint-Petersburg State Polytechnic University, Department of Thermal Physics, Polytechnicheskaya 29, Saint-Petersburg 195251, Russia
[2] Minia University, Faculty of Engineering, Department of Mechanical Power Engineering and Energy, 61111, Egypt

E-mail: mohkoraiem@mu.edu.eg

Abstract. Pyrolysis the waste plastic into fuel is a promising way to protect the environment. In this study, we study the thermal decomposition processes of polymer Poly (methyl methacrylate) PMMA. This work used the microscale combustion calorimeter (MCC) as a tool to study the degradation. MCC is an applicable method to calculate the material combustion parameters by using milligram samples. The experiments were carried out at four heating rates 0.25, 0.5, 0.75, and 1 K/s. The temperature range in which the pyrolysis reaction occurs is ≈ 300-470 oC. We applied several isoconversional methods such as Friedman, Ozawa, Flynn-Wall-Ozawa (FWO), and Kissinger-Akahira-Sunose (KAS) to estimate the apparent activation energy of thermal degradation of Poly(methyl methacrylate). Determine the activation energy from each method evaluated against the experimental data. With investigation, the FWO method shown much higher values for activation energy for PMMA polymer more than the other two methods, and the average value much higher than the Friedman and KAS methods. On the other hand, as a differential method, Friedman showed more accurate results Ea= 216.65kJ/mol, and KAS method 228.61 kJ/mol. Our investigation may give a few experiences to how to more readily portray trial results with hypothetical expectations and essential data for calculating the activation energy of potential pyrolysis reactors.

1. Introduction

The simulation of degradation polymers is essential and exciting from viewpoints. It helps understand the degradation process of polymers, which is complex in typical cases. It assists design the reactors during the pyrolysis or the incineration or recovery of chemicals from recycling processes. One of the requirements to simulate degradation polymers is activation energy affected by the method used to determine the activation energy. Those methods are divided into mains groups, model fitting, and model-free methods. Specific parameters characterize every method to determine the activation energy, and then we find the model-free need to proceeding the pyrolysis at different heating rates to perform the analysis[1,2].

Poly (methyl methacrylate) (PMMA) is considered an inexpensive and economical thermo-plastic. PMMA is widely used, as it has good thermal stability. PMMA is characterized as a combustible polymer with a low oxygen index [3,4].

The analyzing thermal degradation of the original sample can proceed as a single-step or multi-step reaction. For analyzing non-isothermal solid-state kinetic data, any method tries to analyze and determine the kinetic triplet. These methods can be divided into two types model-fitting and model-free (isoconversional) methods[5–7].

The model-free methods can be differential iso-conversional methods or integral iso-conversional methods. one of the differential iso-conversional methods is Friedman which is characterized as an accurate method, but it rarely used in determine the activation energy, this due to it relatively complicated computations and the sensitivity to measurements can because more errors. On another hand there are

Content from this work may be used under the terms of the Creative Commons Attribution 3.0 licence. Any further distribution of this work must maintain attribution to the author(s) and the title of the work, journal citation and DOI.
Published under licence by IOP Publishing Ltd

IOP Conf. Series: Materials Science and Engineering 1100 (2021) 012053 doi:10.1088/1757-899X/1100/1/012053

integral isoconversional methods like Kissinger–Akahira–Sunose (KAS) and Ozawa–Flynn–Wall (FWO). Using the FWO method is widely used even it less accurate method but this refers to the simplicity of using it[8].

Micro combustion calorimetry (MCC) measured the heat release rate and the other important parameter of the pyrolysis and combustion process by using a milligrams sample and measuring the consumed oxygen. MCC as an instrument divided into two parts, first pyrolizer where the sample rapidly heated in an inert (nitrogen) atmosphere, this allows the sample to release the gas continues, then it followed by a combustor where the oxidization pyrolyzed gas in high temperature in excess oxygen[9,10] MCC method has been established as a standard ASTM test method (ASTM 7309)[11].

In this study employ and analyze the isoconversional methods as the estimated values of the activation energy using more than one isoconversional method can give an area for values for every value, where can predict true values of Ea. For that we will use the three-model free method to predicate the activation energy cause it important parameter in modeling degradation for the PMMA (Friedman, FWO, and KAS), with considering the limitation of every method. The results are evaluated and reliability against the experimental data out using MCC.

2. Materials and methods

In this work, the methodology is described in [1], PMMA as a flammable material with molecular weight 35 kg/mol, it supplied by Scientific Polymer Products Inc. The mass of the sample varied from 2-3 mg was placed into a ceramic crucible, which was put into a pyrolysis chamber. The specimen was subjected to a controlled heating (at a constant heating rate) in the nitrogen flow. Run the MCC at choosing heat rate β=0.25, 0.5, 0.75, and 1 K/s with temperature of combustor 900°C and maximum temperature pyrolizer is 750 °C. The measurements were preceded in accordance with in ASTM D7309-07a[11], and every experiment repeat the measurement to performed 3 to 6 times to ensure the repeatability of heat release tests, then it averaging the results data and calculate the standard deviation as shown by vertical bars in Fig. 1.

Kinetic Model

Determine the heat release rate of the oxidation of volatiles by measure the oxygen consumption by using the following equation

$$\dot{q} = -\Delta q_{O_2}\sigma_{O_2}\frac{m_{O_2}}{m_0} = -\Delta q\frac{\dot{m}}{m_0} \quad 1$$

where m_0 is the initial sample mass, σ_{O_2} is the stoichiometric oxygen to fuel mass ratio, $\dot{m}$ is the sample mass loss rate, and $\Delta q_{O_2}\sigma_{O_2} = \Delta q$ is the heat of combustion per unit mass of volatiles.

Due to complete volatile oxidation and synchronization of oxygen consumption rate with the sample temperature, the dependence $\dot{q}(T)$ measured in MCC characterizes the rate of sample decomposition, similar to the mass loss rate $\dot{m}(T)$ measured in TGA. Both dependencies provide the same information about the pyrolysis rate if the heat of combustion, Δq, does not change [12].

$$\overline{\Delta q'} = \int_0^{\infty} \Delta q' dt \quad (2) \qquad \text{where heat rate } \beta = \frac{dT}{dt}$$

Heat release based global conversion

$$\alpha = \frac{1}{\overline{\Delta q'}}\int_0^t \dot{q}(t)dt \int_0^{\infty} \Delta q' dt \quad (3) \qquad \text{With integration reaction rate RR} = \frac{d\alpha}{dt} = \frac{\dot{q}}{\Delta q'} = \quad (4)$$

The rate of pyrolysis from solid to volatile is described by Arrhenius equation.

$$\frac{d\alpha}{dt} = Af(\alpha)exp\left(-\frac{E_a(\alpha)}{RT}\right) = RR \quad 5$$

where E_a is the activation energy, A pre-exponential factor and R is the gas constant 8.314 J/K. mol.

2.1 Isoconversional method

For the calculation of the activation energies all heating rates have been used and they were estimated using the Friedman, Kissinger–Akahira–Sunose and Flynn–Wall–Ozawa methods for comparison reasons all mathematical equations for three methods are describe in Table 1.

Table 1. Mathematical equations of Friedman, FWO and KAS methods.

Method	Expression	Plot	Slop equal	
Friedman	$ln\,RR = -\frac{E_a}{\mathcal{R}T} + ln\,Af(\alpha)$	$\ln RR$ against $\frac{1}{T}$	$-\frac{E_a}{\mathcal{R}}$	[13,14]
FWO	$\ln\beta = -1.0516\frac{E_a}{\mathcal{R}T} + constant$	$\ln\beta$ against $\frac{1}{T}$	$-1.5016\frac{E_a}{\mathcal{R}}$	[15,16]
KAS	$\ln\frac{\beta}{T_\alpha{}^2} = \ln(\frac{AR}{E_a g(\alpha)}) - \frac{E_a}{\mathcal{R}T}$	$\ln\frac{\beta}{T_\alpha{}^2}$ against $\frac{1}{T}$	$-\frac{E_a}{\mathcal{R}}$	[17–19]

3. Results and discussion

The Figure demonstrates an increase in the heat release rate with an increase in the heating rate for PMMA samples. This follows from the fact that increasing the heating rate decreases the residence time to reach the peak temperature. The temperature range, in which the pyrolysis reaction occurs, for the samples of the investigated materials: ≈ 300-470 °C.

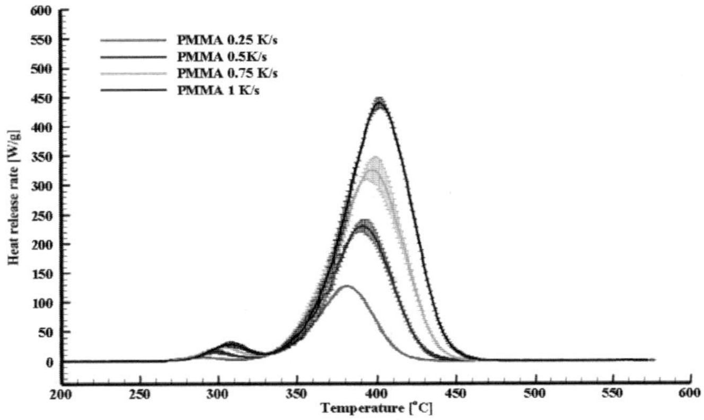

Figure 1. MCC measurements Heat release rate for PMMA (solid lines with vertical bars), b) Heat release rate for polymer blends (solid lines with vertical bars).

The calculated activation energy for PMMA samples is employed at four heating rates 0.25, 05, 0.75, and 1 K/s, during the conversion range 0.1 to 0.9 by using the model-free methods (Friedman, FWO, and KAS). As mention above every method determines the Ea by the assumption and the slop of its linear plots, this shown in Figures 2, 3, and 4. As shown in every figure the plots of Friedman (ln RR), FWO (ln β), and KAS (ln (β/T^2) versus 1/T at selected conversion. KSA and FWO plots, its shows a parallel for every conversion against the others, and this gives a reason to consider and assume the reaction order is the first-order reaction.

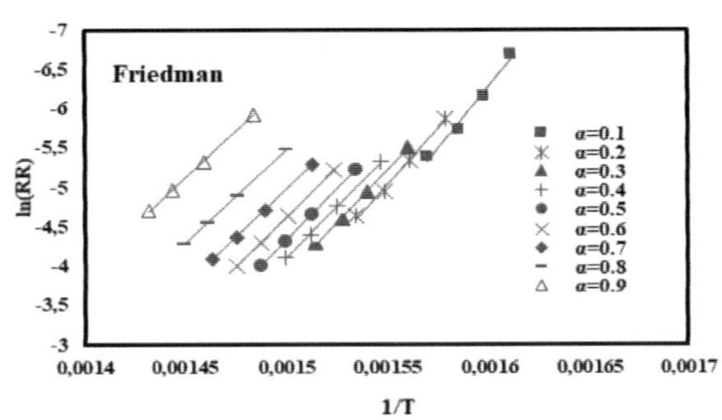

Figure 2. Friedman linear plots for a PMMA at α 0.1-0.9

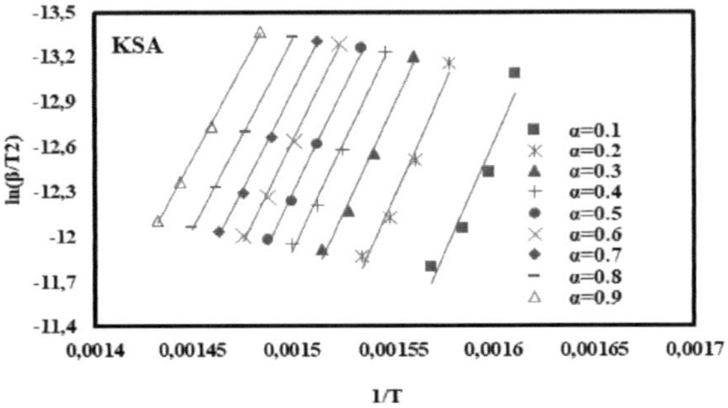

Figure 3. KSA linear plots for a PMMA at α 0.1-0.9

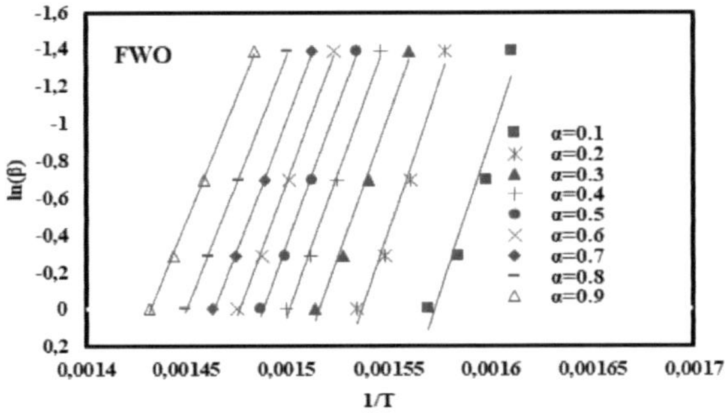

Figure 4. FWO linear plots for a PMMA at α 0.1-0.9

Ea via α is important because it shows how the reaction going on, and the order of the reaction. As the Ea higher variate this refers to its complex kinetic process and can't apply the single-step reaction equation. So, shown Fig. 5 the FWO shows a higher activation energy than KAS, as mention in ICTAC FWO is wide used but it has very low accuracy and limited to linear heating rate conditions method [20]. On another hand, Friedman method is more accurate, this because its a differential method and there is not any assumption, it supposed to show more accuracy and more sensitivity to measurement. According to ICTAC the chosen average Ea should be less than 20% difference between lower and higher Ea. Ea average activation energy using Friedman method (differential method) about 216.56 kJ/mol, and KAS 228.61 kJ/mol., the difference between them is about 7%, but FWO average Ea equal 260.2 kJ/mol. And the difference between FWO and Friedman is about 9.5%.

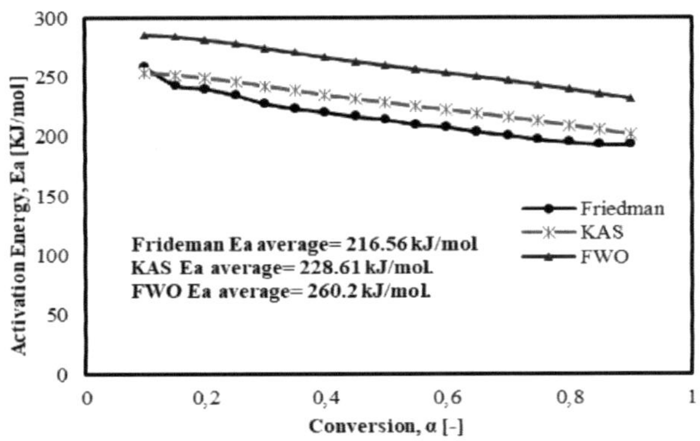

Figure 5. Estimated activation energy E_a for PMMA using Friedman, FWO and KAS methods on the conversion (α)

4. Conclusions

In this study, the kinetic analysis of PMMA pyrolysis using microscale combustion calorimeter (MCC) carried out at four heating rates 0.25, 0.5, 0.75, and 1 K/s. The MCC is confirmed to be a tool for determining the kinetic parameters for the polymer. For that, we able the modal-free method by using the measurements MCC for PMMA to determine activation energy. The activation energy was calculated by the iso-conversional methods (Friedman, Kissinger– Akahira–Sunose, Flynn–Wall–Ozawa). Finally, the degradation of PMMA is a single-step reaction according to the dependency of activation energy and conversional. The calculated activation energy by Friedman shows more accurate than KAS and FWO.

Acknowledgment

This work is supported by the RSF grant 16-49-02017. Support by the Supercomputer center "Polytechnic" (SPbPU) is gratefully acknowledged.

References

[1] Snegirev A Y, Handawy M K, Stepanov V V. and Talalov V A 2019 Pyrolysis and combustion of polymer mixtures: Exploring additivity of the heat release rate *Polym. Degrad. Stab.* **161** 245–59
[2] Handawy M K, Snegirev A Y, Stepanov V V and Talalov V A 2020 Kinetic Modeling of Pyrolysis and Combustion of High-Density Polyethylene-Based Polymer Mixtures *J. Phys. Conf. Ser.* **1565** 012097
[3] Korobeinichev O P, Paletsky A A, Gonchikzhapov M B, Glaznev R K, Gerasimov I E, Naganovsky Y K, Shundrina I K, Snegirev A Y and Vinu R 2019 Kinetics of thermal decomposition of PMMA at different heating rates and in a wide temperature range *Thermochim. Acta* **671** 17–25
[4] Acosta R I, Gross K C and Perram G P 2015 Thermal degradation of Poly(methyl methacrylate) with a 1.064 μm Nd:YAG laser in a buoyant flow *Polym. Degrad. Stab.* **121** 78–89
[5] Chrissafis K 2009 Kinetics of thermal degradation of polymers : Complementary use of isoconversional and model-fitting methods *J. Therm. Anal. Calorim.* **95** 273–83
[6] Slopiecka K, Bartocci P and Fantozzi F 2012 Thermogravimetric analysis and kinetic study of poplar wood pyrolysis *Appl. Energy* **97** 491–7
[7] Khedri S and Elyasi S 2016 Kinetic analysis for thermal cracking of HDPE: A new isoconversional approach *Polym. Degrad. Stab.* **129** 306–18
[8] Vyazovkin S 2015 *Isoconversional kinetics of thermally stimulated processes*
[9] Yang C Q, He Q, Lyon R E and Hu Y 2010 Investigation of the flammability of different textile fabrics using micro-scale combustion calorimetry *Polym. Degrad. Stab.* **95** 108–15
[10] Lyon R E and Janssens M L 2005 Polymer Flammability Report 82
[11] Anon 2007 *ASTM D7309: Standard Test Method for Determining Flammability Characteristics of Plastics and Other Solid Materials Using Microscale Combustion Calorimetry*
[12] Snegirev A Y 2014 Generalized approach to model pyrolysis of flammable materials *Thermochim. Acta* **590** 242–50
[13] Friedman H L 1964 Kinetics of thermal degradation of char-forming plastics from thermogravimetry. Application to a phenolic plastic *J. Polym. Sci. Part C Polym. Symp.* **6** 183–95
[14] Snegirev A Y, Talalov V A, Stepanov V V. and Harris J N 2012 Formal kinetics of polystyrene pyrolysis in non-oxidizing atmosphere *Thermochim. Acta* **548** 17–26
[15] Takeo O 1965 A New Method of Analyzing Thermogravimetric Data *Bull. Chem. Soc. Jpn.* **38** 1881–6
[16] Flynn J H and Wall L A 1966 A quick, direct method for the determination of activation energy from thermogravimetric data *J. Polym. Sci. Part B Polym. Lett.* **4** 323–8
[17] Kissinger H E 1957 Reaction Kinetics in Differential Thermal Analysis *Anal. Chem.* **29** 1702–6
[18] Kissinger H E 1956 Variation of peak temperature with heating rate in differential thermal analysis *J. Res. Natl. Bur. Stand. (1934).* **57** 217
[19] Akahira, T. and Sunose T 1971 Joint Convention of Four Electrical Institutes *Sci. Educ. Technol.* **16** 22–31
[20] Vyazovkin S, Burnham A K, Criado J M, Pérez-Maqueda L A, Popescu C and Sbirrazzuoli N 2011 ICTAC Kinetics Committee recommendations for performing kinetic computations on thermal analysis data *Thermochim. Acta* **520** 1–19

IOP Conf. Series: Materials Science and Engineering 1100 (2021) 012054 doi:10.1088/1757-899X/1100/1/012054

Training of engineers in demand in the engineering industry at PSEDA

V G Lizunkov, E V Politsinskaya, D V Valuev and E O Zhilin

Tomsk Polytechnic University, Tomsk, Russia,
E-mail: vladeslave@rambler.ru, valuevden@rambler.ru

Abstract. The article identified the criteria and levels of formation of the supra-professional competencies (team competency) which are essential for future engineers of technical universities, according to the stakeholders' feedback on the priority social and economic development areas. The authors defined the formation levels of team competency among students of technical areas of training at the Research Tomsk Polytechnic University. They described and investigated theoretical and empirical research methods that allowed formulating specific pedagogical recommendations to solve the problem identified in the article.

1. Introduction

Today the engineering activity has an increasingly innovative character and focuses on developing and creating new equipment and technologies brought to the type of marketable products that provide a new social and economic effect, and therefore demanded and competitive. To develop the economy, Russian Federation creates territories with a special legal status relative to the rest of the country and favorable economic conditions for entrepreneurs.

Resident enterprises have modern knowledge-intensive production, which entails a complication of professional activities and requires from educational institutions fundamentally new approaches to training engineers. Today, competitive production needs engineering personnel with a new consciousness, with the necessary competencies and capable of giving a positive impetus to the development of the industry. The very innovative development of an enterprise predetermines its ability to create new competencies. Nowadays a technical university graduate should be able to plan, design, produce and apply complex engineering objects, processes and systems with added value in modern teamwork conditions.

To improve the quality of engineering education, Russian institutions of higher education are moving to federal state educational standards FSES 3++, focused on quality training of graduates to perform labor functions through active interaction between higher education and labor. The universities are modernizing the main professional educational programs according to the new standards.

Team competency according to FSES 3++ is included in the list of general cultural competencies of a graduate. A modern university graduate besides having professional knowledge should have teamwork skills and the ability to realize himself in various social roles.

2 Subject and Methods of Research

2.1 Analysis of psychological and pedagogical literature

Our analysis of psychological and pedagogical literature, both in Russia and abroad, shows that there are scientific and educational prerequisites to solve the problem of forming a team competency that meets the modern requirements of engineering activity. The most important works that reflect the issues of forming the competencies of team interaction include the works of such authors as E.V. Bolgarina [1], Yu.V. Vasilyeva [2], K. Cruz, J. Pinto [3] et al. The works that describe different pedagogical technologies of forming this competency among students include the work of V.S. Okuneva [4], A.D. Nikolaeva, A.D. Malysheva [5], L.M. Harding [6], Zi-Yu Liu, Zaffar Ahmed Shaikh, Farida Gazizova [7], Nor Kamaliana

Content from this work may be used under the terms of the Creative Commons Attribution 3.0 licence. Any further distribution of this work must maintain attribution to the author(s) and the title of the work, journal citation and DOI.
Published under licence by IOP Publishing Ltd

Khamis, Amirul Mukhlish Abdul Azam. [8] et al. To create a more complete picture of forming teamwork competencies, we used information provided by scientists who study the value systems of different social groups: C.S.Burke, S.M.Fiore, E.Salas [9] et al.

However, despite the valuable contribution of available research, we should note that the professional pedagogy does not yet pay due attention to the problem of forming team competency in the training of such an economically important specialty as a mining engineer, and there is no correlation with the peculiarities of their training.

From the analysis of various scientific works, we will understand team competency as the ability to work effectively on a designated problem in various project teams. It includes the ability to intelligently build communication within a team, the ability to take responsibility for personal and team results, and to participate in resolving intra-group conflicts.

The article aims to reveal the formation level of team competency of future engineers in the transition to FSES 3++ and to substantiate the effectiveness of implementing the collaborative approach to form team competency in the educational process of the university.

2.2 Research Methodology

We tested the CDIO Syllabus training results assessment model on specialists of industrial enterprises. The study involved 146 managers at various levels.

The method of collecting information is a formalized interview. We used a 5-point scale to assess the required competencies and, consequently, the results of training future specialists. The table shows the assessment based on the average value of received employers' answers.

We monitored the development of team competency among students through observation of students' ability to work in a team, questionnaires and interviews. The research involved 135 students of 3-4 years of technical training from the National Research Tomsk Polytechnic University.

The monitoring was conducted in the first semester of 2019/2020 academic year.

3 Survey Results Analysis

3.1 Formation levels team competency

In our research we monitored the formation of team competency among engineering students at the National Research Tomsk Polytechnic University. To study the development of team competency, we identified assessment criteria and levels of its formation considering the standards and analysis of various researchers' work (Table 2).

Table 1. Formation levels of team competency

Criteria	Formation levels		
	Low	Average	High level
Communication skills	A student cannot build effective communication with team members. A student does not know how to listen, clearly express their thoughts.	A student attempts to build an effective system of communication with team members, and not always can clearly articulate his proposals, ideas and present them.	A student clearly and precisely expresses his thoughts, ideas, is able to listen attentively, convincingly gives arguments, can defend his point of view.
Teamwork activity	Personal interests always dominate the interests of the team. A student refuses to perform the tasks assigned to him, cannot negotiate. A specialist generates a conflict situation within a team.	Personal interests tend to dominate the team ones. A specialist performs the assigned tasks only with urgent requests. Loner. A student is reluctant to perform his part of the work, not interested in the overall team result.	The interests of the team are above their own interest. A student shows initiative, puts forward ideas and is ready to help. A student fully fulfills the tasks assigned to him.
Responsibility for the team's results	A student makes mistakes and inaccuracies while performing the tasks. A student does not fit within	A student periodically allows inaccuracies in its activities, thus limiting the effectiveness of the team.	A student fulfills his tasks exactly in time.

	the time frameset.

Based on the selected criteria, the authors studied to identify the level of formation of team competency among students of RTPU. Totally 135 students took part in the research. The results are shown in Figure 1.

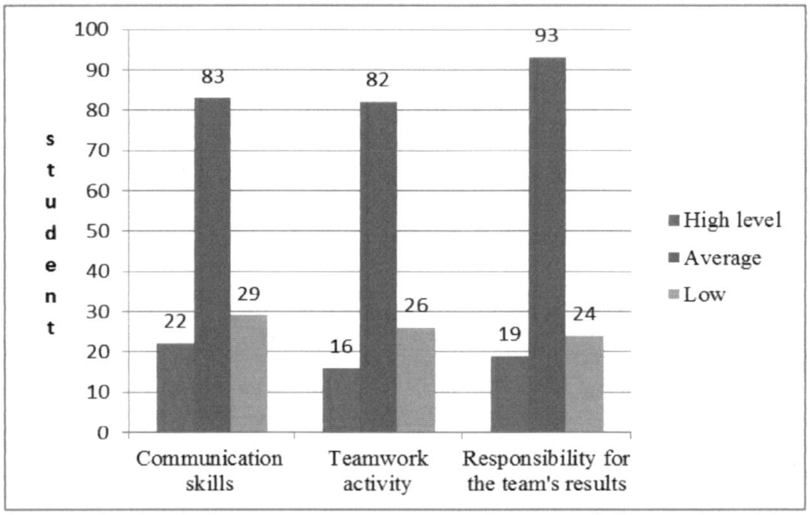

Figure 1. Formation level of team competency among future specialists graduated from RTPU

The results of the study show that the majority of students have mainly the average level of formation of team competency. In other words, the model of education even in competitive universities such as RTPU has difficulties in forming competencies according to the requirements of FSES 3+++. In this regard, we can conclude that this problem can be overcome by using a collaborative approach to learning.

3.2 Collaborative approach to learning

Collaborative learning integrates knowledge and creates "natural learning" situations (as an opposition to a trained result from a highly structured learning situation). Working in a team, students achieve high results due to the synergy effect, i.e. when they achieve together a result that cannot be achieved alone.

Interest in the collaborative approach is growing both in Russia and abroad, as evidenced by several publications [10,11,12,13]. However, the authors' analysis shows that there is a shortage of designed methods for developing teamwork skills, including training in mining specialties.

The process of forming team competency of future engineers should include the following stages:

1. Creation of professional significance and personal value of teamwork among students.

After leaving the university, the future engineer usually still needs a lot of time to adapt to the professional environment.

The professional value orientation of the graduate ensures his conscientious attitude to the chosen profession, causes the desire and striving to engage in professional activities as soon as possible. It becomes possible if the educational process is close to real professional activity. Consequently, all tasks performed by students in special subjects should be related to specific production problems or have an applied nature.

In real life, a mining engineer has to face all kinds of challenges, including extreme ones. Some tasks can be easily solved according to the previously studied algorithm, but there may arise tasks that require analysis, synthesis and so on, that is the problem of operational solution at the level of a large team. In most cases, when students have a choice between group work and individual work, they usually choose individual work because they consider they can do the work much faster and better than in the group. However, some goals cannot all be achieved individually.

It all depends on the specific task and the level of responsibility for the result. Students should clearly see the synergy effect of working together on a simple example. It is possible to offer students a task with the choice of a joint activity or an individual task with the definition of time parameters for the task. The result will be obvious. Students that team up and distribute tasks among the team members will cope with this task faster and with the least effort. It is necessary to discuss the results with students. The role of the teacher is to show all the advantages of teamwork on a concrete example. It is to help the student realize that his future professional activity, his professional growth is constant collaboration, interaction with the team.

While working in a team on practice-oriented tasks and issues, considering the specifics of the professional activity, in the future the adaptation period will pass faster, they will form a value attitude to the future professional activity, which ensures their competitiveness and mobility in the labor market.

2. Consideration of individual personal qualities and value orientations of each team member when organizing teamwork.

Students experience strong emotional experiences during teamwork. They arise at the moment of task distribution, in the process of selecting various alternatives to solve the problem, in preparation for project protection. All members are responsible for the result of teamwork, and confidence in future success unites the participants. In the course of their work, students have to face the individual uniqueness of each group member, show respect and tolerance for each other, and provide support. One condition for the achievement of the set tasks is the presence of psychological and emotional comfort in the team.

An engineer sometimes has to work overtime. The reason may be narrow terms of project delivery or an unforeseen situation that needs to be quickly eliminated. It is unacceptable to have non-constructive interpersonal conflicts for efficient, fast, well-coordinated teamwork. Thus, to organize effective and well-coordinated work, all team members should have a common point of view in understanding the norms of social behavior, morality, acceptability of certain methods, i.e. have common values.

Students work as a team on a project and interact with each other not only on a learning issue but also on a number of other issues. Therefore, the existence of common value orientations is very important in achieving an educational goal. Diagnostics of value orientations allows the teacher to correctly realize what roles students are capable of playing in a team, who will succeed in the role of leader, and who is an effective performer.

In addition, teamwork effectively influences the formation of value orientations. The influence of group values leads to the generalization of individual orientations. Teamwork results in the formation of a positive value attitude among individual students towards their future professional activities.

A group without cohesion, unity of opinion, mutual trust, mutual assistance cannot be called a team. Teamwork is built according to the following principles: effective interpersonal communication; the ability to work in harmony; group relations are open, and the group is ready to face any difficulties and obstacles in the way of effective work. This results from common value concepts.

Thus, when forming teams to maximize the effectiveness and coherence of their work, the teacher should consider the value orientations of students and their personal characteristics [15].

3. Role of a teacher as an organizer who guides students to make decisions by realizing the potential of the group as a whole and each of its participants individually.

When organizing teamwork, the role of the instructor is more of a mentor. His task is to support and guide the team to achieve the planned result. In addition, the role of the teacher involves the ability to choose technological, organizational, socio-psychological capabilities to achieve maximum pedagogical results. The organization of student teamwork allows teachers to get to know students better and maintain effective independent, group work of students [14].

The teacher's intervention should be kept to a minimum. The initial stages of teamwork may require more support from the teacher, but in the future, he should become an observer from a team organizer and guide the team toward the goal only if it is necessary. Once the team has achieved its goals, the teacher should encourage students to reflect on the process and the results of the teamwork. This will help to determine the mistakes and obstacles that prevented the achievement of the set goal. The result of teamwork largely depends on properly built in communication between team members. There are several rules for students to know:

- listen carefully to the team member until the end;
- start speaking only after the previous team member has finished speaking;
- ask questions if everyone understood what you wanted to say;
- make suggestions and ask others to Express their opinion and add;
- openly express their opinion and be interested in the opinion of others;
- discuss proposals and ideas collectively;
- come to a common opinion together;
- sum up the results together.

Teamwork is a very effective way to learn how to solve problems. The process of problem solving forms such important personal qualities of the future specialist as a creative, non-standard approach to solving professional problems, creativity, independence, ability and willingness to self-development, self-actualization, ability to set goals independently, to put forward ideas, ability to plan their activities and analyze their results [16,17].

4.Conclusion

Our research resulted in the following conclusions: it is possible to solve the mentioned problem and to form a high level of team competency of future specialists when implementing a collaborative approach to education in the educational process of the university. Firstly, the relevance of the collaborative approach is that it orients future engineers to work in an innovative mode, motivating them to find a non-standard solution. Secondly, teamwork develops the ability to listen and understand others, to express a point of view, to find a compromise, which meets the requirements of the new federal state educational standards and makes the future engineer quite competitive specialist in the labor market, demanded in priority development areas.

Acknowledgment

The article was carried out with the financial grant support of the Russian Science Foundation (project No. 19-78-00018)

References

[1] Bolgarina E V 2015 Formation of the competency for teamwork readiness in the process of teaching the discipline *J."Software Engineering". Higher school pedagogy*, 2: 27-29.

[2] Vasilyeva Yu V 2016 Formation of competencies of the team professionally-oriented interaction of the future engineer *J. Bulletin of the Orenburg State University*, 10 (198): 3-12.

[3] Cruz K, Pinto J 2019 Team focus in focus: Its implications for real teams and their members J*ournal of Work and Organizational Psychology*, 35 (2): P. 123-13

[4] Okuneva V . 2012 Project activity in forming students' ability to work in a team. *Siberian Pedagogical Journal* 5: 96-102

[5] Malysheva A D 2017 Ability to work in a team as a general cultural competency of university students *J.Modern problems of science and education*, 2: 151

[6] Harding, L M 2017 Students of a Feather "Flocked" Together: A Group Assignment Method for Reducing Free-Riding and Improving Group and Individual Learning Outcomes *Journal of marketing education* 40 (2): 117-127

[7] Zi-Yu Liu, Zaffar Ahmed Shaikh, Farida Gazizova Using the Concept of Game-Based Learning in Education *International Journal of Emerging Technologies in Learning* 15 (14): 53-64

[8] Nor Kamaliana Khamis, Amirul Mukhlish Abdul Azam Learning Aids of Risk Assessment Apps For Practical Engineering Student: A Case Study *International Journal of Emerging Technologies in Learning* 14 (24): 81-95

[9] Burke C S, Fiore S M, Salas E 2003 The role of shared cognition in enabling shared leadership and team adaptability. Shared Leadership: Reframing the Hows and Whys of Leadership P. 103-122.

[10] Marjan Laal, Azadeh Sadat Naser, Mozhgan Laal, Zhina Khattami-Kermanshahi 2013 What do we achieve from learning in collaboration? *J. Procedia-Social and Behavioral Sciences* 93: 1427 – 1432

[11] Medaille Ann, Usinger Janet 2018 "That's going to be the hardest thing for me": tensions experienced by quiet students during collaborative learning situations *J. Educational studies*, 46 (2): 240-257

[12] Nefyodova L A, Mironova D A 2016 "The Role of Collaborative Methods of Teaching in Forming Translator's Language Personality".Humanities Research. Humanitates, 2 (2): 184-193.

[13] PervazIqbal M, Velan G M, O'Sullivan, A J, Balasooriya C 2020 The collaborative learning development exercise (CLeD-EX): an educational instrument to promote key collaborative learning behaviors in medical students. BMC medical education, 20 (1), article number: 62

[14] Nousiainen T, Kangas M, Rikala J Vesisenaho M 2018 Teacher competencies in game-based pedagogy *J. Teaching and Teacher Education* 74: 85-97. https://doi.org/10.1016/j.tate.2018.04.012

[15] Lizunkov V, Politsinskaya E, Malushko E, Pavlov A 2020 Modelling as the Basis for Building a Competency Model of a Specialist Demanded by Industrial Enterprises in Priority Social and

Economic Development Area (PSEDA) *J. International journal of emerging technologies in learning* 15 (13): 321-326
[16] Lizunkov V, Politsinskaya E, Gazin K, The Architecture of Project-Based Learning in the Supplementary Vocational Education System in a Higher Education *J. International journal of emerging technologies in learning* 15 (4): 227-234
[17] Lizunkov Vladislav, Politsinskaya Ekaterina, Yagova Evgenia 2021 Concept of Preparing in-Demand Staff for the Priority Social and Economic Development Areas in Further Education. *J. International Journal of Engineering Pedagogy (iJEP)* 11(1): 95-105. DOI: https://doi.org/10.3991/ijep.v11i1.15013

IOP Conf. Series: Materials Science and Engineering 1100 (2021) 012055 doi:10.1088/1757-899X/1100/1/012055

Structure and properties of a cobalt-chromium-molybdenum alloy obtained by selective laser melting

A A Saprykin[1,a], Yu P Sharkeev[1,2,b], N A Saprykina[1,c], M A Khimich[2,3,d] and E A Ibragimov[1,e]

[1]National Research Tomsk Polytechnic University, Tomsk, Russia
[2]Institute of Strength Physics and Materials Science SB RAS
[3]National Research Tomsk State University, Tomsk, Russia

[a]sapraa@tpu.ru, [b]sharkeev@ispms.tsc.ru, [c]saprikina@tpu.ru, [d]khimich@ispms.tsc.ru, [e]egor83@list.ru

Abstract. Cobalt-based heat-resistant alloys have unique properties that are used in corrosive environments and at high temperatures. The addition of chromium improves resistance to hot corrosion, while molybdenum improves corrosion resistance and strength. When machining a workpiece of a heat-resistant alloy, up to half of the material goes into chips and goes to processing. Modern trends in the development of additive technologies are aimed at saving material, introducing new materials and improving the physical and mechanical characteristics of the resulting products. Selective laser melting (SLM) is one of the additive methods for obtaining bulk metal products and is one of the most promising directions in the development of science and technology in recent years. This paper describes the porosity (the average value was 14 ± 5%), the structure and properties of alloy samples obtained by selective laser melting in the following modes: laser power - 100 W, scanning speed - 350 mm / s, thickness of the applied powder layer – 25 μm, scanning step – 100 μm, frequency of pulse-periodic laser action (modulations) - 2.5 kHz from a spherical powder of the CoCrMo alloy (66 wt% Co, 28 wt% Cr, 6 wt% Mo).

1. Introduction
Alloys of the Co-Cr-Mo system are widely known in the world for their high physical and mechanical properties, corrosion resistance and resistance to aggressive media and critical temperature conditions. Alloys are well studied in domestic and foreign literature. The cobalt structure exists in two crystalline modifications: a low-temperature ε-phase with a hexagonal close-packed lattice and a high-temperature γ-phase with a face-centered cubic lattice. Studies have shown that the value of the hardness of the alloy directly depends on the amount of the low-temperature phase. The laser melting method allows to reduce the time of exposure of the laser beam to the powder composition by increasing the power and speed of the laser movement. Due to the high thermal conductivity of the metal alloy, rapid crystallization and cooling occur, which contributes to an increase in the proportion of the ε-martensite phase in the alloy and an increase in the hardness and wear resistance of parts. Obtaining products from alloys of this system by selective laser melting makes it possible to form the structure and phase composition, and, consequently, a complex of physical and mechanical properties, which are impossible or very problematic to obtain using traditional methods of manufacturing alloys and products from them. Powder compositions used in selective laser melting must meet certain requirements: spherical shape of particles, particle size, phase and elemental composition, fluidity, flowability, etc. This paper describes the porosity (the average value was 14 ± 5%), the structure and properties of alloy samples obtained by selective laser melting in the following modes: laser power - 100 W, scanning speed - 350 mm / s, thickness of the applied powder layer - 25 μm, scanning step - 100 μm, frequency of pulse-periodic laser action (modulations) - 2.5 kHz from a spherical powder of the CoCrMo alloy (66 wt% Co, 28 wt% Cr, 6 wt% Mo).

2. Results and discussion

Content from this work may be used under the terms of the Creative Commons Attribution 3.0 licence. Any further distribution of this work must maintain attribution to the author(s) and the title of the work, journal citation and DOI.
Published under licence by IOP Publishing Ltd

The produced powder is particles of the required chemical composition (66 wt.% Co, 28 wt.% Cr, 6 wt.% Mo), particle size distribution with a particle size in the range of 20-70 μm, phase composition (stabilized high-temperature γ-phase) with spherical or spherical-like shape. Such a set of characteristics is obtained by spheroidization methods through melting and rapid crystallization of powder particles. In recent years, metal powders, including pure iron, stainless steel, tool steel, titanium alloys, cobalt-chromium alloys, nickel-based superalloys, alloys based on copper, aluminum, etc., have been used in additive technologies [1]. A part produced by laser melting layer-by-layer consists of cured tracks and combined layers. In the process of local melting of a metal powder by a laser beam, rapid crystallization and cooling, it is possible to obtain unique features of the microstructure inherent only to this method. A promising direction is the creation of a martensitic structure. All existing studies were carried out on samples obtained from commercially available Co28Cr6Mo alloy powder and on industrial commercial equipment. A large number of works are devoted to the determination of laser melting conditions for obtaining samples from CoCrMo alloy powder [2, 3]. An increase in the martensitic structure in samples of CoCrMo alloy powder obtained by selective laser melting after aging is described in [4]. The samples with the highest microhardness were obtained by aging at 900 °C for 10 hours. In these studies, a relationship was established between the heat treatment process, microstructure and microhardness. In articles [5], the preparation of an alloy in the process of laser melting from a powder mixture of a CoCrW and Cu alloy was investigated and the mechanical properties of the obtained samples were studied. Before laser melting, the two powders were uniformly mixed together for 30 minutes. Barucca et al. [6] studied the structure of Co24Cr5Mo samples prepared by selective laser melting and confirmed the phase transition from β (cubic face-centered lattice) to α (hexagonal close-packed lattice) and carbide formation by electron microscopy and found unexpectedly high values hardness exceeding the values of a sample of this alloy, made by traditional methods. In Russia, in STANKIN, A. Nazarov. [7] developed a technological process for the manufacture of complex-profile parts from a heat-resistant cobalt alloy by the method of selective laser melting. The relationships between the parameters of selective laser melting, the structure and phase composition of the heat-resistant cobalt alloy are determined. Large-scale research in the field of selective laser melting from a CoCrMo alloy is carried out at the Massachusetts Institute of Technology (Boston, USA) [8]. In Russia, the team of Professor I. Shishkovsky (Skolkovo) is engaged in selective laser sintering (melting) of products. [9] in cooperation with the National School of Engineering Saint-Etienne (France) and the "Laboratory of innovative additive technologies" MSTU "STANKIN". Equipment for the manufacture of medical implants is produced by several companies in the world. Concept Laser GmbH (Germany) manufactures Mlab cusing and R Mlab cusing installations, which can work with titanium alloys, titanium powder and cobalt-chromium powders [10]. SLM Solutions GmbH (Germany) manufactures SLM® 500 HL units that can work with titanium and cobalt-chromium alloys [11].

The experiment on the formation of bulk samples was carried out on a VARISKAF-100MVS selective laser melting facility. The melting process of powder materials was carried out in a chamber filled with an inert Ar gas after preliminary evacuation. The samples were formed on a VT1-0 titanium substrate preheated to a temperature of 190-200 °C. Analysis of the literature data and the available experience of working with refractory metal powders made it possible to preliminarily estimate the ranges of variable parameters of the installation. The scanning scheme is line-by-line by areas, the direction of shading is mutually perpendicular from layer to layer. The experience of previous studies has shown that with the proposed strategy for the formation of a single layer, a decrease in the scan length is accompanied by a decrease in residual stresses [12, 13].

The samples obtained had the shape of a parallelepiped with a square base, the side of which was 10 mm. The height of the obtained samples reached 5 mm. The surface of all the obtained samples after the completion of the melting process had developed morphology, relief, and roughness. To assess the structure and phase composition, the obtained samples were subjected to mechanical grinding in sections parallel and perpendicular to the substrate, polishing with diamond pastes, and chemical etching in a $HCl:H_2O_2$ mixture in a 5:1 ratio for 15 s [14]. The volumetric porosity of the obtained samples was estimated by the metallographic method. The study of the microstructure and elemental composition was carried out by the methods of optical metallography, scanning electron microscopy, and energy dispersive microanalysis. To assess the phase composition, studies were carried out using X-ray diffraction methods. Specimens prepared from spherical powder had varying porosity. Its average value was 14 ± 5%. The pores were irregular in shape with rounded edges. Most of the pores in the volume of the sample are interconnected and the pores form open porosity, which affects the strength and toughness of the alloy of the formed products. In the longitudinal and cross sections after etching, closed gas pores were observed in the samples, in the vicinity of which there were areas of dark contrast. A similar picture was observed by the authors of [15], who investigated the $\gamma \rightarrow \varepsilon$ transformation in alloys of the CoCrMo system, initiated by an increase in the defectiveness of the material under multiple plastic deformation by rolling. In their work, they identify such areas of dark contrast as the ε-phase based on the low-temperature cobalt phase. In some areas, grains were observed whose shape was close to equiaxial, and the average size did not exceed 10 μm. Both open and closed porosity are observed in the raster images of the obtained sample in longitudinal and cross sections, Figure 1. The pores had an irregular shape. The closed pore size did not exceed 25 μm. In the vicinity of the pores, areas of dark contrast were observed. In individual images, areas of molten material were observed inside the pores. The study by the method of energy dispersive microanalysis showed a uniform distribution of the components of cobalt, chromium and molybdenum over the cross section of the sample. Analysis of the chemical composition over the entire cross-sectional area showed an integral ratio of the values of the concentrations of cobalt, chromium and molybdenum, corresponding to the declared alloy, i.e. 66 wt. % Co, 28 wt. % Cr, 6 wt. % Mo. The results of point microanalysis showed that the concentration of cobalt, chromium and molybdenum in certain areas of the surface, in the vicinity of the pores and in the areas of molten material inside the pores corresponds to 66, 28, and 6 wt. %, respectively. Oxygen is present in the alloy. Its mass fraction did not exceed 2 wt. % and indicates the formation of an oxide layer on the surface, which is formed during storage and transportation of samples. The elemental composition in dark and light areas is identical. The single-phase state of the studied samples is identified by X-ray diffraction methods, Figure 2. All reflections present in the profile correspond to reflections from the planes of the ε-phase. Broadened X-ray peaks indicate the small size of the coherent scattering regions, which indirectly indicates the small size of the structure elements, as well as a high level of internal residual stresses. The lattice parameters calculated from the profiles turned out to be equal: $a = 0.2535 \pm 0.0001$ nm, $c = 0.4114 \pm 0.0002$ nm, which exceeds the reference values of the ε-cobalt lattice parameters ($a = 0.2507$ nm, $c = 0.40695$ nm [16]). This difference in lattice parameters is a consequence of the deformation of the crystal lattice of the base, in this case, cobalt, due to the incorporation of molybdenum and chromium atoms into the crystal lattice, an increase in defectiveness, as well as thermal stresses that accompany the SLM process. A significant redistribution of intensities on the X-ray profiles of the alloy obtained should be noted. The predominant growth is observed in the crystallographic direction <002>. This direction is perpendicular to the plane of the substrate, parallel to the direction of growth of the samples, and is traditional for the samples obtained by SLS. Thus, the resulting alloy is a solid solution based on a hexagonal low-temperature phase of cobalt - ε, in which cobalt, chromium and molybdenum are uniformly distributed, with the predominant crystallographic growth direction being <002>. It should be noted that the ε-phase is thermodynamically stable in the alloy and increases its strength, workability at low temperatures, and plasticity [17].

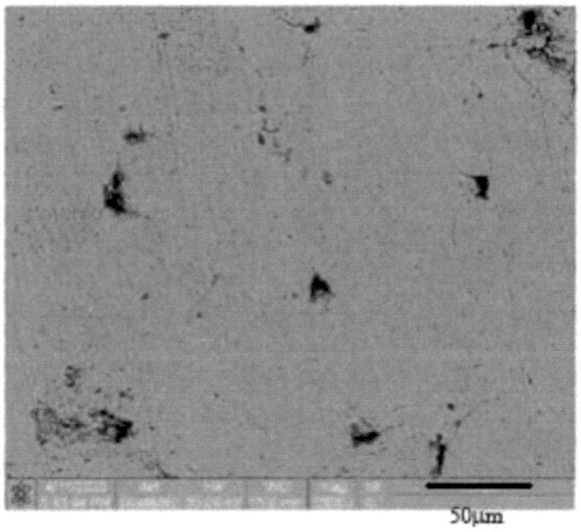

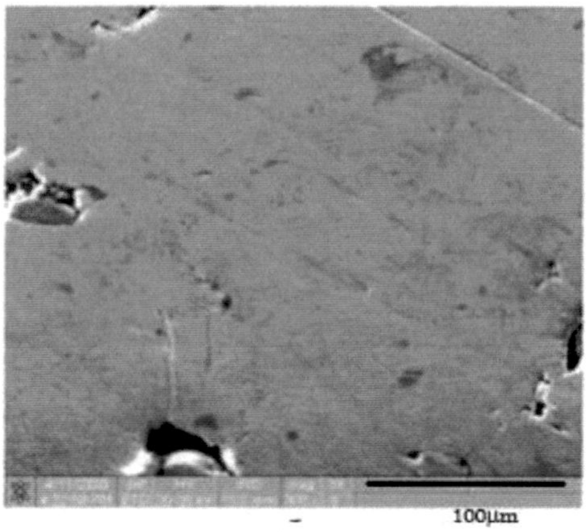

Figure 1. SEM images of the polished surfaces of an SLM sample of a Co-Cr-Mo alloy made of a spherical powder in longitudinal (a) and transverse (b) sections.

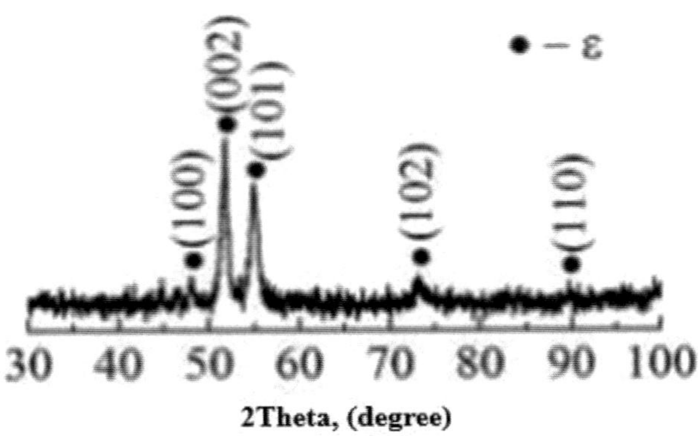

Figure 2. X-ray profile of a spherical powder SLM sample

The microhardness of the SLM samples obtained from the powder was studied in sections parallel and perpendicular to the substrate. Its value in the section parallel to the substrate was 4800 ± 400 MPa; perpendicular – 5600 ± 310 MPa. Such a difference in the values in the longitudinal and cross sections can be associated with the applied strategy of selective laser melting [18].

Conclusions

This article discusses the possibility of obtaining bulk samples from a spherical powder of the CoCrMo alloy (66 wt% Co, 28 wt% Cr, 6 wt% Mo) by selective laser melting. The results of the influence of the selective laser melting mode (laser radiation power – 100 W, scanning speed – 350 mm/s, thickness of the applied powder layer – 25 µm, scanning step – 100 µm, frequency of pulse-periodic laser action (modulations) - 2.5 kHz are presented.) on the porosity of the samples (the average value was 14 ± 5%), changes in the structure of the material, microhardness (in the section parallel to the substrate, 4800 ± 400 MPa; in the perpendicular section, 5600 ± 310 MPa).

Acknowledgments

«The reported study was funded by RFBR and Tomsk region according to the research project №19-48-700022»

References

[1] Zhou, X., Li, K., Zhang, D., Liu, X., Ma, J., Liu, W., Shen, Z., 2015. Textures formed in a CoCrMo alloy by selective laser melting. J. Alloys Compd. 631, 153–164. https://doi.org/10.1016/j.jallcom.2015.01.096

[2] Saprykin, A.A., Sharkeev, Y.P., Saprykina, N.A., Ibragimov, E.A., 2020. Surface formation mechanisms in selective laser melting of cobalt-chromium-molybdenum powder. Key Engineering Materials, , 839 KEM, 73–78

[3] Saprykin, A.A., Sharkeev, Y.P., Saprykina, N.A., Ibragimov, E.A., The mechanism of forming coagulated particles in selective laser melting of cobalt-chromium-molybdenum powder. Key Engineering Materials, 2020, 839 KEM, 79–85

[4] Zhang, M., Yang, Y., Song, C., Bai, Y., Xiao, Z., 2018. An investigation into the aging behavior of CoCrMo alloys fabricated by selective laser melting. J. Alloys Compd. 750, 878–886. https://doi.org/10.1016/j.jallcom.2018.04.054

[5] Lu, Y., Ren, L., Xu, X., Yang, Y., Wu, S., Luo, J., Yang, M., Liu, L., Zhuang, D., Yang, K., Lin, J., 2018. Effect of Cu on microstructure, mechanical properties, corrosion resistance and cytotoxicity of CoCrW alloy fabricated by selective laser melting. J. Mech. Behav. Biomed. Mater. 81, 130–141. https://doi.org/10.1016/j.jmbbm.2018.02.026

[6] Barucca, G., Santecchia, E., Majni, G., Girardin, E., Bassoli, E., Denti, L., Gatto, A., Iuliano, L., Moskalewicz, T., Mengucci, P., 2015. Structural characterization of biomedical Co-Cr-Mo components produced by direct metal laser sintering. Mater. Sci. Eng. C 48, 263–269. https://doi.org/10.1016/j.msec.2014.12.009

[7] Nazarov A. P. Development of the technological process for manufacturing complex-profile parts and heat-resistant cobalt alloy by selective laser melting: dis. candidate of technical Sciences, Moscow, 2013
[8] Benjamin M. Wua, Scott W. Borlandb, Russell A. Giordanob, Linda G. Cimac, Emanuel M. Sachsd, Michael J. Cima Solid free-form fabrication of drug delivery devices. Journal of Controlled Release Volume 40, Issues 1–2, June 1996, Pages 77–87.
[9] I. I. Zhuravleva, T. V. Rodchenko, A. L. Petrov, V. I. Shcherbakov, A. I. Snarev, and I. V. Shishkovsky Modeling and studying the properties of gradient filtering elements using the synthesized SLS method. Laser engineering and technology 2006 with 419-428
[10] I. V. Shishkovskii, I. A. Yadroitsev, I. Yu. Smurov Selective laser sintering/melting of nitinol–hydroxyapatite composite for medical applications. Powder Metallurgy and Metal Ceramics September 2011, Volume 50, Issue 5-6, pp 275-283.
[11] Nazarov A. P., Okunkova A. A. Typical samples of products obtained by selective laser sintering. Bulletin of the Saratov state technical University No. 1 (67) / volume 3 / 2012 p. 76-82
[12] Sharkeev Yu P, Eroshenko A Yu, Kovalevskaya Zh G, Saprykin A A, Ibragimov E A, Glukhov I A, Khimich M A, Uvarkin P V and Babakova E V 2016 Structural and Phase State of Ti–Nb Alloy at Selective Laser Melting of the Composite Powder Russian Physics Journal 59(3) pp 430-434
[13] Saprykin A A, Sharkeev Yu P, Ibragimov E A, Babakova E V, Kovalevskaya Zh G, Eroshenko A Yu, Khimich M A, Uvarkin P V, Glukhov I A 2016 Synthesizing conditions and structural-phase state of Ti-Nb alloy when selective laser melting ECCM 2016 - Proceeding of the 17th European Conference on Composite Materials
[14] F. Z. Hassani, M. Ketabchi, S. Bruschi, A. Ghiotti Effects of carbide precipitation on the microstructural and tribological properties of Co–Cr–Mo–C medical implants after thermal treatment. J. Mater. Sci. 2016. 51. 4495–4508. Doi: 10.1007/s10853-016- 9762-5.
[15] M. Mori, K. Yamanaka, S. Sato, S. Tsubaki, K. Satoh, M. Kumagai, M. Imafuku, T. Shobu, A. Chiba Tuning strain-induced γ-to-ε martensitic transformation of biomedical Co-Cr-Mo alloys by introducing parent phase lattice defects. Journal of the Mechanical Behavior of Biomedical Materials. 90. (2019). 523– 529. Doi: https://doi.org/10.1016/j.jmbbm.2018.10.038
[16] https://www.webelements.com/cobalt/crystal_structure.html
[17] M. Béreš, C.C. Silva, P.W.C. Sarvezuk, L. Wu, L.H.M. Antunes, A.L. Jardini, A.L.M. Feitosa, J. Žilková, H.F.G. de Abreu, R.M. Filho Mechanical and phase transformation behavior of biomedical CoCr-Mo alloy fabricated by direct metal laser sintering. Materials Science & Engineering A. 714. (2018). 36–42. Doi: https://doi.org/10.1016/j.msea.2017.12.087
[18] Mechanical and phase transformation behaviour of biomedical Co-Cr-Mo alloy fabricated by direct metal laser sintering. M Béreš, CC Silva, PWC Sarvezuk, L Wu, LHM Antunes, AL Jardini, Materials Science and Engineering: A 714, 36-42, 2018. 9. 2018

AUTHOR INDEX

Institute of Physics
Dirac House, Temple Back
Bristol BS1 6BE UK

ISSN: 1757-8981
ISBN 978-1-7138-4598-0

CURRAN ASSOCIATES INC.
proceedings.com